ALGEBRAIC NUMBER THEORY AND FERMAT'S LAST THEOREM

FOURTH EDITION

ALGEBRAIC NUMBER THEORY AND FERMAT'S LAST THEOREM

FOURTH EDITION

Ian Stewart
University of Warwick
United Kingdom

David Tall
University of Warwick
United Kingdom

CRC Press
Taylor & Francis Group
Boca Raton London New York

CRC Press is an imprint of the
Taylor & Francis Group, an **informa** business
A CHAPMAN & HALL BOOK

CRC Press
Taylor & Francis Group
6000 Broken Sound Parkway NW, Suite 300
Boca Raton, FL 33487-2742

First issued in paperback 2020

© 2016 by Taylor & Francis Group, LLC
CRC Press is an imprint of Taylor & Francis Group, an Informa business

No claim to original U.S. Government works

ISBN-13: 978-1-4987-3839-2 (hbk)
ISBN-13: 978-0-367-65871-7 (pbk)

Library of Congress Cataloging-in-Publication Data

Stewart, Ian, 1945-
  Algebraic number theory and Fermat's last theorem / Ian Stewart and David Tall.
-- Fourth edition.
      pages cm
  "A CRC title."
  Includes bibliographical references and index.
  ISBN 978-1-4987-3839-2 (alk. paper)
  1. Algebraic number theory. 2. Fermat's last theorem. I. Tall, David Orme. II. Title.

QA247.S76 2016
512.7'4--dc23   2015023199

Visit the Taylor & Francis Web site at
http://www.taylorandfrancis.com

and the CRC Press Web site at
http://www.crcpress.com

Contents

Preface to the Third Edition

The title of this book indicates a dual purpose. Our first aim is to introduce fundamental ideas of algebraic numbers. The second is to tell one of the most intriguing stories in the history of mathematics—the quest for a proof of Fermat's Last Theorem. We use this celebrated theorem to motivate a general study of the theory of algebraic numbers, from a reasonably concrete point of view. The range of topics that we cover is selected to allow students to make early progress in understanding the necessary concepts.

'Algebraic Number Theory' can be read in two distinct ways. One is the theory of numbers viewed algebraically, the other is the study of algebraic numbers. Both apply here. We illustrate how basic notions from the theory of algebraic numbers may be used to solve problems in number theory. However, our main focus is to extend properties of the natural numbers to more general number structures: algebraic number fields, and their rings of algebraic integers. These structures have most of the standard properties that we associate with ordinary whole numbers, but some subtle properties concerning primes and factorization sometimes fail to generalize.

A Diophantine equation (named after Diophantus of Alexandria, who—it is thought—lived around 250 and whose book *Arithmetica* systematized such concepts) is a polynomial equation, or a system of polynomial equations, that is to be solved in integers or rational numbers. The central problem of this book concerns solutions of a very special Diophantine equation:

$$x^n + y^n = z^n$$

where the exponent n is a positive integer. For $n = 2$ there are many integer solutions—in fact, infinitely many—which neatly relate to the theorem of

Pythagoras. For $n \geq 3$, however, there appear to be no integer solutions. It is this assertion that became known as Fermat's Last Theorem. (It is equivalent to there being no rational solutions—try to work out why.)

One method of attack might be to imagine the equation $x^n + y^n = z^n$ as being situated in the complex numbers, and to use the complex nth root of unity $\zeta = e^{2\pi i/n}$ to obtain the factorization (valid for odd n)

$$x^n + y^n = (x + y)(x + \zeta y) \ldots (x + \zeta^{n-1} y).$$

This approach entails introducing algebraic ideas, including the notion of factorization in the ring $\mathbf{Z}[\zeta]$ of polynomials in ζ. This promising line of attack was pursued for a time in the 19th century, until it was discovered that this particular ring of algebraic numbers does not possess all of the properties that it 'ought to'. In particular, factorization into 'primes' is not unique in this ring. (It fails, for instance, when $n = 23$, although this is not entirely obvious.) It took a while for this idea to be fully understood and for its consequences to sink in, but as it did so, the theory of algebraic numbers was developed and refined, leading to substantial improvements in our knowledge of Diophantine equations. In particular, it became possible to prove Fermat's Last Theorem in a whole range of special cases. Subsequently, geometric methods and other approaches were introduced to make further gains, until, at the end of the 20th century, Andrew Wiles finally set the last links in place to establish the proof after a three hundred year search.

To gain insight into this extended story we must assume a certain level of algebraic background. Our choice is to start with fundamental ideas that are usually introduced into algebra courses, such as commutative rings, groups and modules. These concepts smooth the way for the modern reader, but they were not explicitly available to the pioneers of the theory. The leading mathematicians in the 19th and early 20th centuries developed and used most of the basic results and techniques of linear algebra—for perhaps a hundred years—without ever defining an abstract vector space. There is no evidence that they suffered as a consequence of this lack of an explicit theory. This historical fact indicates that abstraction can be built only on an already existing body of specific concepts and relationships. This indicates that students will profit from direct contact with the manipulation of examples of number-theoretic concepts, so the text is interspersed with such examples. The algebra that we introduce—which is what we consider necessary for grasping the essentials of the struggle to prove Fermat's Last Theorem—is therefore not as 'abstract' as it might be. We believe that in mathematics it is important to 'get your hands dirty'. This requires struggling with calculations in specific contexts, where the elegance of polished theory may disguise the essential nature of the mathematics. For instance, factorization into primes in specific number fields displays the tendency of mathematical objects to take on a life of their own. In some situations something works, in others it does not, and the reasons why are often far from obvious. Without experiencing the struggle in person, it is quite impossible to understand why the pioneers in algebraic number theory had such difficulties. Of such frustrating yet stimulating stuff is the mathematical fabric woven.

We therefore do not begin with later theories that have proved to be of value in a wider range of problems, such as Galois theory, valuation rings, Dedekind domains, and the like. Our purpose is to get students involved in performing calculations that will enable them to build a platform for understanding the theory. However, *some* algebraic background is necessary. We assume a working knowledge of a variety of topics from algebra, reviewed in detail in Chapter 1. These include commutative rings and fields, ideals and quotient rings, factorization of polynomials with real coefficients, field extensions, symmetric polynomials, modules, and free abelian groups. Apart from these concepts we assume only some elementary results from the theory of numbers and a superficial comprehension of multiple integrals.

For organizational reasons rather than mathematical necessity, the book is divided into four parts. Part I develops the basic theory from an algebraic standpoint, introducing the ring of integers of a number field and exploring factorization within it. Quadratic and cyclotomic fields are investigated in more detail, and the Euclidean imaginary fields are classified. We then consider the notion of factorization and see how the notion of a 'prime' p can be pulled apart into two distinct ideas. The first is the concept of being 'irreducible' in the sense that p has no factors other than 1 and p. The second is what we now call 'prime': that if p is a factor of the product ab (possibly multiplied by units—invertible elements) then it must be a factor of either a or b. In this sense, a prime must be irreducible, but an irreducible need not be prime. It turns out that factorization into irreducibles is not always unique in a number field, but useful sufficient conditions for uniqueness may be found. The factorization theory of ideals in a ring of algebraic integers is more satisfactory, in that every ideal is a unique product of prime ideals. The extent to which factorization is not unique can be 'measured' by the group of ideal classes (fractional ideals modulo principal ones).

Part II emphasizes the power of geometric methods arising from Minkowski's theorem on convex sets relative to a lattice. We prove this key result geometrically by looking at the torus that appears as a quotient of Euclidean space by the lattice concerned. As illustrations of these ideas we prove the two- and four-squares theorems of classical number theory; as the main application we prove the finiteness of the class group.

Part III concentrates on applications of the theory thus far developed, beginning with some slightly *ad hoc* computational techniques for class numbers, and leading up to a special case of Fermat's Last Theorem that exemplifies the development of the theory by Kummer, prior to the final push by Wiles.

Part IV describes the final breakthrough, when—after a long period of solitary thinking—Wiles finally put together his proof of Fermat's Last Theorem. Even this tale is not without incident. His first announcement in a lecture series in Cambridge turned out to contain a subtle unproved assumption, and it took another year to rectify the error. However, the proof is finally in a form that has been widely accepted by the mathematical community. In this text we cannot give the full proof in all its glory. Instead we discuss the new ingredients that make the proof possible: the ideas of elliptic curves and elliptic integrals, and the link that shows that the existence of a counterexample to Fermat's Last Theorem would lead to a mathematical construction involving elliptic integrals. The proof of the theorem rests upon showing that such a construction cannot exist. We end with a brief survey of later developments, new conjectures, and open problems.

There follow two appendices which are of importance in algebraic number theory, but do not contribute directly to the proof of Fermat's Last Theorem. The first deals with quadratic residues and the quadratic reciprocity theorem of Gauss. It uses straightforward computational techniques (deceptively so: the ideas are very clever). It may be read at an early stage—for example, right at the beginning, or alongside Chapter 3 which is rather short: the two together would provide a block of work comparable to the remaining chapters in the first part of the book. The second appendix proves the Dirichlet Units Theorem, again a beacon in the development of algebraic number theory, but not directly required in the proof of Fermat's Last Theorem.

A preliminary version of Parts I–III of the book was written in 1974 by Ian Stewart at the University of Tübingen, under the auspices of the Alexander von Humbolt Foundation. This version was used as the basis of a course for students in Warwick in 1975; it was then revised in the light of that experience, and was published by Chapman and Hall. That edition also benefited from the subtle comments of a perceptive but anonymous referee; from the admirable persistence of students attending the course; and from discussions with colleagues. The book has been used by successive generations of students, and a second edition in 1986 brought the story up to date—at that time—and corrected typographical and computational errors.

In the 1980s a proof of Fermat's Last Theorem had not been found. In fact, graffiti on the wall of the Warwick Mathematics Institute declared 'I have a proof that Fermat's Last Theorem is equivalent to The Four Colour Theorem, but this wall is too small for me to write it.' Since that time, both Fermat's Last Theorem and the Four Colour Theorem have fallen, after centuries of effort by the mathematical community. The final conquest of Fermat's Last Theorem required a new version that would give a reasonable idea of the story behind the complete saga. This new version, brought out with a new publisher, is the result of further work to bring the book up to date for the 21st century. It involved substantial rewriting of much of the material, and two new chapters on elliptic curves and elliptic functions. These topics, not touched upon in previous editions, were required to complete the final solution of the most elusive conundrum in pure mathematics of the last three hundred years.

Coventry, February 2001.   Ian Stewart
                                                                                          David Tall

Preface to the Fourth Edition

There are three main changes to this fourth edition.

We provide up-to-date information on what is known about unique prime factorization for real quadratic number fields, especially Malcolm Harper's proof that $\mathbb{Z}(\sqrt{14})$ is Euclidean.

We have added one very important new result: Preda Mihăilescu's stunning proof of the Catalan Conjecture of 1844. This states the only non-trivial consecutive integer powers are 8 ($= 2^3$) and 9 ($= 3^2$). We discuss the history of this problem and sketch the current version of the proof, which is an extensive technical application of cyclotomic integers $\mathbb{Z}(\zeta)$ where ζ is a complex root of unity.

Chapter 14 of the previous edition has been split into two separate chapters for reasons of length. Chapter 14 now covers classical ideas about modular functions. Chapter 15 sketches the new ideas of Frey, Wiles and others that led to the long-sought proof of Fermat's Last Theorem. Section 15.4 on recent developments has been updated.

We have also corrected known typographical errors, extended and corrected the index, improved several figures, updated the bibliography and the further reading list, clarified a few historical remarks, and made many small stylistic changes, usually to conform to current practice. Among them is the replacement of boldface symbols such as $\mathbf{R}$ by 'blackboard bold' symbols $\mathbb{R}$.

Coventry and Kenilworth, May 2015.

Ian Stewart
David Tall

Index of Notation

(q/p)	Legendre symbol
$\mathbb{Z}$	Integers
$\mathbb{Q}$	Rationals
$\mathbb{R}$	Reals
$\mathbb{C}$	Complex numbers
$\mathbb{N}$	Natural numbers
$\mathbb{Z}_n$	Integers modulo n
R/I	Quotient ring
ker f	Kernel of f
im f	Image of f
$\langle X \rangle$	Ideal generated by X
$\langle x_1, \ldots, x_n \rangle$	Ideal generated by $x_1, \ldots, x_n$
$R[t]$	Ring of polynomials over R in t
∂p	Degree of polynomial p
$b|a$	b divides a
$\mathrm{D}f$	Formal derivative of f
$L : K$	Field extension
$[L : K]$	Degree of field extension
$K(\alpha_1, \ldots, \alpha_n)$	Field obtained by adjoining $\alpha_1, \ldots, \alpha_n$ to K
$R(\alpha_1, \ldots, \alpha_n)$	Ring generated by R and $\alpha_1, \ldots, \alpha_n$
$s_r(t_1, \ldots, t_n)$	rth elementary symmetric polynomial in $t_1, \ldots, t_n$
N/M	Quotient module
$\langle X \rangle_R$	R-submodule generated by X
det (A)	Determinant of A
(a_{ij})	Matrix
$\mathbb{Z}^n$	Set of n-tuples with integer entries
$\tilde{A}$	Adjoint of matrix A

$\lvert X \rvert$	Cardinality of set x
$\mathbb{A}$	Algebraic numbers
$f_\alpha(t)$	Field polynomial of α
$p_\alpha(t)$	Minimum polynomial of α
ω	$\frac{1}{2}(-1 + i\sqrt{3})$
$\Delta[\alpha_1, \ldots, \alpha_n]$	Discriminant of a basis
$\mathbb{B}$	Algebraic integers
$\mathfrak{O}$	Ring of integers of number field
$\mathfrak{O}_K$	Ring of integers of number field K
$\mathrm{N}_K(\alpha)$	Norm of α
$\mathrm{T}_K(\alpha)$	Trace of α
$\mathrm{N}(\alpha)$	Norm of α
$\mathrm{T}(\alpha)$	Trace of α
Δ_G	Discriminant of $\alpha_1, \ldots, \alpha_n$ when this generates G
$\begin{pmatrix} j \\ i \end{pmatrix}$	Binomial coefficient
$U(R)$	Groups of units of R
$\mathfrak{a}, \mathfrak{b}, \mathfrak{c}, \mathfrak{p}$, etc.	Ideals
$\mathfrak{a}^{-1}$	Inverse of a fractional ideal
$\mathfrak{a} \mid \mathfrak{b}$	$\mathfrak{a}$ divides $\mathfrak{b}$: equivalently, $\mathfrak{a} \supseteq \mathfrak{b}$
$\mathrm{N}(\mathfrak{a})$	Norm of $\mathfrak{a}$
$B_r[x]$	Closed ball centre x, radius r
$\lVert x - y \rVert$	Distance from x to y in $\mathbb{R}^n$
$\mathbb{S}$	Circle group
$\mathbb{T}^n$	$\mathbb{N}^n / \mathbb{Z}^n$, the n-dimensional torus
$v(X)$	Volume of X
v	Natural homomorphism $\mathbb{R}^N \to \mathbb{T}^n$
$\mathbb{L}^{st}$	$\mathbb{R}^s \times \mathbb{C}^t$
s	Number of real monomorphisms $K \to \mathbb{C}$
t	Half number of complex monomorphisms $K \to \mathbb{C}$
σ	Map $K \to \mathbb{L}^{st}$
$\mathcal{F}$	Group of fractional ideals
$\mathcal{P}$	Group of principal ideals
$\mathcal{H}$	Class-group $\mathcal{F} \mid \mathcal{P}$
$h(\mathfrak{O})$	Class-number
h	Class-number
$\mathfrak{a} \sim \mathfrak{b}$	Equivalence of fractional ideals modulo principal ideals
$[\mathfrak{a}]$	Equivalence class of $\mathfrak{a}$
Δ	Discriminant of K
M_{st}	Minkowski constant $(\frac{4}{\pi})^t (s + 2t)^{-s-2t} (s + 2t)!$
$\mathfrak{J}$	Ideal of $\mathbb{Z}(\zeta)$ generated by $1 - \zeta$ where $\zeta = e^{2\pi i/p}$

λ	$1 - \zeta$
$\bar{z}$	Complex conjugate of z
B_k	kth Bernouilli number
l	Map $\mathbb{L}^{st} \to \mathbb{R}^{s+l}$
U	Group of units of $\mathfrak{O}$
$\phi(x)$	Euler function
$\mathbb{RP}^2$	Real projective plane
$\mathcal{P}$	The plane $\{(x, y, z) : z = 1\}$
$\mathcal{Q}$	The plane $\{(x, y, z) : z = 0\}$
$\mathbb{CP}^2$	Complex projective plane
$\sim$	Equivalence relation for homogeneous coordinates
g_2, g_3	Coefficients in Weierstrass normal form of a cubic
$\mathbb{O}$	Specific rational point on an elliptic curve
$\mathcal{G}$	Set of rational points on an elliptic curve
$P * Q$	Geometric construction on elliptic curve
$P + Q$	Group operation on elliptic curve
$F(k, v)$	Elliptic integral of the first kind
sn u	Elliptic function
cn u	Elliptic function
dn u	Elliptic function
ω_1, ω_2	Periods of an elliptic function
$\mathcal{L}_{\omega_1, \omega_2}$	Lattice generated by ω_1, ω_2
$\wp$	Weierstrass $\wp$-function
$P \oplus Q$	Renaming of $P + Q$ for clarity
$\mathbb{C} \cup \{\infty\}$	Riemann sphere
$\mathbb{SL}_2(\mathbb{Z})$	Special linear group
$\mathbb{PSL}_2(\mathbb{Z})$	Projective special linear group
$\mathbb{H}$	Upper half-plane in $\mathbb{C}$
$\mathbb{D}$	Modular domain
$X_0(N)$	Modular curve of level N
$\mathcal{F}$	Frey elliptic curve
$P(N)$	Power function of N
$P(A, B, C)$	Power of (A, B, C)

The Origins of
Algebraic Number Theory

Numbers have fascinated the human race for millennia. The Pythagoreans studied many properties of the natural numbers $1, 2, 3, \ldots$, and the famous theorem of Pythagoras, though geometrical, has a pronounced number-theoretic content. Earlier Babylonian civilizations had noted empirically many so-called Pythagorean triples, such as 3, 4, 5 and 5, 12, 13. These are natural numbers a, b, c such that

$$a^2 + b^2 = c^2. \tag{1}$$

A clay tablet from about 1500 BC includes the triple 4961, 6480, 8161, demonstrating the sophisticated techniques of the Babylonians.

The Ancient Greeks, though concentrating on geometry, continued to take an interest in numbers. Around 250 AD, Diophantus of Alexandria wrote a highly influential treatise on polynomial equations which studied solutions in fractions. Particular cases of these equations with natural number solutions have been called *Diophantine* equations to this day.

The study of algebra developed over the centuries, too. Indian and Chinese mathematicians dealt with increasing confidence with negative numbers and zero. Meanwhile the Rashidun Caliphate conquered Alexandria in the 7th century, sweeping across north Africa and Spain. The ensuing civilization brought an enrichment of mathematics with Muslim ingenuity grafted on to Greek and Hindu influence. The word 'algebra' itself derives from the Arabic title 'al jabr w'al muqābalah' (literally 'restoration and equivalence') of a book written by the Persian Al-Khowarizmi in about 825. By the 13th century, peaceful coexistence of Islam and Christianity led to most Greek and Arabic classics being available in Latin translations.

In the 16th century, Girolamo Cardano used negative and imaginary solutions in his famous book *Ars Magna* (The Great Art), and in succeeding centuries complex numbers were used with greater understanding and flexibility.

Meanwhile the theory of natural numbers was not neglected. One of the greatest number theorists of the 17th century was Pierre de Fermat (1601–1665). His fame rests on his correspondence with other mathematicians, for he published very little. He would set challenges in number theory based on his own calculations; and at his death he left a number of theorems whose proofs were known, if at all, only to himself. The most notorious of these was a marginal note in his own personal copy of Diophantus, written in Latin, which translates:

> To resolve a cube into [the sum of] two cubes, a fourth power into fourth powers, or in general any power higher than the second into two of the same kind, is impossible; of which fact I have found a remarkable proof. The margin is too small to contain it.

More precisely, Fermat asserted that, in contrast to the case of Pythagorean triples, the equation

$$x^n + y^n = z^n \tag{2}$$

has no integer solutions x, y, z (other than the trivial ones with one or more of x, y, z equal to zero).

In the years following Fermat's death, almost all of his stated results were furnished with a proof. An exception was his claim that $F_n = 2^{2^n} + 1$ is prime for all positive integers n. In a letter to Pierre de Carcavi in 1659 he claimed a proof of this conjecture, but it was subsequently shown that he was wrong: for instance, F_5 is divisible by 641. Even the great Fermat could make mistakes. But one by one, his other assertions were furnished with proofs, until by the mid-19th century only one elusive jewel remained. A proof of his statement about the non-existence of solutions of (2) for $n \geq 3$ exceeded the powers of all 19th century mathematicians. This beguiling and infuriating assertion, so simple to state, yet so subtle in its labyrinthine complexity, became known as 'Fermat's Last Theorem'. This romantic epithet is in fact doubly inappropriate for, without a proof, it was not a 'theorem', neither was it the last result that Fermat studied—only the last to remain unproved by other mathematicians.

Given that a proof is so elusive, is it really credible that Fermat could have possessed a genuine proof—a clever way of looking at the problem, which eluded later generations? Or had he made a subtle error, which passed unnoticed, so that his 'theorem' had no proof at all? No one knows for sure, but there is a strong consensus that if he did have what he thought was a proof, it would not survive modern scrutiny. Consensus and certainty are not the same thing, however.

Be that as it may, during the late 19th and early 20th centuries the name stuck, with its glow of romanticism—somehow lacking in the more appropriate title 'Fermat Conjecture'. It has the two classic ingredients of a problem that can capture the imagination of a wider public—a simple statement that can be widely understood, but a proof that defeats the greatest intellects.

Another classic problem of this type—the impossibility of trisecting an angle using only ruler and compasses—took two thousand years to be solved. This problem was posed by the Greeks in their study of geometry; it was solved in the early 19th century using algebraic techniques. In the same way the advancement in the solution of Fermat's Last Theorem has moved away from the original domain, the theory of natural numbers, to a different area of mathematical study, algebraic numbers. By the 19th century the developing theory of algebra had matured to a state where it could usefully be applied in number theory.

As it happened, Fermat's Last Theorem was not the main problem being attacked by number theorists at the time; for example, when Kummer made the all-important breakthrough that we are to describe in this text, he was working on a different problem: 'higher reciprocity laws'. At this stage it is worth making a minor diversion to look at this subject, for it was here that algebraic numbers entered number theory in the work of Carl Friedrich Gauss. In 1796 the eighteen-year-old Gauss had given the first proof of a remarkable fact observed empirically by Leonhard Euler in 1783. Euler had investigated when an integer q is congruent to a perfect square modulo a prime p,

$$x^2 \equiv q \pmod{p}.$$

If so, q is a *quadratic residue* of p. Euler concentrated on the case when p, q are distinct odd primes and noted: if at least one of the odd primes p, q is of the form $4r + 1$, then q is a quadratic residue of p if and only if p is a quadratic residue of q; on the other hand, if both p, q are of the form $4r + 3$, then precisely one is a quadratic residue of the other. However, he failed to find a proof.

Because of the reciprocal nature of the relationship between p and q, this result was known as the *quadratic reciprocity law*. Adrien-Marie Legendre attempted a proof in 1785 but assumed that certain arithmetic series contain infinitely many primes—a theorem whose proof turned out to be far deeper than the quadratic reciprocity law itself. Legendre also introduced the symbol

$$(q/p) = \begin{cases} 1 & \text{if } q \text{ is a quadratic residue of } p \\ -1 & \text{if not,} \end{cases}$$

in terms of which the law becomes

$$(q/p)(p/q) = (-1)^{(p-1)(q-1)/4}.$$

We now call this the Legendre symbol. It is commonly written

$$\left(\frac{q}{p}\right)$$

but (q/p) is more convenient typographically.

Gauss gave the first proof of the law of quadratic reciprocity in 1796, but he was dissatisfied because his method did not seem a natural way to attack so seemingly simple a theorem. He went on to give several more proofs, two of which appeared in his book *Disquisitiones Arithmeticae* (1801), a definitive text on number theory which still remains in print, Gauss [32]. His second proof depends on a numerical criterion that he discovered, and we give a computational proof depending on this criterion in Appendix A.

Between 1808 and 1832 Gauss continued to look for similar laws for powers higher than squares. This entailed looking for relationships between p and q so that q is a cubic residue of p ($x^3 \equiv q \pmod{p}$) or a biquadratic residue ($x^4 \equiv q \pmod{p}$), and so on. He found some partial results about higher reciprocity laws, and in doing so he discovered that his calculations were simplified by working over the Gaussian integers $a + bi$ ($a, b \in \mathbb{Z}, i = \sqrt{-1}$), rather than the integers alone. This led him to develop a theory of prime factorization for Gaussian integers. He proved that decomposition into primes is unique in that context, and from that he developed a law of biquadratic reciprocity. In the same way, he considered cubic reciprocity by using numbers of the form $a + b\omega$ where $\omega = e^{(2\pi i)/3}$. These higher reciprocity laws do not have the same striking simplicity as quadratic reciprocity, and we shall not study them in this text. But Gauss's use of these new types of number is of fundamental importance for Fermat's Last Theorem, and the study of their factorization properties is a deep and fruitful source of methods and problems.

The numbers concerned are all examples of a particular type of complex number, namely one that is a solution of a polynomial equation

$$a_n x^n + \ldots + a_1 x + a_0 = 0$$

where all the coefficients a_j are integers. Such a complex number is said to be *algebraic*; if further $a_n = 1$ it is called an *algebraic integer*. Examples of algebraic integers include i (which satisfies $x^2 + 1 = 0$), $\sqrt{2}$ ($x^2 - 2 = 0$) and more complicated examples, such as the roots of $x^7 - 265x^3 + 7x^2 - 2x + 329 = 0$. The number $\frac{1}{2}i$ (satisfying $4x^2 + 1 = 0$) is algebraic but not an integer. On the other hand, there are complex numbers which are not algebraic, such as e or π, although proofs of those statements are difficult.

In the wider setting of algebraic integers, we can factorize a solution of Fermat's equation $x^n + y^n = z^n$ (if one exists) by introducing a complex nth root of unity $\zeta = e^{2\pi i/n}$ and writing (2) as

$$(x + y)(x + \zeta y) \ldots (x + \zeta^{n-1} y) = z^n. \tag{3}$$

If $\mathbb{Z}[\zeta]$ denotes the set of algebraic integers of the form $a_0 + a_1 \zeta + \ldots + a_r \zeta^r$ where each a_r is an ordinary integer, then this factorization takes place in the ring $\mathbb{Z}[\zeta]$.

In 1847 the French mathematician Gabriel Lamé announced a proof of Fermat's Last Theorem. In outline his proposal was to show that only the case where x, y have no common factors need be considered, and then deduce that in this case $x + y, x + \zeta y, \ldots, x + \zeta^{n-1} y$ have no common factors, that is, they are relatively prime. He then argued that a product of relatively prime numbers in (3) can equal an nth power only if each factor is an nth power. So

$$
\begin{aligned}
x + y &= u_1^n \\
x + \zeta y &= u_2^n \\
&\vdots \\
x + \zeta^{n-1} &= u_n^n
\end{aligned}
\tag{4}
$$

On this basis Lamé derived a contradiction.

Announcing a proof does not imply that it is one. Joseph Liouville immediately pointed out that the deduction of (4) from (3) assumes uniqueness of factorization in a subtle manner. Liouville's fears were confirmed when he later received a letter from Ernst Kummer, who had shown that uniqueness of factorization fails in some cases, the first being $n = 23$. Over the summer of 1847 Kummer went on to devise his own proof of Fermat's Last Theorem for certain exponents n, surmounting the difficulties of non-uniqueness of factorization by introducing the theory of 'ideal' complex numbers. In retrospect this theory can be viewed as introducing numbers from outside $\mathbb{Z}[\zeta]$ to use as factors when factorizing elements within $\mathbb{Z}[\zeta]$. These 'ideal factors' restore a version of unique factorization.

Subsequently the theory began to take on a different form from that in which Kummer had left it, but the key concept of an 'ideal'—a reformulation by Richard Dedekind of Kummer's 'ideal number'—gave the theory a major boost. By using his theory of ideal numbers, Kummer proved

Fermat's Last Theorem for a wide range of prime powers—the so-called 'regular' primes. He also evolved a powerful machine with applications to many other problems in mathematics. In fact a large part of classical number theory can be expressed in the framework of algebraic numbers. This point of view was urged most strongly by David Hilbert in his *Zahlbericht* (Number Report) of 1897, which had an enormous influence on the development of number theory, see Reid [63].

As a result, algebraic number theory today is a flourishing and important branch of mathematics, with deep methods and insights, and—most significantly—applications not only to number theory, but also to group theory, algebraic geometry, topology, and analysis. It was these wider links that eventually led to the final proof of Fermat's Last Theorem, establishing it once and for all as a theorem, not a conjecture. The eventual proof was made possible by various significant inroads, which were made using techniques from elliptic functions, modular forms, and Galois representations.

The breakthrough, as indicated above, was made by Andrew Wiles. As a teenager, fascinated by the simplicity of the statement of the theorem, Wiles had begun a long and mostly solitary journey in search of a proof. The event that triggered his final push was a conjecture put forward by two Japanese mathematicians, Yutaka Taniyama and Goro Shimura, who hypothesized a link between elliptic curves and modular forms. Their ideas were later refined by André Weil. This proposal became known as the Taniyama–Shimura–Weil Conjecture, and it was discovered that if this conjecture could be proved, then Fermat's Last Theorem could be deduced from it. At this point, Wiles leaped into action. He worked in solitude for seven years before he convinced himself that he had proved a special case of the Taniyama–Shimura–Weil Conjecture that was strong enough to imply Fermat's Last Theorem. He announced his result in a lecture in Cambridge on 23 June 1993.

When his proof was being checked, a query from a colleague revealed a gap, and Wiles accepted that some details required attention. It took him so long to do this that some questioned whether he had ever been close to the proof at all. However, in the autumn of 1994, working with his former student Richard Taylor, he finally realised that he could complete the proof satisfactorily. He released the proof for scrutiny in October 1994 and it was published in May 1995.

Fermat's Last Theorem probably has the distinction of being the theorem with the greatest number of false 'proofs', so the proof was scrutinized very carefully. However, this time the ideas fitted together so tightly that experts in the mathematical community agreed that all was well. In the ensuing period nothing has happened to change this opinion: Fermat's Last Theorem has at last been declared true. However, the proof uses techniques far beyond what would have been available to Fermat. So when he stated that he had found a proof that could not be fitted into the margin of his book, had he truly found a perceptive insight that has been missed by mathematicians for over 350 years? Or was it, as observed by the historian Dirk Struik [81], that 'even the great Fermat slept sometimes'?

I

Algebraic Methods

1

Algebraic Background

Fermat's Last Theorem is a special problem in the general theory of Diophantine equations—integer solutions of polynomial equations. To place the problem in context, we move to the wider realm of algebraic numbers, which arise as the real or complex solutions of polynomials with integer coefficients; we focus particularly on algebraic integers, which are solutions of polynomials with integer coefficients where the leading coefficient is 1. For example, the equation $x^2 - 2 = 0$ has no integer solutions, but it has two real solutions, $x = \pm\sqrt{2}$. The leading coefficient of the polynomial $x^2 - 2$ is 1, so $\pm\sqrt{2}$ are algebraic integers.

To operate with such numbers, it is useful to work in subsystems of the complex numbers that are closed under the usual operations of arithmetic. Such subsystems include subrings (which are closed under addition, subtraction and multiplication) and subfields (closed under all four arithmetic operations including division). Thus along with $\pm\sqrt{2}$ we consider the ring of all numbers $a + b\sqrt{2}$ for $a, b \in \mathbb{Z}$, and the field of all numbers $p + q\sqrt{2}$ for $p, q \in \mathbb{Q}$.

In this chapter we lay the foundations for algebraic number theory by considering some fundamental facts about rings, fields, and other algebraic structures, including abelian groups and modules, which are relevant to our theoretical development. We expect the reader to be acquainted with elementary properties of groups, rings and fields, and to have a basic knowledge of linear algebra over an arbitrary field, up to simple properties of determinants. Familiar results at this level will be stated without proof; results that we think may be less familiar to some readers are proved in full or in outline as appropriate. References are given for results not proved in full, in case the reader wishes to pursue them in greater depth. Useful general references on abstract algebra, with emphasis on rings and fields, are Fenrick [27], Fraleigh [28], Jacobson [44], Lang [47], Sharpe [74], and Stewart and Tall [79]. For group theory, see Burn [12], Humphreys [42], Macdonald [49], Neumann *et al.* [61], and Rotman [69].

First we set up the ring-theoretic language, in particular the notion of an ideal, which proves to be so important. Then we consider factorization of polynomials over a ring, which in this book is often a subring of the complex numbers. Topics of central importance at this stage are factorization of a polynomial over an extension field, and the theory of elementary symmetric polynomials. Module-theoretic language helps us to clarify certain points later. Results concerning finitely generated abelian groups are proved because they are vital in describing the additive group structure of the subrings of the complex numbers that occur.

1.1   Rings and Fields

Unless explicitly stated to the contrary, the term *ring* in this book will always mean a commutative ring R with identity element 1 (or 1_R). If such a ring has no zero-divisors (so that in R, $a \neq 0$, $b \neq 0$ implies $ab \neq 0$), and if $1 \neq 0$ in it, then it is called a *domain*. (Another common term is *integral domain*, but we omit 'integral' throughout.) An element a in a ring R is called a *unit* if there exists $b \in R$ such that $ab = 1$. Suppose $ab = ac = 1$. Then $c = 1c = abc = acb = 1b = b$. The unique b such that $ab = 1$ is denoted by a^{-1}, and ca^{-1} is also denoted by c/a. If $1 \neq 0$ in R and every non-zero element in R is a unit, then R is called a *field*.

We use standard notation $\mathbb{N}$ for the set of natural numbers $0, 1, 2, \ldots$, $\mathbb{Z}$ for the integers, $\mathbb{Q}$ for the rationals, $\mathbb{R}$ for the reals and $\mathbb{C}$ for the complex numbers. Under the usual operations $\mathbb{Q}, \mathbb{R}, \mathbb{C}$ are fields, $\mathbb{Z}$ is a domain, and $\mathbb{N}$ is not even a ring. For $n \in \mathbb{N}$, $n > 0$, we denote the ring of integers modulo n by $\mathbb{Z}_n$. If n is composite, then $\mathbb{Z}_n$ has zero divisors, but for n prime, then $\mathbb{Z}_n$ is a field; see Fraleigh [28] p. 217.

Our convention is that a subring S of a ring R is required to contain 1_R. We can check that S is a subring by demonstrating that $1_R \in S$, and if $s, t \in S$ then $s + t, -s, st \in S$. The subset S then forms a ring in its own right under the operations restricted from R. In the same way, if K is a field, then a subfield F of K is a subset that is a field under the operations restricted from K. We can check that F is a subfield of K by demonstrating that $1_k \in F$, and if $s, t \in F$ ($s \neq 0$) then $s + t, -s, st, s^{-1} \in F$.

The concept of an *ideal* is of central importance in this text. Recall that an ideal is a non-empty subset I of a ring R such that if $r, s \in I$, then

$r - s \in I$, and if $r \in R$, $s \in I$ then $rs \in I$. We also require the concept of the *quotient ring* R/I of R by an ideal I. The elements of R/I are cosets $I + r$ of the additive group of I in R, with addition and multiplication defined by

$$(I + r) + (I + s) = I + (r + s)$$
$$(I + r)(I + s) = I + rs$$

for all $r, s \in R$. For example, if $n\mathbb{Z}$ is the set of integer multiples of $n \in \mathbb{Z}$, then $\mathbb{Z}/n\mathbb{Z}$ is isomorphic to $\mathbb{Z}_n$.

A *homomorphism* $f : R_1 \to R_2$, where R_1 and R_2 are rings, is a function such that

$$f(1_{R_1}) = 1_{R_2}$$
$$f(r + s) = f(r) + f(s)$$
$$f(rs) = f(r)f(s)$$

for all $r, s \in R_1$. A *monomorphism* is an injective (1–1) homomorphism and an *isomorphism* is a bijective (1–1 and onto) homomorphism.

The *kernel* and *image* of a homomorphism f are defined in the usual way:

$$\ker f = \{r \in R_1 \mid f(r) = 0\}$$
$$\operatorname{im} f = \{f(r) \in R_2 \mid r \in R_1\}.$$

The kernel is an ideal of R_1; the image is a subring of R_2; and the *isomorphism theorem* states that there is an isomorphism from $R_1/\ker f$ to $\operatorname{im} f$. For details, see Fraleigh [28], Jacobson [44], or Sharpe [74].

If X and Y are subsets of a ring R we write $X + Y$ for the set of all elements $x + y$ ($x \in X, y \in Y$), and XY for the set of all finite sums $\Sigma x_i y_i$ ($x_i \in X, y_i \in Y$). When X and Y are both ideals, so are $X + Y$ and XY.

The sum $X + Y$ of two subsets can be generalized to an arbitrary collection $\{X_i\}_{i \in I}$ by defining $\Sigma_{i \in I} X_i$ to be the set of all finite sums $x_{i_1} + \ldots + x_{i_n}$ of elements $x_{i_j} \in X_{i_j}$.

We make the customary compression of notation with regard to $\{x\}$ and x, writing for example xY for $\{x\}Y$, $x + Y$ for $\{x\} + Y$, and 0 for $\{0\}$.

The ideal *generated* by a subset X of R is the smallest ideal of R containing X; we denote this by $\langle X \rangle$. If $X = \{x_1, \ldots, x_n\}$, we write $\langle X \rangle$ as $\langle x_1, \ldots, x_n \rangle$. (Some writers use (X) where we have written $\langle X \rangle$, but then the last-mentioned simplification of notation would reduce to the notation for an n-tuple $(x_1, \ldots, x_n)$, so $\langle X \rangle$ is to be preferred.)

A simple calculation shows that

$$\langle X \rangle = XR = \sum_{x \in X} xR.$$

The identity element 1_R is crucial in this equation. In a commutative ring without identity we would also have to add on to $\Sigma_{x \in X}\, xR$, and to XR, the additive group generated by X.

If there exists a finite subset $X = \{x_1, \ldots, x_n\}$ of R such that $I = \langle X \rangle$, then we say that I is *finitely generated* as an ideal of R. If $I = \langle x \rangle$ for an element $x \in R$ we say that I is the *principal ideal* generated by x.

Example 1.1. Let $R = \mathbb{Z}$, $X = \{4, 6\}$. Then $\langle 4, 6 \rangle$ is finitely generated. In fact $\langle 4, 6 \rangle$ contains $2 \cdot 4 - 6 = 2$ and it easily follows that $\langle 4, 6 \rangle = \langle 2 \rangle$, so this ideal is principal. More generally, every ideal of $\mathbb{Z}$ is of the form $\langle n \rangle$ for some $n \in N$, hence principal.

Example 1.2. Let R be the set $\mathbb{Q}$ under the usual operation of addition, but define a non-standard multiplication on R by setting $xy = 0$ for all $x, y \in R$. The ideal $\langle X \rangle$ for a subset $X \subseteq R$ is then equal to the abelian group generated by X under addition. Now R is an ideal of R, but is not finitely generated. To see this, suppose that R is generated as an abelian group by elements $p_1/q_1, \ldots, p_n/q_n$. Then the only primes dividing the denominators of elements of R will be those dividing $q_1, \ldots, q_n$, which is a contradiction.

If K is a field and R is a subring of K then R is a domain. Conversely, every domain D can be embedded in a field L; and there exists such an L consisting only of elements d/e where $d, e \in D$ and $e \neq 0$. Such an L, which is unique up to isomorphism, is called the *field of fractions* or *field of quotients* of D. See Fraleigh [28] Theorem 26.1 p. 239.

Theorem 1.3. *Every finite domain is a field.*

Proof: Let D be a finite domain. Since $1 \neq 0$, then D has at least 2 elements. For $0 \neq x \in D$ the elements xy, as y runs through D, are distinct; for if $xy = xz$ then $x(y - z) = 0$ and so $y = z$ since D has no zero-divisors. Hence, by counting, the set of all elements xy is D. Thus $1 = xy$ for some $y \in D$, so D is a field. $\square$

Every field has a unique minimal subfield, the *prime subfield*, and this is isomorphic either to $\mathbb{Q}$ or to $\mathbb{Z}_p$ where p is a prime number: see Fraleigh [28] Theorem 29.7 p. 260. Correspondingly, we say that the *characteristic* of the field is 0 or p. In a field of characteristic p we have $px = 0$ for every element x, where as usual we write

$$px = (1 + 1 + \ldots + 1)x$$

where there are p summands 1; moreover, p is the smallest positive integer with this property. In a field of characteristic zero, if $nx = 0$ for some non-zero element x and integer n, then $n = 0$. Our major concern in the sequel will be subfields of $\mathbb{C}$ (the complex numbers), which of course have characteristic zero, but fields of prime characteristic arise naturally from time to time.

We use without further comment the fact that $\mathbb{C}$ is *algebraically closed*: given any polynomial p over $\mathbb{C}$ there exists $x \in \mathbb{C}$ such that $p(x) = 0$. For a proof of this see Stewart [78] p. 22–25. Different proofs using analysis or topological considerations are in Hardy [37] p. 492 and Titchmarsh [85] p. 118.

1.2 Factorization of Polynomials

Later we consider factorization in a more general context. Here we concentrate on factorizing polynomials. First, a few general remarks.

In a ring S, if we can write $a = bc$ for $a, b, c \in S$, then we say that b, c are *factors* of a. We also say that 'b divides a', and write

$$b \mid a.$$

For any unit $e \in S$ we can always write

$$a = e(e^{-1}a),$$

so, trivially, a unit is a factor of all elements in S. If $a = bc$ where neither b nor c is a unit, then b and c are called *proper factors* and a is said to be *reducible*. In particular $0 = 0 \cdot 0$ is reducible.

Note that if a is itself a unit and $a = bc$, we have

$$1 = aa^{-1} = bca^{-1},$$

so b and c are both units. A unit cannot have a proper factorization. We therefore concentrate on factorization of non-units. A non-unit $a \in S$ is said to be *irreducible* if it has no proper factors.

Now we turn our attention to the case $S = R[t]$, the ring of polynomials in an indeterminate t with coefficients in a ring R. The elements of $R[t]$ are expressions

$$r_n t^n + r_{n-1} t^{n-1} + \ldots + r_1 t + r_0$$

where $r_0, r_1, \ldots, r_n \in R$ and addition and multiplication are defined in the obvious way. (For a formal treatment of polynomials, and why not to use it, see Fraleigh [28] pp. 263–265.)

Given a non-zero polynomial

$$p = r_n t^n + \ldots + r_0,$$

define the *degree* of p to be the largest value of n for which $r_n \neq 0$, and write it ∂p. Polynomials of degree $0, 1, 2, 3, 4, 5, \ldots$, are often referred to as *constant, linear, quadratic, cubic, quartic, quintic,...*, polynomials respectively. In particular a constant polynomial is just a (non-zero) element of R.

If R is a domain, then

$$\partial pq = \partial p + \partial q$$

for non-zero p, q so $R[t]$ is also a domain. If $p = aq$ in $R[t]$, then $\partial p = \partial a + \partial q$ implies that

$$\partial q \leq \partial p.$$

When R is not a field, then it is perfectly possible to have a non-trivial factorization in which $\partial p = \partial q$. For example

$$3t^2 + 6 = 3(t^2 + 2)$$

in $\mathbb{Z}[t]$, where neither 3 nor $t^2 + 2$ is a unit. This is because of the existence of non-units in R. However, if R is a field, then all (non-zero) constants in $R[t]$ are units and so if q is a proper factor of p for polynomials over a field, then $\partial q < \partial p$.

Concentrate first on polynomials over a field K. Here we have the *division algorithm* which states that if $p, q \neq 0$ then

$$p = qs + r$$

where either $r = 0$ or $\partial r < \partial q$. The proof is by induction on ∂p and in practice is no more than long division of p by q leaving remainder r, which is either zero (in which case $q \mid p$) or of degree lower than q.

The division algorithm is used repeatedly in the *Euclidean algorithm*, which is a particularly efficient method for finding the *highest common factor* d of non-zero polynomials p, q. This is defined by the properties:

(a) $d \mid p, d \mid q$,

(b) If $d' \mid p$ and $d' \mid q$ then $d' \mid d$.

These define d uniquely up to non-zero constant multiples. To calculate d we first suppose that p, q are named so that $\partial p \geq \partial q$; then divide q into p to get

$$p = qs_1 + r_1 \qquad \partial r_1 < \partial q \leq \partial p,$$

and continue in the following way:

$$q = r_1 s_2 + r_2 \qquad \partial r_2 < \partial r_1$$
$$r_1 = r_2 s_3 + r_3 \qquad \partial r_3 < \partial r_2$$
$$\vdots$$
$$r_{n-2} = r_{n-1} s_n + r_n \quad \partial r_n < \partial r_{n-1}$$

until we arrive at a zero remainder:

$$r_{n-1} = r_n s_{n+1}.$$

The last non-zero remainder r_n is the highest common factor. (From the last equation $r_n \mid r_{n-1}$, and working back successively, r_n is a factor of $r_{n-2}, \ldots, r_1, p, q$, verifying (a). If $d' \mid p, d' \mid q$, then from the first equation, d' is a factor of $r_1 = p - q s_1$, and successively working down the equations, d' is a factor of $r_2, r_3, \ldots, r_n$, so $d' \mid r_n$, verifying (b).) Beginning with the first equation, and substituting in those which follow, we find that $r_i = a_i p + b_i q$ for suitable $a_i, b_i \in K[t]$, and in particular the highest common factor $d = r_n$ is of the form

$$d = ap + bq \quad \text{for suitable } a, b \in K[t]. \tag{1.1}$$

A useful special case is when $d = 1$, when p, q are called *coprime* and (1.1) gives

$$ap + bq = 1 \quad \text{for suitable } a, b \in K[t].$$

This technique for calculating the highest common factor can also be used to find the polynomials a, b.

Example 1.4. $p = t^3 + 1, q = t^2 + 1 \in \mathbb{Q}[t]$.
  Then

$$t^3 + 1 = t(t^2 + 1) + (-t + 1),$$
$$t^2 + 1 = (-t - 1)(-t + 1) + 2,$$
$$-t + 1 = (-\tfrac{1}{2}t + \tfrac{1}{2})2.$$

The highest common factor is 2, or up to a constant factor, 1, so p and q are coprime, and substituting back from the second equation,

$$1 = \tfrac{1}{2}(t^2 + 1) + \tfrac{1}{2}(t + 1)(-t + 1).$$

Then substituting for $-t + 1$ using the first equation:

$$1 = \tfrac{1}{2}(t^2 + 1) + \tfrac{1}{2}(t + 1)((t^3 + 1) - t(t^2 + 1))$$
$$= (-\tfrac{1}{2}t^2 - \tfrac{1}{2}t + \tfrac{1}{2})(t^2 + 1) + (\tfrac{1}{2}t + \tfrac{1}{2})(t^3 + 1).$$

Factorizing a single polynomial p is by no means as straightforward as finding the highest common factor of two. It is known that every non-zero polynomial over a field K is a product of finitely many irreducible factors, and these are unique up to the order in which they are multiplied and up to constant factors: see Fraleigh [28] Theorem 31.8 p.284; Stewart [78] Theorem 3.12 p. 37 and Theorem 3.16 p. 38. Finding these factors is very much an *ad hoc* matter. Linear factors are easiest, since $(x - \alpha) \mid p$ if and only if $p(\alpha) = 0$.

If $p(\alpha) = 0$, then α is called a *zero* of p. If $(t - \alpha)^m \mid p$ where $m \geq 2$, then α is a *repeated zero* and the largest such m is the *multiplicity* of α.

To detect repeated zeros we use a method which, like much in this chapter, was more familiar at the turn of the century than it is now. Given a polynomial

$$f = \sum_{i=0}^{n} r_i t^i$$

over a ring R we define

$$Df = \sum_{i=0}^{n} i r_i t^{i-1},$$

called for obvious reasons the *formal derivative* of f. It is not hard to check directly that

$$
\begin{aligned}
D(f + g) &= Df + Dg \\
D(fg) &= (Df)g + f(Dg).
\end{aligned}
$$

This enables us to check for repeated factors. A factor q of a polynomial p is *repeated* if $q^r \mid p$ for some $r \geq 2$. In particular q is repeated if its square divides p.

Theorem 1.5. *Let K be a field of characteristic zero. A non-zero polynomial f over K is divisible by the square of a polynomial of degree > 0 if and only if f and Df have a common factor of degree > 0.*

Proof: First suppose $f = g^2 h$. Then

$$Df = g^2 Dh + 2g(Dg)h$$

and so f and Df have g as a common factor.

Now suppose that f has no squared irreducible factor. For any irreducible factor g of f we have

$$f = gh$$

where g and h are coprime, otherwise g would be a factor of h and would occur as a squared factor in gh. If f and Df have a common factor g, which we take to be irreducible, then on differentiating formally we obtain

$$Df = (Dg)h + g(Dh)$$

So g is a factor of $(Dg)h$, hence of Dg because g and h are coprime. But Dg is of lower degree than g, hence it can only have g as a factor if $Dg = 0$. Since K has characteristic zero, by direct computation, this implies g is constant, so f and g can have no non-trivial common factor. $\square$

If the field has characteristic $p > 0$, then the first part of Theorem 1.5, that f having a squared factor implies f and Df have a common factor, is still true, and the proof is the same as above.

A result we need later is:

Corollary 1.6. *An irreducible polynomial over a subfield K of $\mathbb{C}$ has no repeated zeros in $\mathbb{C}$.*

Proof: Suppose f is irreducible over K. Then f and Df must be coprime, because a common factor would be a squared factor of f by (1.5), but f is irreducible. Thus there exist polynomials a, b over K such that $af + bDf = 1$, and the same equation interpreted over $\mathbb{C}$ shows f and Df to be coprime over $\mathbb{C}$. By Theorem 1.5, f has no repeated zeros. $\square$

We often consider factorization of polynomials over $\mathbb{Q}$. When such a polynomial has integer coefficients it turns out that we need consider only factors with integer coefficients. This fact is enshrined in a result of Gauss:

Lemma 1.7. (Gauss's Lemma.) *Let $p \in \mathbb{Z}[t]$ and suppose that $p = gh$ where $g, h \in \mathbb{Q}[t]$. Then there exists $\lambda \in \mathbb{Q}, \lambda \neq 0$, such that $\lambda g, \lambda^{-1}h \in \mathbb{Z}[t]$.*

Proof: Multiplying by the product of the denominators of the coefficients of g, h we rewrite $p = gh$ as

$$np = g'h'$$

where g', h' are rational multiples of g, h respectively, $n \in \mathbb{Z}$ and $g', h' \in \mathbb{Z}[t]$. Therefore n divides the coefficients of the product $g'h'$. We now divide the equation successively by the prime factors of n. We shall establish that if k is a prime factor of n, then k divides all the coefficients of g' or all those of h'. Whichever it is, we can divide that particular polynomial by k

to give another polynomial with integer coefficients. After dividing in this way by all the prime factors of n, we are left with

$$p = \overline{g}\overline{h}$$

where $\overline{g}, \overline{h} \in \mathbb{Z}[t]$ are rational multiples of g, h respectively. Putting $\overline{g} = \lambda g$ for $\lambda \in \mathbb{Q}$, we obtain $\overline{h} = \lambda^{-1} h$ and the result will follow.

It remains to prove that if

$$\begin{aligned} g' &= g_0 + g_1 + \ldots + g_r t^r \\ h' &= h_0 + h_1 + \ldots + h_s t^s \end{aligned}$$

and a prime k divides all the coefficients of $g'h'$, then k must divide all the g_i or all the h_j. But if a prime k does not divide all the g_i and all the h_j, we can choose the *first* of each set of coefficients, say g_m, h_q, that are not divisible by k. Then the coefficient of t^{m+q} in $g'h'$ is

$$g_0 h_{m+q} + g_1 h_{m+q-1} + \ldots + g_m h_g + \ldots g_{m+q} h_0.$$

Since every term in this expression is divisible by k except $h_q g_m$, the whole coefficient is not divisible by k, a contradiction.   □

We need methods to prove irreducibility for various specific polynomials over $\mathbb{Z}$. The first of these is *Eisenstein's criterion*, named after Gotthold Eisenstein:

Theorem 1.8. (Eisenstein's Criterion.) *Let*

$$f = a_0 + a_1 t + \ldots + a_n t^n$$

be a polynomial over $\mathbb{Z}$. Suppose there is a prime q such that

(a) $q \nmid a_n$,
(b) $q \mid a_i$, $(i = 0, 1, \ldots, n-1)$,
(c) $q^2 \nmid a_0$.

Then, apart from constant factors, f is irreducible over $\mathbb{Z}$, and hence irreducible over $\mathbb{Q}$.

Proof: By Lemma 1.7 it is enough to show that f can have only constant factors over $\mathbb{Z}$.

If not, then $f = gh$ where

$$\begin{aligned} g &= g_0 + g_1 t + \ldots + g_r t^r \\ h &= h_0 + h_1 t + \ldots + h_s t^s \end{aligned}$$

with all $g_i, h_j \in \mathbb{Z}$ and $r, s > 1, r + s = n$.

Now $g_0 h_0 = a_0$, so (b) implies q divides one of g_0, h_0 while (c) implies that it cannot divide both. Without loss of generality, suppose q divides g_0 but not h_0. Not all g_i are divisible by q because this would imply that q divides a_n, contrary to (a). Let g_m be the first coefficient of g not divisible by q. Then

$$a_m = g_0 h_m + \ldots + g_m h_0$$

where $m \leq r < n$. All the summands on the right are divisible by q except the last, which means that a_m is not divisible by q, contradicting (b). □

A second useful method is *reduction modulo n*, as follows. Suppose $0 \neq p \in \mathbb{Z}[t]$, with p reducible: say $p = qr$. The natural homomorphism $\mathbb{Z} \to \mathbb{Z}_n$ gives rise to a homomorphism $\mathbb{Z}[t] \to \mathbb{Z}_n[t]$. Using bars to denote images under this map, we have $\bar{p} = \bar{q}\bar{r}$. If $\partial \bar{p} = \partial p$, then clearly $\partial \bar{q} = \partial q$, $\partial \bar{r} = \partial r$, and $\bar{p}$ is also reducible. This proves:

Theorem 1.9. *If $p \in \mathbb{Z}[t]$ and its image $\bar{p} \in \mathbb{Z}_n[t]$ is irreducible, with $\partial \bar{p} = \partial p$, then p is irreducible as an element of $\mathbb{Z}[t]$.* □

In practice we take n to be prime, though this is not essential. The point of reducing modulo n is that $\mathbb{Z}_n$ being finite, there are only a finite number of possible factors of $\bar{p}$ to be considered.

Examples.

(1) The polynomial $t^2 - 2$ satisfies Eisenstein's criterion with $q = 2$.

(2) The polynomial $t^{11} - 7t^6 + 21t^5 + 49t - 56$ satisfies Eisenstein's criterion with $q = 7$.

(3) The polynomial $t^5 - t + 1$ does not satisfy Eisenstein's criterion for any q. Instead we try reduction modulo 5. There is no linear factor since none of 0, 1, 2, 3, 4 yield 0 when substituted for t, so the only possible way to factorize is

$$t^5 - t + 1 = (t^2 + \alpha t + \beta)(t^3 + \gamma t^2 + \delta t + \epsilon)$$

where $\alpha, \beta, \gamma, \delta, \epsilon$ take values 0, 1, 2, 3 or 4 (mod 5). This gives a system of equations on comparing coefficients: there are only a finite number of possibilities all of which are easily eliminated. Hence the polynomial is irreducible mod 5, so irreducible over $\mathbb{Z}$.

1.3   Field Extensions

When finding the zeros of a polynomial p over a field K, it is often necessary
to pass to a larger field L that contains K. In these circumstances, L is
called a *field extension* of K. For example, $p(t) = t^2 + 1$ has no zeros in $\mathbb{R}$,
but considering p as a polynomial over $\mathbb{C}$, it has zeros $\pm i$ and a factorization

$$p(t) = (t + i)(t - i).$$

Field extensions often arise in a slightly more general context as a
monomorphism $j : K \to L$ where K and L are fields. It is customary
in these cases to identify K with its image $j(K)$, which is a subfield of L;
then a field extension is a pair of fields (K, L) where K is a subfield of L.
We talk of the extension

$$L : K$$

of K. Most field extensions we encounter involve two subfields of $\mathbb{C}$.

If $L : K$ is a field extension, then L has a natural structure as a vector
space over K (where vector addition is addition in L and scalar multiplication of $\lambda \in K$ on $v \in L$ is just $\lambda v \in L$). The dimension of this vector space
is called the *degree* of the extension, or the *degree of L over K*, written

$$[L : K]$$

The degree has an important multiplicative property:

Theorem 1.10. *If $H \subseteq K \subseteq L$ are fields, then*

$$[L : H] = [L : K][K : H].$$

Proof: We sketch this. Details are in Stewart [78], Theorem 6.4 p. 68. Let
$\{a_i\}_{i \in I}$ be a basis for L over K, and $\{b_j\}_{j \in J}$ a basis for K over H. Then
$\{a_i b_j\}_{(i,j) \in I \times J}$ is a basis for L, over H.   □

If $[L : K]$ is finite we say that L is a *finite extension* of K.

Given a field extension $L{:}K$ and an element $\alpha \in L$, there may or may
not exist a polynomial $p \in K[t]$ such that $p(\alpha) = 0$, $p \neq 0$. If not, we
say that α is *transcendental* over K. If such a p exists, we say that α
is *algebraic* over K. If α is algebraic over K, then there exists a unique
monic polynomial q of minimal degree subject to $q(\alpha) = 0$, and q is called
the *minimum polynomial* of α over K. (A *monic* polynomial is one with
highest coefficient 1.) It is easy to prove that the minimum polynomial of
α is irreducible over K, see Stewart [78] p. 58–62.

If $\alpha_1, \ldots \alpha_n \in L$, we write

$$K(\alpha_1, \ldots, \alpha_n)$$

for the smallest subfield of L containing K and the elements $\alpha_1, \ldots, \alpha_n$.

In an analogous way, if S is a subring of a ring R and $\alpha_1, \ldots, \alpha_n \in R$, we write

$$S[\alpha_1, \ldots, \alpha_n]$$

for the smallest subring of R containing S and the elements $\alpha_1, \ldots, \alpha_n$. Clearly $S[\alpha_1, \ldots, \alpha_n]$ consists of all polynomials in $\alpha_1, \ldots, \alpha_n$ with coefficients in S. For instance $S[\alpha]$ consists of all polynomials

$$s_0 + s_1\alpha + \ldots + s_m\alpha^m \qquad (s_i \in S).$$

The structure of the field $K(\alpha)$ depends on α in an interesting way. If α is transcendental over K, then for $k_m \neq 0$ we have

$$k_0 + k_1\alpha + \ldots + k_m\alpha^m \neq 0 \qquad (k_i \in K).$$

In this case $K(\alpha)$ must include all rational expressions

$$\frac{s_0 + s_1\alpha + \ldots + s_n\alpha^n}{k_0 + k_1\alpha + \ldots + k_m\alpha^m} \qquad (s_j, k_i \in K, k_m \neq 0)$$

and clearly consists precisely of these elements.

However, for α algebraic, we have:

Theorem 1.11. *If $L : K$ is a field extension and $\alpha \in L$, then α is algebraic over K if and only if $K(\alpha)$ is a finite extension of K. In this case, $[K(\alpha){:}K] = \partial p$ where p is the minimum polynomial of α over K, and $K(\alpha) = K[\alpha]$.*

Proof: Once more we sketch the proof, given in full in Stewart [78] Theorem 5.12 p. 62. If $[K(\alpha) : K] = n < \infty$ then the powers, $1, \alpha, \alpha^2, \ldots, \alpha^n$ are linearly dependent over K, whence α is algebraic. Conversely, suppose α algebraic with minimum polynomial p of degree m. We claim that $K(\alpha)$ is the vector space over K spanned by $1, \alpha, \ldots, \alpha^{m-1}$. This space, call it V, is certainly closed under addition, subtraction, and multiplication by α; for the last statement note that $\alpha^m = -p(\alpha) + \alpha^m = q(\alpha)$ where $\partial q < m$. Hence V is closed under multiplication, and so forms a ring. All we need prove now is that if $0 \neq v \in V$ then $1/v \in V$. Now $v = h(\alpha)$ where $h \in K[t]$ and $\partial h < m$. Since p is irreducible, p and h are coprime, so there exist $f, g \in K[t]$ such that

$$f(t)p(t) + g(t)h(t) = 1.$$

Then
$$1 = f(\alpha)p(\alpha) + g(\alpha)h(\alpha) = g(\alpha)h(\alpha)$$

so that $1/v = g(\alpha) \in V$ as required. But it follows at once that $[K(\alpha) : K] = \dim_K V = m$. $\qquad\square$

If we specify in advance K and an irreducible monic polynomial $p(t) \in K[t]$ then there exists up to isomorphism a unique extension field L such that L contains an element α with minimum polynomial p, and $L = K(\alpha)$. This can be constructed as $K[t]/\langle p \rangle$. It is customary to express this construction by the phrase 'adjoin to K an element α with $p(\alpha) = 0$' and to write $K(\alpha)$ for the resulting field. This, and much else, is discussed in Stewart [78] Chapter 17, p. 177–190.

1.4   Symmetric Polynomials

Let $R[t_1, t_2, \ldots, t_n]$ denote the ring of polynomials in indeterminates $t_1, t_2, \ldots, t_n$ with coefficients in a ring R. Let S_n denote the symmetric group of permutations on $\{1, 2, \ldots, n\}$. For any permutation $\pi \in S_n$ and any polynomial $f \in R[t_1, \ldots, t_n]$ define the polynomial f^π by

$$f^\pi(t_1, \ldots, t_n) = f(t_{\pi(1)}, \ldots, t_{\pi(n)}).$$

For example if $f = t_1 + t_2 t_3$ and π is the cycle (123) then $f^\pi = t_2 + t_3 t_1$. The polynomial f is *symmetric* if $f^\pi = f$ for all $\pi \in S_n$. For example $t_1 + \ldots + t_n$ is symmetric. More generally we have the *elementary symmetric polynomials*

$$s_r(t_1, \ldots, t_n) \qquad (1 \leq r \leq n)$$

defined to be the sum of all possible distinct products of r distinct t_i's. Thus

$$
\begin{aligned}
s_1(t_1, \ldots, t_n) &= t_1 + t_2 + \ldots + t_n, \\
s_2(t_1, \ldots, t_n) &= t_1 t_2 + t_1 t_3 + \ldots + t_2 t_3 + \ldots + t_{n-1} t_n, \\
&\;\;\vdots \\
s_n(t_1, \ldots, t_n) &= t_1 t_2 \ldots t_n.
\end{aligned}
$$

These arise in the following circumstances. Consider a polynomial of degree n over a subfield K of $\mathbb{C}$:

$$f = a_n t^n + \ldots + a_0.$$

Resolve it into linear factors over $\mathbb{C}$:

$$f = a_n(t - \alpha_1) \ldots (t - \alpha_n).$$

Expand the product:

$$f = a_n(t^n - s_1 t^{n-1} + \ldots + (-1)^n s_n),$$

where s_i denotes $s_i(\alpha_1, \ldots, \alpha_n)$.

A polynomial in $s_1, \ldots, s_n$ can clearly be rewritten as a symmetric polynomial in $t_1, \ldots, t_n$. The converse is also true, a fact first proved by Isaac Newton:

Theorem 1.12. *Let R be a ring. Then every symmetric polynomial in $R[t_1, \ldots, t_n]$ is expressible as a polynomial with coefficients in R in the elementary symmetric polynomials $s_1, \ldots, s_n$.*

Proof: We demonstrate a specific technique for reducing a symmetric polynomial into elementary ones. First we order the monomials $t_1^{\alpha_1} \ldots t_n^{\alpha_n}$ by a 'lexicographic' order in which $t_1^{\alpha_1} \ldots t_n^{\alpha_n}$ precedes $t_1^{\beta_1} \ldots t_n^{\beta_n}$ if the first nonzero $\alpha_i - \beta_i$ is positive. Given a polynomial $p \in R[t_1, \ldots, t_n]$ we order its terms lexicographically. If p is symmetric, then for every monomial $a t_1^{\alpha_1} \ldots t_n^{\alpha_n}$ occurring in p, there occurs a similar monomial with the exponents permuted. Let α_1 be the highest exponent occurring in monomials of p: then there is a term containing $t_1^{\alpha_1}$. The leading term of p in lexicographic ordering contains $t_1^{\alpha_1}$, and among all such monomials we select the one with the highest occurring power of t_2 and so on. In particular, the leading term of a symmetric polynomial is of the form $a t_1^{\alpha_1} \ldots t_n^{\alpha_n}$ where $\alpha_1 \geq \ldots \geq \alpha_n$. For example, the leading term of

$$s_1^{k_1} \ldots s_n^{k_n} = (t_1 + \ldots + t_n)^{k_1} \ldots (t_1 \ldots t_n)^{k_n}$$

is

$$t_1^{k_1 + \ldots + k_n} t_2^{k_2 + \ldots + k_n} \ldots t_n^{k_n}.$$

By choosing $k_1 = \alpha_1 - \alpha_2, \ldots, k_{n-1} = \alpha_{n-1} - \alpha_n$, $k_n = \alpha_n$ (which is possible because $\alpha_1 \geq \ldots \geq \alpha_n$) we can make this the same as the leading term of p. Then

$$p - a s_1^{\alpha_1 - \alpha_2} \ldots s_{n-1}^{\alpha_{n-1} - \alpha_n} s_n^{\alpha_n}$$

has a lexicographic leading term

$$b t_1^{\beta_1} \ldots t_n^{\beta_n} \qquad (\beta_1 \geq \ldots \geq \beta_n)$$

which comes after $at_1^{\alpha_1} \dots t_n^{\alpha_n}$ in the ordering. But only a finite number of monomials $t_1^{\gamma_1} \dots t_n^{\gamma_n}$ satisfying $\gamma_1 \geq \dots \geq \gamma_n$ follow $t_1^{\alpha_1} \dots t_n^{\alpha_n}$ lexicographically, and so a finite number of repetitions of the given process reduce p to a polynomial in $s_1, \dots, s_n$. $\qquad\square$

Example 1.13. The symmetric polynomial

$$p = t_1^2 t_2 + t_1^2 t_3 + t_1 t_2^2 + t_1 t_3^2 + t_2^2 t_3 + t_2 t_3^2$$

is written lexicographically. Here $n = 3$, $\alpha_1 = 2$, $\alpha_2 = 1$, $\alpha_3 = 0$ and the method tells us to consider

$$p - s_1 s_2.$$

This simplifies to give

$$p - s_1 s_2 = 3 t_1 t_2 t_3.$$

The polynomial $3 t_1 t_2 t_3$ is visibly $3 s_3$, but the method, using $\alpha_1 = \alpha_2 = \alpha_3 = 1$, also leads us to this conclusion.

This result about symmetric functions proves to be extremely useful in the following instance:

Corollary 1.14. *Suppose that L is an extension of the field K, $p \in K[t]$, $\partial p = n$ and the zeros of p are $\theta_1, \dots, \theta_n \in L$. If $h(t_1, \dots, t_n) \in K[t_1, \dots, t_n]$ is symmetric, then $h(\theta_1, \dots, \theta_n) \in K$.*

1.5 Modules

Let R be a ring. By an R-*module* we mean an abelian group M (written additively), together with a function $\alpha : R \times M \to M$, for which we write $\alpha(r, m) = rm$ $(r \in R, m \in M)$, satisfying

(a) $\quad (r + s)m \quad = \quad rm + sm,$
(b) $\quad r(m + n) \quad = \quad rm + rn,$
(c) $\quad\quad r(sm) \quad = \quad (rs)m,$
(d) $\quad\quad\quad 1m \quad = \quad m$

for all $r, s \in R$, $m, n \in M$.

Although (d) is always obligatory in this text, be warned that in other parts of mathematics it may not be required to be so.

The function α is called an R-*action* on M.

If R is a field K, then an R-module is the same thing as a vector space over K. In this sense we can think of an R-module as a generalization of a vector space, but because division need not be possible in R, many of the results and techniques of vector space theory do not carry over unchanged to R-modules. The basic theory of modules may be found in Fraleigh [28] section 37.2, p. 338. In particular we define an *R-submodule* of M to be a subgroup N of M (under addition) such that if $n \in N$, $r \in R$, then $rn \in N$. We may then define the *quotient module* M/N to be the corresponding quotient group, with R-action

$$r(N + m) = N + rm \qquad (r \in R, m \in M).$$

If $X \subseteq M, Y \subseteq R$, we define YX to be the set of all finite sums $\sum_i y_i x_i$ where $y_i \in Y, x_i \in X$.

The submodule of M *generated by* X, which we write

$$\langle X \rangle_R,$$

is the smallest submodule containing X. This is equal to RX. If $N = \langle x_1, \ldots, x_n \rangle_R$ then we say that N is a *finitely generated R-module*.

A $\mathbb{Z}$-module is nothing more than an abelian group M (written additively), and conversely, given an additive abelian group M we can make it into a $\mathbb{Z}$-module by defining

$$0m = 0, \quad 1m = m \quad (m \in M)$$

then inductively

$$(n + 1)m = nm + m \quad (n \in \mathbb{Z}, n > 0)$$

and

$$(-n)m = -nm \quad (n \in \mathbb{Z}, n > 0).$$

We discuss this case further in the next section.

More generally there are several natural ways in which R-modules can arise, of which we distinguish three:

(1) Suppose that R is a subring of a ring S. Then S is an R-module with action

$$\alpha(r, s) = rs \quad (r \in R, s \in S)$$

where the product is just that of elements in S.

(2) Suppose that I is an ideal of the ring R. Then I is an R-module under

$$\alpha(r, i) = ri \quad (r \in R, i \in I)$$

where the product is that in R.

(3) Suppose that $J \subseteq I$ is another ideal. Then J is also an R-module. The quotient module I/J has the action

$$r(J + i) = J + ri \quad (r \in R, i \in I).$$

1.6   Free Abelian Groups

The study of algebraic numbers involves subfields and subrings of $\mathbb{C}$. A typical instance is the subring

$$\mathbb{Z}[i] = \{a + bi \in \mathbb{C} \mid a, b \in \mathbb{Z}\}.$$

The additive group of $\mathbb{Z}[i]$ is isomorphic to $\mathbb{Z} \times \mathbb{Z}$. More generally the additive groups of those subrings of $\mathbb{C}$ that we study are usually isomorphic to the direct product of a *finite* number of copies of $\mathbb{Z}$. In this section we study such abelian groups, for later use.

Let G be an abelian group. In this section we use additive notation for G, so the group operation is denoted by $+$, the identity by 0, the inverse of g by $-g$ and powers of g by $2g, 3g, \ldots$. In later chapters we encounter cases where multiplicative notation is more appropriate.

If G is finitely generated as a $\mathbb{Z}$-module, so there exist $g_1, \ldots, g_n \in G$ such that every $g \in G$ is a sum

$$g = m_1 g_1 + \ldots + m_n g_n \qquad (m_i \in \mathbb{Z})$$

then G is a *finitely generated abelian group.*

Generalizing the notion of linear independence in a vector space, we say that elements $g_1, \ldots, g_n$ in an abelian group G are *linearly independent* (over $\mathbb{Z}$) if any equation

$$m_1 g_1 + \ldots + m_n g_n = 0$$

with $m_1, \ldots, m_n \in \mathbb{Z}$ implies $m_1 = \ldots = m_n = 0$. A linearly independent set that generates G is a *basis* ($\mathbb{Z}$-basis for emphasis). If $\{g_1, \ldots, g_n\}$ is a basis, every $g \in G$ has a unique representation in the form

$$g = m_1 g_1 + \ldots + m_n g_n \qquad (m_i \in \mathbb{Z})$$

because an alternative expression

$$g = k_1 g_1 + \ldots + k_n g_n \qquad (k_i \in \mathbb{Z})$$

implies

$$(m_1 - k_1)g_1 + \ldots + (m_n - k_n)g_n = 0$$

and linear independence implies that $m_i - k_i = 0$, that is, $m_i = k_i (1 \leq i \leq n)$.

If $\mathbb{Z}^n$ denotes the direct product of n copies of the additive group of integers, it follows that a group with a basis of n elements is isomorphic to $\mathbb{Z}^n$.

To show that two different bases of G have the same number of elements, let $2G$ be the subgroup of G consisting of all elements of the form $g + g$ ($g \in G$). If G has a basis of n elements, then $G/2G$ is a group of order 2^n. Since the definition of $2G$ does not depend on any particular basis, every basis must have the same number of elements.

An abelian group with a basis of n elements is called a *free abelian group* of *rank n*. If G is free abelian of rank n and $\{x_1, \ldots, x_n\}$, $\{y_1, \ldots, y_n\}$ are both bases, then there exist integers a_{ij}, b_{ij} such that

$$y_i = \sum_j a_{ij} x_j \qquad x_i = \sum_j b_{ij} y_j.$$

If we consider the matrices

$$A = (a_{ij}) \qquad B = (b_{ij})$$

then $AB = I_n$, the identity matrix. Hence

$$\det(A)\det(B) = 1$$

and since $\det(A)$ and $\det(B)$ are integers,

$$\det(A) = \det(B) = \pm 1.$$

A square matrix over $\mathbb{Z}$ with determinant ± 1 is *unimodular*. We have:

Lemma 1.15. *Let G be a free abelian group of rank n with basis $\{x_1, \ldots, x_n\}$. Suppose (a_{ij}) is an $n \times n$ matrix with integer entries. Then the elements*

$$y_i = \sum_j a_{ij} x_j$$

form a basis of G if and only if (a_{ij}) is unimodular.

Proof: The 'only if' part has already been dealt with. Now suppose $A = (a_{ij})$ is unimodular. Since $\det(A) \neq 0$ the y_j are linearly independent. We have

$$A^{-1} = (\det(A))^{-1} \tilde{A}$$

where $\tilde{A}$ is the adjoint matrix and has integer entries. Hence $A^{-1} = \pm \tilde{A}$ has integer entries. Putting $B = A^{-1} = (b_{ij})$ we obtain $x_i = \sum_j b_{ij} y_j$, demonstrating that the y_j generate G. Thus they form a basis. $\square$

The central result in the theory of finitely generated free abelian groups concerns the structure of subgroups:

Theorem 1.16. *Every subgroup H of a free abelian group G of rank n is free of rank $s \leq n$. Moreover there exists a basis $u_1, \ldots, u_n$ for G and positive integers $\alpha_1, \ldots, \alpha_s$ such that $\alpha_1 u_1, \ldots, \alpha_s u_s$ is a basis for H.*

Proof: We use induction on the rank n of G. For $n = 1$, G is infinite cyclic and the result is a consequence of the subgroup structure of the cyclic group. If G has rank n, pick any basis $w_1, \ldots, w_n$ of G. Every $h \in H$ is of the form

$$h = h_1 w_1 + \ldots + h_n w_n.$$

Either $H = \{0\}$, in which case the theorem is trivial, or there exist non-zero coefficients h_i for some $h \in H$. From all such coefficients, let $\lambda(w_1, \ldots, w_n)$ be the least positive integer occurring. Now choose the basis $w_1, \ldots, w_n$, to make $\lambda(w_1, \ldots, w_n)$ minimal. Let α_1 be this minimal value, and number the w_i in such a way that

$$v_1 = \alpha_1 w_1 + \beta_2 w_2 + \ldots + \beta_n w_n$$

is an element of H in which α_1 occurs as a coefficient. Let

$$\beta_i = \alpha_1 q_i + r_i \qquad (2 \leq i \leq n)$$

where $0 \leq r_i < \alpha_1$, so that r_i is the remainder on dividing β_i by α_1. Define

$$u_1 = w_1 + q_2 w_2 + \ldots + q_n w_n.$$

Then it is easy to verify that $u_1, w_2, \ldots, w_n$ is another basis for G. (The appropriate matrix is clearly unimodular.) With respect to the new basis,

$$v_1 = \alpha_1 u_1 + r_2 w_2 + \ldots + r_n w_n.$$

By the minimality of $\alpha_1 = \lambda(w_1, \ldots, w_n)$ for *all* bases we have

$$r_2 = \ldots = r_n = 0.$$

Hence

$$v_1 = \alpha_1 u_1.$$

With respect to the new basis, let

$$H' = \{m_1 u_1 + m_2 w_2 + \ldots + m_n w_n \mid m_1 = 0\}.$$

Clearly $H' \cap V_1 = \{0\}$, where V_1 is the subgroup generated by v_1. We claim that $H = H' + V_1$. For if $h \in H$ then

$$h = \gamma_1 u_1 + \gamma_2 w_2 + \ldots + \gamma_n w_n$$

and putting

$$\gamma_1 = \alpha_1 q + r_1 \qquad (0 \le r_1 < \alpha_1)$$

it follows that H contains

$$h - qv_1 = r_1 u_1 + \gamma_2 w_2 + \ldots + \gamma_n w_n$$

and the minimality of α_1 once more implies that $r_1 = 0$. Hence $h - qv_1 \in H'$. It follows that H is isomorphic to $H' \times V_1$ and H' is a subgroup of the group G' which is free abelian of rank $n - 1$ with generators $w_2, \ldots, w_n$. By induction, H' is free of rank $\le n - 1$, and there exist bases $u_2, \ldots, u_n$ of G' and $v_2, \ldots, v_s$ of H' such that $v_i = \alpha_i u_i$ for positive integers α_i. $\square$

From the above two results we deduce a useful theorem about orders of quotient groups. In its statement we use $|X|$ to denote the cardinality of the set X, and $|x|$ to denote the absolute value of the real number x. No confusion need arise.

Theorem 1.17. *Let G be a free abelian group of rank r, and H a subgroup of G. Then G/H is finite if and only if the ranks of G and H are equal. If this is the case, and if G and H have $\mathbb{Z}$-bases $x_1, \ldots, x_r$ and $y_1, \ldots, y_r$ with $y_i = \sum_j a_{ij} x_j$, then*

$$|G/H| = |\det(a_{ij})|.$$

Proof: Let H have rank s. Use Theorem 1.16 to choose $\mathbb{Z}$-bases $u_1, \ldots, u_r$ of G and $v_1, \ldots, v_s$ of H with $v_i = \alpha_i u_i$ for $1 \le i \le s$. Clearly G/H is the direct product of finite cyclic groups of orders $\alpha_1, \ldots, \alpha_s$ and $r - s$ infinite cyclic groups. Hence $|G/H|$ is finite if and only if $r = s$, and in that case

$$|G/H| = \alpha_1 \ldots \alpha_r.$$

Now

$$u_i = \sum_j b_{ij} x_j \qquad v_i = \sum_j c_{ij} u_j \qquad y_i = \sum_j d_{ij} v_j$$

where the matrices $(b_{ij}) = B$ and $(d_{ij}) = D$ are unimodular by Lemma 1.15, and

$$C = (c_{ij}) = \begin{bmatrix} \alpha_1 & & & \\ & \alpha_2 & & \mathbf{0} \\ & & \ddots & \\ \mathbf{0} & & & \alpha_r \end{bmatrix}$$

Clearly if $A = (a_{ij})$ then $A = BCD$, so

$$\det(A) = \det(B)\det(C)\det(D).$$

Therefore

$$|\det(A)| = |\pm 1||\det(C)||\pm 1| = |\alpha_1 \ldots \alpha_r| = |G/H|$$

as claimed.   □

For example, if G has rank 3 and $\mathbb{Z}$-basis x, y, z, and if H has $\mathbb{Z}$-basis

$$3x + y - 2z,$$
$$4x - 5y + z,$$
$$x \qquad + 7z,$$

then $|G/H|$ is the absolute value of

$$\begin{vmatrix} 3 & 1 & -2 \\ 4 & -5 & 1 \\ 1 & 0 & 7 \end{vmatrix},$$

namely 142.

Suppose now that G is a finitely generated group, and let its generators be $w_1, \ldots, w_n$ where the latter need not be independent. Then we can define a map $f : \mathbb{Z}^n \to G$ by:

$$f(m_1, \ldots, m_n) = m_1 w_1 + \ldots + m_n w_n.$$

This is surjective, so G is isomorphic to $\mathbb{Z}^n/H$ where H is the kernel of f. We can use Theorem 1.16 to choose a new basis $u_1, \ldots, u_n$ of $\mathbb{Z}^n$ so that $\alpha_1 u_1, \ldots, \alpha_s u_s$ is a basis for H. Let A be the subgroup of $\mathbb{Z}^n$ generated by $u_1, \ldots, u_s$ and B be the subgroup generated by $u_{s+1}, \ldots, u_n$. Clearly G is isomorphic to $(A/H) \times B$, so is the direct product of a finite abelian group A/H and a free group B on $n - s$ generators. Putting $n - s = k$, we have proved:

Proposition 1.18. *Every finitely generated abelian group with n generators is the direct product of a finite abelian group and a free group on k generators where $k \leq n$.*   □

If K is any subgroup of a finitely generated abelian group G, then writing $G = F \times B$ where F is finite and B is finitely generated and free, we find $K \cong (F \cap K) \times H$ where $H \subseteq B$. Then $F \cap K$ is finite and (by

Theorem 1.16) H is finitely generated and free, so we find K is finitely generated. Hence we have:

Proposition 1.19. *A subgroup of a finitely generated abelian group is finitely generated.* □

Of course the results in this section are not the best possible that can be proved in finitely generated abelian group theory. Refinements may be found in Fraleigh [39] Chapter 9 pp. 86–93. The results we have established are ample for our needs, so we can now make a start on the substance of algebraic number theory.

1.7 Exercises

1. Show that Theorem 1.3 becomes false if the word 'finite' is omitted from the hypotheses.

2. Which of the following polynomials over $\mathbb{Z}$ are irreducible?
   (a) $x^2 + 3$
   (b) $x^2 - 169$
   (c) $x^3 + x^2 + x + 1$
   (d) $x^3 + 2x^2 + 3x + 4$

3. Write down some polynomials over $\mathbb{Z}$ and factorize them into irreducibles.

4. Does Theorem 1.5 remain true over a field of characteristic $p > 0$?

5. Find the minimum polynomial over $\mathbb{Q}$ of
   (i) $(1 + i)/\sqrt{2}$
   (ii) $i + \sqrt{2}$
   (iii) $e^{2\pi i/3} + 2$

6. Find the degrees of the following field extensions:
   (a) $\mathbb{Q}(\sqrt{7}) : \mathbb{Q}$
   (b) $\mathbb{C}(\sqrt{7}) : \mathbb{C}$
   (c) $\mathbb{Q}(\sqrt{5}, \sqrt{7}, \sqrt{35}) : \mathbb{Q}$
   (d) $\mathbb{R}(\theta) : \mathbb{R}$ where $\theta^3 - 7\theta + 6 = 0$ and $\theta \notin \mathbb{R}$
   (e) $\mathbb{Q}(\pi) : \mathbb{Q}$ (*Hint:* You may assume that π is transcendental.)

7. Let K be the field generated by the elements $e^{2\pi i/n}$ $(n = 1, 2, \ldots)$. Show that K is an algebraic extension of $\mathbb{Q}$, but that $[K : \mathbb{Q}]$ is not finite. (*Hint:* It may help to show that the minimum polynomial of $e^{2\pi i/p}$ for p prime is $t^{p-1} + t^{p-2} + \ldots + 1$.)

8. Express the following polynomials in terms of elementary symmetric polynomials, where possible.
   - (a) $t_1^2 + t_2^2 + t_3^2$ $(n = 3)$
   - (b) $t_1^3 + t_2^3$ $(n = 2)$
   - (c) $t_1 t_2^2 + t_2 t_3^2 + t_3 t_1^2$ $(n = 3)$
   - (d) $t_1 + t_2^2 + t_3^3$ $(n = 3)$

9. A polynomial belonging to $\mathbb{Z}[t_1, \ldots, t_n]$ is *antisymmetric* if it is invariant under even permutations of the variables, but changes sign under odd permutations. Let

$$\Delta = \prod_{i<j} (t_i - t_j).$$

Show that Δ is antisymmetric. If f is any antisymmetric polynomial, prove that f is expressible as a polynomial in the elementary symmetric polynomials, together with Δ. (*Hint:* Show that Δ divides f and consider f/Δ.)

10. Find the orders of the groups G/H where G is free abelian with $\mathbb{Z}$-basis x, y, z and H is generated by:
   - (a) $2x$, $3y$, $7z$
   - (b) $x + 3y - 5z$, $2x - 4y$, $7x + 2y - 9z$
   - (c) x
   - (d) $41x + 32y - 999z$, $16y + 3z$, $2y + 111z$
   - (e) $41x + 32y - 999z$

11. Let K be a field. Show that M is a K-module if and only if it is a vector space over K. Show that the submodules of M are precisely the vector subspaces. Do these statements remain true if we do not use convention (d) of Section 1.5 for modules?

12. Let $\mathbb{Z}$ be a $\mathbb{Z}$-module with the obvious action. Find all the submodules.

13. Let R be a ring, and let M be a finitely generated R-module. Is it true that M necessarily has only finitely many distinct R-submodules? If not, is there an extra condition on R which will lead to this conclusion?

14. An abelian group G is said to be *torsion-free* if $g \in G$, $g \neq 0$ and $kg = 0$ for $k \in \mathbb{Z}$ implies $k = 0$. Prove that a finitely generated torsion-free abelian group is a finitely generated free group.

15. By examining the proof of Theorem 1.16 carefully, or by other means, prove that if H is a subgroup of a free group G of rank n then there exists a basis $u_1, \ldots, u_n$ for G and a basis $v_1, \ldots, v_s$ for H where $s \leq n$ and $v_i = \alpha_i u_i$ $(1 \leq i \leq s)$ where the α_i are positive integers and α_i divides α_{i+1} $(1 \leq i \leq s - 1)$.

2

Algebraic Numbers

In this chapter we introduce the algebraic numbers as solutions of polynomial equations with integer coefficients. Among these numbers, the major players are the solutions of equations with integer coefficients whose leading coefficient is 1. These are the algebraic integers. We develop a theory of factorization of algebraic integers, analogous to factorization of whole numbers. In many ways the theories are alike, but in at least one essential way—uniqueness of factorization—there are differences.

The most important of these is that in many rings of algebraic integers, factorization into irreducibles is not unique. We postpone discussion of this issue until Chapter 4.

Here we observe a simpler issue: factorization into irreducibles depends on the ring in which factorization is performed. In $\mathbb{Z}$ the number 5 is irreducible. The only ways to write it as a product are trivial: multiply ± 5 and ± 1. However, in $\mathbb{Z}[\sqrt{5}]$ it can be written as the non-trivial product $5 = \sqrt{5} \cdot \sqrt{5}$; moreover, it turns out that $\sqrt{5}$ cannot be further factorized in this ring. Thus 5 is irreducible in $\mathbb{Z}$, yet reducible in $\mathbb{Z}[\sqrt{5}]$. It is therefore essential to specify the ring in which factorization is carried out.

The natural context is a ring of algebraic integers, contained in its associated algebraic number field. We begin with algebraic number fields that obey a finiteness condition: they are finite-dimensional as vector spaces over the rationals. We prove that such a field is of the form $\mathbb{Q}[\theta]$ for a single algebraic number θ.

We introduce the conjugates of an algebraic number and the discriminant of a basis for $\mathbb{Q}[\theta]$ over $\mathbb{Q}$, using the conjugates of θ to show that the discriminant is always a non-zero rational number. Algebraic integers are defined and shown to form a ring. The ring of algebraic integers in a number field is shown to have an integral basis whose discriminant is an integer. This integer is independent of the choice of integral basis and is called the discriminant of the number field.

Finally, we introduce the norm and trace of an algebraic number, which are ordinary integers when the algebraic number is an algebraic integer. Using the norm and trace in later chapters we can translate statements about algebraic integers into statements about ordinary integers, which are easier to handle.

2.1   Algebraic Numbers

A complex number α is *algebraic* if it is algebraic over $\mathbb{Q}$, that is, it satisfies a non-zero polynomial equation with coefficients in $\mathbb{Q}$. Equivalently, clearing out denominators, we may assume the coefficients are in $\mathbb{Z}$. Let $\mathbb{A}$ denote the set of algebraic numbers. In fact $\mathbb{A}$ is a field:

Theorem 2.1. *The set $\mathbb{A}$ of algebraic numbers is a subfield of the complex field $\mathbb{C}$.*

Proof: We use Theorem 1.11, which in this case says that α is algebraic if and only if $[\mathbb{Q}(\alpha) : \mathbb{Q}]$ is finite. Suppose that α, β are algebraic. Then

$$[\mathbb{Q}(\alpha, \beta) : \mathbb{Q}] = [\mathbb{Q}(\alpha, \beta) : \mathbb{Q}(\alpha)][\mathbb{Q}(\alpha) : \mathbb{Q}]$$

Since β is algebraic over $\mathbb{Q}$ it is certainly algebraic over $\mathbb{Q}(\alpha)$, so the first factor on the right is finite; and the second factor is also finite. Hence $[\mathbb{Q}(\alpha, \beta) : \mathbb{Q}]$ is finite. But each of $\alpha + \beta$, $\alpha - \beta$, $\alpha\beta$, and (for $\beta \neq 0$) α/β belongs to $\mathbb{Q}(\alpha, \beta)$. So all of these are in $\mathbb{A}$, and the theorem is proved. $\square$

The whole field $\mathbb{A}$ is not as interesting, for us, as certain of its subfields. We define a *number field* to be a subfield K of $\mathbb{C}$ such that $[K : \mathbb{Q}]$ is finite. This implies that every element of K is algebraic, so $K \subseteq \mathbb{A}$. The trouble with $\mathbb{A}$ is that $[\mathbb{A} : \mathbb{Q}]$ is not finite (see Chapter 1, Exercise 7, or Stewart [78], Exercise 4.8, p. 55). If K is a number field then $K = \mathbb{Q}(\alpha_i, \ldots, \alpha_n)$ for finitely many algebraic numbers $\alpha_1, \ldots, \alpha_n$ (for instance, a basis for K as vector space over $\mathbb{Q}$). We can strengthen this observation considerably:

Theorem 2.2. *If K is a number field then $K = \mathbb{Q}(\theta)$ for some algebraic number θ.*

Proof: Arguing by induction, it is sufficient to prove that if $K = K_1(\alpha, \beta)$ where K_1 is a subfield of K, then $K = K_1(\theta)$ for some θ. Let p and q respectively be the minimum polynomials of α, β over K_1, and suppose that over $\mathbb{C}$ these factorize as

$$
\begin{aligned}
p(t) &= (t - \alpha_1) \dots (t - \alpha_n), \\
q(t) &= (t - \beta_1) \dots (t - \beta_m),
\end{aligned}
$$

where we choose the numbering so that $\alpha_1 = \alpha$, $\beta_1 = \beta$. By Corollary 1.6 the α_i are distinct, as are the β_j. Hence for each i and each $k \neq 1$ there is at most one element $x \in K_1$ such that

$$
\alpha_i + x\beta_k = \alpha_1 + x\beta_1.
$$

Since there are only finitely many such equations, we may choose $c \neq 0$ in K_1, not equal to any of these x's, and then

$$
\alpha_i + c\beta_k \neq \alpha_1 + c\beta_1
$$

for $1 \leq i \leq n$, $2 \leq k \leq m$. Define

$$
\theta = \alpha + c\beta.
$$

We prove that $K_1(\theta) = K_1(\alpha, \beta)$. Obviously $K_1(\theta) \subseteq K_1(\alpha, \beta)$, and it suffices to prove that $\beta \in K_1(\theta)$ since $\alpha = \theta - c\beta$.

Observe that

$$
p(\theta - c\beta) = p(\alpha) = 0.
$$

Define the polynomial

$$
r(t) = p(\theta - ct) \in K_1(\theta)[t]
$$

Now β is a zero of both $q(t)$ and $r(t)$ as polynomials over $K_1(\theta)$. These polynomials have only one common zero, for if $q(\xi) = r(\xi) = 0$ then ξ is one of $\beta_1, \dots, \beta_m$ and also $\theta - c\xi$ is one of $\alpha_1, \dots, \alpha_n$. Our choice of c forces $\xi = \beta$. Let $h(t)$ be the minimum polynomial of β over $K_1(\theta)$. Then $h(t) \mid q(t)$ and $h(t) \mid r(t)$. Since q and r have just one common zero in $\mathbb{C}$ we have $\partial h = 1$, so

$$
h(t) = t + \mu
$$

for $\mu \in K_1(\theta)$. Now $0 = h(\beta) = \beta + \mu$ so that $\beta = -\mu \in K_1(\theta)$ as required. $\qquad\square$

Example 2.3. $\mathbb{Q}(\sqrt{2}, \sqrt[3]{5})$.

We have
$$\alpha_1 = \sqrt{2}, \alpha_2 = -\sqrt{2},$$
$$\beta_1 = \sqrt[3]{5}, \beta_2 = \omega\sqrt[3]{5}, \beta_3 = \omega^2\sqrt[3]{5}$$

where
$$\omega = \tfrac{1}{2}(-1 + \sqrt{-3})$$

is a complex cube root of 1. The number $c = 1$ satisfies

$$\alpha_i + c\beta_k \neq \alpha + c\beta$$

for $i = 1, 2$, $k = 2, 3$; since the number on the left is not real in any of the four cases, whereas that on the right is. Hence $\mathbb{Q}(\sqrt{2}, \sqrt[3]{5}) = \mathbb{Q}(\sqrt{2} + \sqrt[3]{5})$.

The expression of K as $\mathbb{Q}(\theta)$ is, of course, not unique; for $\mathbb{Q}(\theta) = \mathbb{Q}(-\theta) = \mathbb{Q}(\theta + 1) = \ldots$ and so on.

2.2   Conjugates and Discriminants

If $K = \mathbb{Q}(\theta)$ is a number field there are, in general, several distinct monomorphisms $\sigma : K \to \mathbb{C}$. For instance, if $K = \mathbb{Q}(i)$ where $i = \sqrt{-1}$ then the possibilities are

$$\sigma_1(x + iy) \;=\; x + iy,$$
$$\sigma_2(x + iy) \;=\; x - iy,$$

for $x, y \in \mathbb{Q}$. The full set of such monomorphisms plays a fundamental role in the theory, so we begin with a description.

Theorem 2.4. *Let $K = \mathbb{Q}(\theta)$ be a number field of degree n over $\mathbb{Q}$. Then there are exactly n distinct monomorphisms $\sigma_i : K \to \mathbb{C}$ ($i = 1, \ldots, n$). The elements $\sigma_i(\theta) = \theta_i$ are the distinct zeros in $\mathbb{C}$ of the minimum polynomial of θ over $\mathbb{Q}$.*

Proof: Let $\theta_1, \ldots, \theta_n$ be the (by Corollary 1.3 distinct) zeros of the minimum polynomial p of θ. Then each θ_i also has minimum polynomial p (it must divide p, and p is irreducible) and so there is a unique field isomorphism $\sigma_i : \mathbb{Q}(\theta) \to \mathbb{Q}(\theta_i)$ such that $\sigma_i(\theta) = \theta_i$. In fact, if $\alpha \in \mathbb{Q}(\theta)$ then $\alpha = r(\theta)$ for a unique $r \in \mathbb{Q}[t]$ with $\partial r < n$, and

$$\sigma_i(\alpha) = r(\theta_i).$$

See Garling [31] Corollary 2 to Theorem 7.4, p. 66 or Stewart [78] Theorem 3.8, p. 43.

Conversely if $\sigma : K \to \mathbb{C}$ is a monomorphism then σ is the identity on $\mathbb{Q}$. Now

$$0 = \sigma(p(\theta)) = p(\sigma(\theta))$$

so $\sigma(\theta)$ is one of the θ_i, hence σ is one of the σ_i. □

Keep this notation, and for each $\alpha \in K = \mathbb{Q}(\theta)$ define the *field polynomial* of α over K to be

$$f_\alpha(t) = \prod_{i=1}^{n}(t - \sigma_i(\alpha)).$$

As it stands, this is in $K[t]$. In fact more is true:

Theorem 2.5. *The coefficients of the field polynomial are rational numbers, so that $f_\alpha(t) \in \mathbb{Q}[t]$.*

Proof: We have $\alpha = r(\theta)$ for $r \in \mathbb{Q}[t]$, $\partial r < n$. The field polynomial takes the form

$$f_\alpha(t) = \prod_{i}(t - r(\theta_i))$$

where the θ_i run through all zeros of the minimum polynomial p of θ, whose coefficients are in $\mathbb{Q}$. It is easy to see that the coefficients of $f_\alpha(t)$ are of the form

$$h(\theta_1, \ldots, \theta_n)$$

where $h(t_1, \ldots, t_n)$ is a symmetric polynomial in $\mathbb{Q}[t_1, \ldots, t_n]$. Now use Corollary 1.14. □

The elements $\sigma_i(\alpha)$ for $i = 1, \ldots, n$ are the *K-conjugates* of α. Although the θ_i are distinct (and are the K-conjugates of θ) it is not always the case that the K-conjugates of α are distinct: for instance $\sigma_i(1) = 1$ for all i. The precise situation is given by:

Theorem 2.6. *With the above notation,*
   (a) The field polynomial f_α is a power of the minimum polynomial p_α.
   (b) The K-conjugates of α are the zeros of p_α in $\mathbb{C}$, each repeated n/m times where $m = \partial p_\alpha$ is a divisor of n.
   (c) The element $\alpha \in \mathbb{Q}$ if and only if all of its K-conjugates are equal.
   (d) $\mathbb{Q}(\alpha) = \mathbb{Q}(\theta)$ if and only if all K-conjugates of α are distinct.

Proof: The main point is (a). Now $q = p_\alpha$ is irreducible and α is a zero of $f = f_\alpha$, so $f = q^s h$ where q and h are coprime and both are monic. (This follows from factorizing f into irreducibles.) We claim that h is constant. If not, some $\alpha_i = \sigma_i(\alpha) = r(\theta_i)$ is a zero of h, where $\alpha = r(\theta)$. Therefore if $g(t) = h(r(t))$ then $g(\theta_i) = 0$. Let p be the minimum polynomial of θ over $\mathbb{Q}$, hence also of each θ_i. Then $p|g$, so that $g(\theta_j) = 0$ for all j, and in particular $g(\theta) = 0$. Therefore, $h(\alpha) = h(r(\theta)) = g(\theta) = 0$, so q divides h, a contradiction. Hence h is constant and monic, so $h = 1$ and $f = q^s$.

(b) is an immediate consequence of (a) by the definition of the field polynomial.

To prove (c), it is clear that $\alpha \in \mathbb{Q}$ implies $\sigma_i(\alpha) \in \mathbb{Q}$. Conversely, if all $\sigma_i(\alpha)$ are equal then since the zeros of $q = p_\alpha$ are distinct and $f_\alpha = q^s$, we have $\partial q = 1$ so $\alpha \in \mathbb{Q}$.

Finally for (d): if all $\sigma_i(\alpha)$ are distinct then $\partial p_\alpha = n$, so $[\mathbb{Q}(\alpha) : \mathbb{Q}] = n = [\mathbb{Q}(\theta) : \mathbb{Q}]$. Thus $\mathbb{Q}(\alpha) = \mathbb{Q}(\theta)$. Conversely if $\mathbb{Q}(\alpha) = \mathbb{Q}(\theta)$ then $\partial p_\alpha = n$ so the $\sigma_i(\alpha)$ are distinct. $\square$

Warning. The K-conjugates of α need not be elements of K. Even the θ_i need not be elements of K. For example, let θ be the real cube root of 2. Then $\mathbb{Q}(\theta)$ is a subfield of $\mathbb{R}$. The K-conjugates of θ, however, are θ, $\omega\theta$, $\omega^2\theta$, where $\omega = \frac{1}{2}(-1 + \sqrt{-3})$. The last two of these are nonreal, hence do not lie in $\mathbb{Q}(\theta)$.

Still with $K = \mathbb{Q}(\theta)$ of degree n, let $\{\alpha_i, \ldots, \alpha_n\}$ be a basis of K (as vector space over $\mathbb{Q}$). Define the *discriminant* of this basis to be

$$\Delta[\alpha_1, \ldots, \alpha_n] = \{\det[\sigma_i(\alpha_j)]\}^2. \tag{2.1}$$

If we pick another basis $\{\beta_1, \ldots, \beta_n\}$ then

$$\beta_k = \sum_{i=1}^n c_{ik}\alpha_i \qquad (c_{ik} \in \mathbb{Q})$$

for $k = 1, \ldots, n$, and

$$\det(c_{ik}) \neq 0.$$

The product formula for determinants, and the fact that the σ_i are monomorphisms, hence the identity on $\mathbb{Q}$, shows that

$$\Delta[\beta_1, \ldots, \beta_n] = [\det(c_{ik})]^2 \Delta[\alpha_1, \ldots, \alpha_n]. \tag{2.2}$$

Theorem 2.7. *The discriminant of any basis for $K = \mathbb{Q}(\theta)$ is rational and non-zero. If all K-conjugates of θ are real then the discriminant of any basis is positive.*

Proof: Pick a basis that makes computations straightforward: the obvious one is $\{1, \theta, \ldots, \theta^{n-1}\}$. If the conjugates of θ are $\theta_1, \ldots, \theta_n$ then

$$\Delta[1, \theta, \ldots, \theta^{n-1}] = (\det \theta_i^j)^2.$$

A determinant of the form $D = \det(t_i^j)$ is called a *Vandermonde* determinant, and has value

$$D = \prod_{1 \le i < j \le n} (t_i - t_j). \tag{2.3}$$

To see this, think of everything as lying inside $\mathbb{Q}[t_1, \ldots, t_n]$. Then for $t_i = t_j$ the determinant has two equal rows, so vanishes. Hence D is divisible by each $(t_i - t_j)$. To avoid repeating such a factor twice we take $i < j$. Then comparison of degrees easily shows that D has no other non-constant factors; comparing coefficients of $t_1 t_2^2 \ldots t_n^n$ gives (2.3).

Hence

$$\Delta = \Delta[1, \theta, \ldots, \theta^{n-1}] = \left(\prod (\theta_i - \theta_j) \right)^2.$$

Now D is antisymmetric in the t_i, so D^2 is symmetric. By the usual argument about symmetric polynomials, Corollary 1.14, Δ is rational. Since the θ_i are distinct, $\Delta \ne 0$.

Now let $\{\beta_1, \ldots, \beta_n\}$ be any basis. Then

$$\Delta[\beta_1, \ldots, \beta_n] = (\det c_{ik})^2 \Delta$$

for certain rational numbers c_{ik}, and $\det(c_{ik}) \ne 0$, so

$$\Delta[\beta_1, \ldots \beta_n] \ne 0$$

and is rational. Clearly if all θ_i are real then Δ is a positive real number, hence so is $\Delta[\beta_1, \ldots, \beta_n]$. $\square$

With the above notation, Δ vanishes if and only if some θ_i is equal to another θ_j. Hence the non-vanishing of Δ lets us 'discriminate' among the θ_i, which motivates calling Δ the discriminant.

2.3 Algebraic Integers

A complex number θ is an *algebraic integer* if there is a *monic* polynomial $p(t)$ with integer coefficients such that $p(\theta) = 0$. In other words,

$$\theta^n + a_{n-1} \theta^{n-1} + \ldots + a_0 = 0$$

where $a_i \in \mathbb{Z}$ for all i.

For example, $\theta = \sqrt{-2}$ is an algebraic integer, since $\theta^2 + 2 = 0$; $\tau = \frac{1}{2}(1 + \sqrt{5})$ is an algebraic integer, since $\tau^2 - \tau - 1 = 0$. But $\phi = 22/7$ is not. It satisfies equations like $7\phi - 22 = 0$, but this is not monic; or like $\phi - 22/7 = 0$, whose coefficients are not integers; but it can be shown without difficulty that ϕ does not satisfy any monic polynomial equation with integer coefficients.

We write $\mathbb{B}$ for the set of algebraic integers. One of our aims is to prove that $\mathbb{B}$ is a subring of $\mathbb{A}$. We prepare for this by proving:

Lemma 2.8. *A complex number θ is an algebraic integer if and only if the additive group generated by all powers $1, \theta, \theta^2, \ldots$ is finitely generated.*

Proof: If θ is an algebraic integer, then for some n

$$\theta^n + a_{n-1}\theta^{n-1} + \ldots + a_0 = 0 \tag{2.4}$$

where the $a_i \in \mathbb{Z}$. We claim that every power of θ lies in the additive group generated by $1, \theta, \ldots, \theta^{n-1}$. Call this group Γ. Then (2.4) shows that $\theta^n \in \Gamma$. Inductively, if $m \geq n$ and $\theta^m \in \Gamma$ then

$$\theta^{m+1} = \theta^{m+1-n}\theta^n = \theta^{m+1-n}(-a_{n-1}\theta^{n-1} - \ldots - a_0) \in \Gamma.$$

This proves that every power of θ lies in Γ, which gives one implication.

For the converse, suppose that every power of θ lies in a finitely generated additive group G. The subgroup Γ of G generated by the powers $1, \theta, \theta^2, \ldots$ must also be finitely generated by Proposition 1.19. Let $v_1, \ldots, v_n$ be generators. Each v_i is a polynomial in θ with integer coefficients, so θv_i is also such a polynomial. Hence there exist integers b_{ij} such that

$$\theta v_i = \sum_{j=1}^{n} b_{ij} v_j.$$

This leads to a system of homogeneous equations for the v_i of the form

$$(b_{11} - \theta)v_1 + b_{12}v_2 + \ldots + b_{1n}v_n = 0$$
$$b_{21}v_1 + (b_{22} - \theta)v_2 + \ldots + b_{2n}v_n = 0$$
$$\vdots$$
$$b_{n1}v_1 + b_{n2}v_2 + \ldots + (b_{nn} - \theta)v_n = 0.$$

Since there exists a solution $v_1, \ldots, v_n \in \mathbb{C}$, not all zero, the determinant

$$\begin{vmatrix} b_{11} - \theta & b_{12} & \cdots & b_{1n} \\ b_{21} & b_{22} - \theta & \cdots & b_{2n} \\ \vdots & \vdots & \ddots & \vdots \\ b_{n1} & b_{n2} & \cdots & b_{nn} - \theta \end{vmatrix}$$

is zero. Expand to see that θ satisfies a monic polynomial equation with integer coefficients. □

Theorem 2.9. *The algebraic integers form a subring of the field of algebraic numbers.*

Proof: Let $\theta, \phi \in \mathbb{B}$. We have to show that $\phi + \theta$ and $\theta\phi \in \mathbb{B}$. By Lemma 2.8 all powers of θ lie in a finitely generated additive subgroup Γ_θ of $\mathbb{C}$, and all powers of ϕ lie in a finitely generated additive subgroup Γ_ϕ. But now all powers of $\theta + \phi$ and of $\theta\phi$ are integer linear combinations of elements $\theta^i\phi^j$ which lie in $\Gamma_\theta\Gamma_\phi \subseteq \mathbb{C}$. But if Γ_θ has generators $v_1, \ldots, v_n$ and Γ_ϕ has generators $w_1, \ldots, w_m$, then $\Gamma_\theta\Gamma_\phi$ is the additive group generated by all v_iw_j for $1 \le i \le n$, $1 \le j \le m$. Hence all powers of $\theta + \phi$ and of $\theta\phi$ lie in a finitely generated additive subgroup of $\mathbb{C}$, so by Lemma 2.8 $\theta + \phi$ and $\theta\phi$ are algebraic integers. Hence $\mathbb{B}$ is a subring of $\mathbb{A}$. □

A simple extension of this technique lets us prove a useful theorem:

Theorem 2.10. *Let θ be a complex number satisfying a monic polynomial equation whose coefficients are algebraic integers. Then θ is an algebraic integer.*

Proof: Suppose that

$$\theta^n + \psi_{n-1}\theta^{n-1} + \ldots + \psi_0 = 0$$

where $\psi_0, \ldots, \psi_{n-1} \in \mathbb{B}$. Then these generate a subring Ψ of $\mathbb{B}$. The argument of Lemma 2.8 shows that all powers of θ lie inside a finitely generated Ψ-submodule M of $\mathbb{C}$, spanned by $1, \theta, \ldots, \theta^{n-1}$. By Theorem 2.9, each ψ_i and all its powers lie inside a finitely generated additive group Γ_i with generators γ_{ij} $(1 \le j \le n_i)$. Therefore M lies inside the additive group generated by all elements

$$\gamma_{1j_1}, \gamma_{2j_2}, \ldots, \gamma_{n-1,j_{n-1}}\theta^k$$

$(1 \le j_i \le n_i, 0 \le i \le n-1, 0 \le k \le n-1)$, which is a finite set. So M is finitely generated as an additive group. □

Theorems 2.9 and 2.10 let us construct many new algebraic integers from known ones. For instance, $\sqrt{2}$ and $\sqrt{3}$ are clearly algebraic integers. Then Theorem 2.9 says that numbers such as $\sqrt{2} + \sqrt{3}$, $7\sqrt{2} - 41\sqrt{3}$, $(\sqrt{2})^5(1 + \sqrt{3})^2$ are also algebraic integers. And Theorem 2.10 says that zeros of polynomials such as

$$t^{23} - (14 + \sqrt[5]{3})t^9 + (\sqrt[3]{2})t^5 - 19\sqrt{3}$$

are algebraic integers. It would not be easy, particularly in the last instance, to compute explicit polynomials over $\mathbb{Z}$ of which these algebraic integers are zeros; although it can in principle be done by using symmetric polynomials. In fact Theorems 2.9 and 2.10 can be proved this way.

The Ring of Integers of a Number Field

We have given this topic its own heading because the ideas involved are absolutely central to the entire book.

For any number field K write

$$\mathfrak{O} = K \cap \mathbb{B},$$

and call $\mathfrak{O}$ the *ring of integers* of K. The symbol '$\mathfrak{O}$' is traditional but confusing. It look a bit like a letter D, but actually it is a Gothic (often called 'Fraktur') capital O (for 'order', the old terminology, in German). In cases where it is not immediately clear which number field is involved, we write more explicitly $\mathfrak{O}_K$. Since K and $\mathbb{B}$ are subrings of $\mathbb{C}$ it follows that $\mathfrak{O}$ is a subring of K. Further $\mathbb{Z} \subseteq \mathbb{Q} \subseteq K$ and $\mathbb{Z} \subseteq \mathbb{B}$ so $\mathbb{Z} \subseteq \mathfrak{O}$.

The following lemma is easy to prove:

Lemma 2.11. *If $\alpha \in K$ then $c\alpha \in \mathfrak{O}$ for some non-zero $c \in \mathbb{Z}$.*

Corollary 2.12. *If K is a number field then $K = \mathbb{Q}(\theta)$ for an algebraic integer θ.*

Proof: By Theorem 2.2, $K = \mathbb{Q}(\phi)$ for an algebraic number ϕ. By Lemma 2.11, $\theta = c\phi$ is an algebraic integer for some $0 \neq c \in \mathbb{Z}$. Clearly $\mathbb{Q}(\phi) = \mathbb{Q}(\theta)$. $\quad\square$

Warning. For $\theta \in \mathbb{C}$, write $\mathbb{Z}[\theta]$ for the set of elements $p(\theta)$ for polynomials $p \in \mathbb{Z}[t]$. If $K = \mathbb{Q}(\theta)$ where θ is an algebraic integer then certainly $\mathfrak{O}$ contains $\mathbb{Z}[\theta]$ since $\mathfrak{O}$ is a ring containing θ. However, $\mathfrak{O}$ need not equal $\mathbb{Z}[\theta]$. For example, $\mathbb{Q}(\sqrt{5})$ is a number field and $\sqrt{5}$ an algebraic integer.

But

$$\frac{1 + \sqrt{5}}{2}$$

is a zero of $t^2 - t - 1$, hence an algebraic integer; and it lies in $\mathbb{Q}(\sqrt{5})$ so it belongs to $\mathfrak{O}$. It does not belong to $\mathbb{Z}[\sqrt{5}]$.

There is a useful criterion, in terms of the minimum polynomial, for a number to be an algebraic integer:

Lemma 2.13. *An algebraic number α is an algebraic integer if and only if its minimum polynomial over $\mathbb{Q}$ has coefficients in $\mathbb{Z}$.*

Proof: Let p be the minimum polynomial of α over $\mathbb{Q}$, and recall that this is monic and irreducible in $\mathbb{Q}[t]$. If $p \in \mathbb{Z}[t]$ then α is an algebraic integer. Conversely, if α is an algebraic integer then $q(\alpha) = 0$ for some monic $q \in \mathbb{Z}[t]$, and $p|q$. By Gauss's Lemma 1.7 it follows that $p \in \mathbb{Z}[t]$, because some rational multiple λp lies in $\mathbb{Z}[t]$ and divides q, and the monicity of q and p implies $\lambda = 1$.   □

To avoid confusion about the word 'integer' we adopt the following convention: a *rational integer* is an element of $\mathbb{Z}$, and a plain *integer* is an algebraic integer. (The aim is to reserve the shorter term for the concept most often encountered.) Any remaining possibility of confusion is eliminated by:

Lemma 2.14. *An algebraic integer is a rational number if and only if it is a rational integer. Equivalently, $\mathbb{B} \cap \mathbb{Q} = \mathbb{Z}$.*

Proof: Clearly $\mathbb{Z} \subseteq \mathbb{B} \cap \mathbb{Q}$. Let $\alpha \in \mathbb{B} \cap \mathbb{Q}$; since $\alpha \in \mathbb{Q}$ its minimum polynomial over $\mathbb{Q}$ is $t - \alpha$. By Lemma 2.13 the coefficients of this are in $\mathbb{Z}$, hence $-\alpha \in \mathbb{Z}$, hence $\alpha \in \mathbb{Z}$.   □

2.4   Integral Bases

Let K be a number field of degree n (over $\mathbb{Q}$). A *basis* (or *$\mathbb{Q}$-basis* for emphasis) of K is a basis for K as a vector space over $\mathbb{Q}$. By Corollary 2.11 we have $K = \mathbb{Q}(\theta)$ where θ is an algebraic integer, so the minimum polynomial p of θ has degree n and $\{1, \theta, \ldots, \theta^{n-1}\}$ is a basis for K.

The ring $\mathfrak{O}$ of integers of K is an abelian group under addition. A $\mathbb{Z}$-basis for $(\mathfrak{O}, +)$ is called an *integral basis* for K (or for $\mathfrak{O}$). Thus

$\{\alpha_1, \ldots, \alpha_s, \}$ is an integral basis if and only if all $\alpha_i \in \mathfrak{O}$ and every element of $\mathfrak{O}$ is *uniquely* expressible in the form

$$a_1\alpha_1 + \ldots + a_s\alpha_s$$

for rational integers $a_1, \ldots, a_s$. It is obvious from Lemma 2.11 that any integral basis for K is a $\mathbb{Q}$-basis, so $s = n$. But we have to verify that integral bases exist. In fact they do, but they are not always what naively we might expect them to be.

For instance, $K = \mathbb{Q}[\theta]$ (which equals $\mathbb{Q}(\theta)$) for an algebraic integer θ by Corollary 2.12, so $\{1, \theta, \ldots, \theta^{n-1}\}$ is a $\mathbb{Q}$-basis for K which consists of integers. However, it does *not* follow that $\{1, \theta, \ldots, \theta^{n-1}\}$ is an integral basis, because some elements in $\mathbb{Q}[\theta]$ with non-integer coefficients may also be (algebraic) integers. As an example, consider $K = \mathbb{Q}(\sqrt{5})$. We saw that $\frac{1}{2} + \frac{1}{2}\sqrt{5}$ satisfies the equation

$$t^2 - t + 1 = 0,$$

so is an integer in $\mathbb{Q}(\sqrt{5})$, but it is not an element of $\mathbb{Z}[\sqrt{5}]$.

Our first problem, therefore, is to show that integral bases exist. That they do is equivalent to the statement that $(\mathfrak{O}, +)$ is a free abelian group of rank n. To prove this we first establish:

Lemma 2.15. *If $\{\alpha_1, \ldots, \alpha_n\}$ is a basis of K consisting of integers, then the discriminant $\Delta[\alpha_1, \ldots, \alpha_n]$ is a rational integer, not equal to zero.*

Proof: By Theorem 2.7, $\Delta = \Delta[\alpha_1, \ldots, \alpha_n]$ is rational. It is an integer since the α_i are. Hence by Lemma 2.14 it is a rational integer. By Theorem 2.7, $\Delta \neq 0$.   □

Theorem 2.16. *Every number field K possesses an integral basis, and the additive group of $\mathfrak{O}$ is free abelian of rank n equal to the degree of K.*

Proof: We have $K = \mathbb{Q}(\theta)$ for θ an integer. Hence there exist bases for K consisting of integers: for example $\{1, \theta, \ldots, \theta^{n-1}\}$. We have already seen that such $\mathbb{Q}$-bases need not be integral bases. However, the discriminant of a $\mathbb{Q}$-basis consisting of integers is always a rational integer (Lemma 2.15), so what we do is to select a basis $\{\omega_1, \ldots, \omega_n\}$ of integers for which

$$|\Delta[\omega_1, \ldots, \omega_n]|$$

is least. We claim that this is in fact an integral basis. If not, there is an integer ω of K such that

$$\omega = a_1\omega_1 + \ldots + a_n\omega_n$$

for $a_i \in \mathbb{Q}$, *not all in* $\mathbb{Z}$. Choose the numbering so that $a_1 \notin \mathbb{Z}$. Then $a_1 = a + r$ where $a \in \mathbb{Z}$ and $0 < r < 1$. Define

$$\psi_1 = \omega - a\omega_1, \qquad \psi_i = \omega_i \qquad (i = 2, \ldots, n).$$

Then $\{\psi_1, \ldots, \psi_n\}$ is a basis consisting of integers. The determinant relevant to the change of basis from the ω's to the ψ's is

$$\begin{vmatrix} a_1 - a & a_2 & a_3 & \cdots & a_n \\ 0 & 1 & 0 & \cdots & 0 \\ 0 & 0 & 1 & \cdots & 0 \\ \vdots & \vdots & \vdots & \ddots & \vdots \\ 0 & 0 & 0 & \cdots & 1 \end{vmatrix} = r,$$

so

$$\Delta[\psi_1, \ldots, \psi_n] = r^2 \Delta[\omega_1, \ldots, \omega_n].$$

Since $0 < r < 1$ this contradicts the choice of $\{\omega_1, \ldots, \omega_n\}$ making $|\Delta[\omega_1, \ldots, \omega_n]|$ minimal.

It follows that $\{\omega_1, \ldots, \omega_n\}$ is an integral basis, and so $(\mathfrak{O}, +)$ is free abelian of rank n. □

This raises the question of finding integral bases in cases such as $\mathbb{Q}(\sqrt{5})$ where the $\mathbb{Q}$-basis $\{1, \sqrt{5}\}$ is not an integral basis. We consider a more general case in the next chapter, but this particular example is worth a brief discussion here.

An element of $\mathbb{Q}(\sqrt{5})$ is of the form $p + q\sqrt{5}$ for $p, q \in \mathbb{Q}$, and has minimum polynomial

$$(t - p - q\sqrt{5})(t - p + q\sqrt{5}) = t^2 - 2pt + (p^2 - 5q^2).$$

Then $p + q\sqrt{5}$ is an integer if and only if the coefficients $2p, p^2 - 5q^2$ are rational integers. Thus $p = \frac{1}{2}P$ where P is a rational integer. For P even, we have p^2 a rational integer, so $5q^2$ is a rational integer also, implying q is a rational integer. For P odd, a straightforward calculation (performed in the next chapter in greater generality) shows $q = \frac{1}{2}Q$ where Q is also an odd rational integer.

From this it follows that $\mathfrak{O} = \mathbb{Z}[\frac{1}{2} + \frac{1}{2}\sqrt{5}]$ and an integral basis is $\{1, \frac{1}{2} + \frac{1}{2}\sqrt{5}\}$.

We can prove this by another route using the discriminant. The two monomorphisms $\mathbb{Q}(\sqrt{5}) \to \mathbb{C}$ are

$$\begin{aligned} \sigma_1(p + q\sqrt{5}) &= p + q\sqrt{5}, \\ \sigma_2(p + q\sqrt{5}) &= p - q\sqrt{5}. \end{aligned}$$

Hence the discriminant $\Delta[1, \frac{1}{2} + \frac{1}{2}\sqrt{5}]$ is

$$\begin{vmatrix} 1 & \frac{1}{2} + \frac{1}{2}\sqrt{5} \\ 1 & \frac{1}{2} - \frac{1}{2}\sqrt{5} \end{vmatrix}^2 = 5.$$

Define a rational integer to be *squarefree* if it is not divisible by the square of a prime. For example, 5 is squarefree, as are 6, 7, but not 8 or 9. Given a $\mathbb{Q}$-basis of K consisting of integers, we compute the discriminant and then we have:

Theorem 2.17. *Suppose that* $\alpha_1, \ldots, \alpha_n \in \mathfrak{O}$ *form a* $\mathbb{Q}$-*basis for* K. *If* $\Delta[\alpha_1, \ldots, \alpha_n]$ *is squarefree then* $\{\alpha_1, \ldots, \alpha_n\}$ *is an integral basis.*

Proof: Let $\{\beta_1, \ldots, \beta_n\}$ be an integral basis. Then there exist rational integers c_{ij} such that $\alpha_i = \Sigma c_{ij}\beta_j$, and

$$\Delta[\alpha_1, \ldots, \alpha_n] = (\det c_{ij})^2 \Delta[\beta_1, \ldots, \beta_n].$$

Since the left-hand side is squarefree, $\det c_{ij} = \pm 1$, so (c_{ij}) is unimodular. Hence by Lemma 1.15 $\{\alpha_1, \ldots, \alpha_n\}$ is a $\mathbb{Z}$-basis for $\mathfrak{O}$, that is, an integral basis for K. $\square$

For example, the $\mathbb{Q}$-basis $\{1, \frac{1}{2} + \frac{1}{2}\sqrt{5}\}$ for $\mathbb{Q}(\sqrt{5})$ consists of integers and has discriminant 5 (calculated above). Since 5 is squarefree, this is an integral basis. Later we show that there exist integral bases whose discriminants are not squarefree, so the converse of Theorem 2.17 is false.

For two integral bases $\{\alpha_1, \ldots, \alpha_n\}$, $\{\beta_1, \ldots, \beta_n\}$ of an algebraic number field K, we have

$$\Delta[\alpha_1, \ldots, \alpha_n] = (\pm 1)^2 \Delta[\beta_1, \ldots, \beta_n] = \Delta[\beta_1, \ldots, \beta_n],$$

because the matrix corresponding to the change of basis is unimodular. Hence the discriminant of an integral basis is independent of which integral basis we choose. This common value is called the *discriminant of* K (or of $\mathfrak{O}$). It is always a non-zero rational integer. Obviously, isomorphic number fields have the same discriminant. The important role played by the discriminant will become apparent as the drama unfolds.

2.5   Norms and Traces

These important concepts often let us transform a problem about algebraic integers into one about rational integers. As usual, let $K = \mathbb{Q}(\theta)$ be a number field of degree n and let $\sigma_1, \ldots, \sigma_n$ be the monomorphisms $K \to \mathbb{C}$. The field polynomial is a power of the minimum polynomial by Theorem 2.6(a), so by Lemma 2.13 and Gauss's Lemma 1.7 it follows that $\alpha \in K$ is an integer if and only if the field polynomial has rational integer coefficients. For any $\alpha \in K$ define the *norm*

$$N_K(\alpha) = \prod_{i=1}^{n} \sigma_i(\alpha)$$

and *trace*

$$T_K(\alpha) = \sum_{i=1}^{n} \sigma_i(\alpha).$$

Where the field K is clear from the context, we abbreviate the norm and trace of α to $N(\alpha)$ and $T(\alpha)$ respectively.

Since the field polynomial is

$$f_\alpha(t) = \prod_{i=1}^{n} (t - \sigma_i(\alpha))$$

The remark above implies that *if α is an integer then the norm and trace of α are rational integers.* Since the σ_i are monomorphisms it is clear that

$$N(\alpha\beta) = N(\alpha)N(\beta) \tag{2.5}$$

and if $\alpha \neq 0$ then $N(\alpha) \neq 0$. If p, q are rational numbers then

$$T(p\alpha + q\beta) = pT(\alpha) + qT(\beta). \tag{2.6}$$

For instance, if $K = \mathbb{Q}(\sqrt{7})$ then the integers of K are $\mathfrak{O} = \mathbb{Z}[\sqrt{7}]$, see Theorem 3.2. The maps σ_i are

$$\begin{aligned} \sigma_1(p + q\sqrt{7}) &= p + q\sqrt{7}, \\ \sigma_2(p + q\sqrt{7}) &= p - q\sqrt{7}. \end{aligned}$$

Hence

$$\begin{aligned} N(p + q\sqrt{7}) &= p^2 - 7q^2, \\ T(p + q\sqrt{7}) &= 2p. \end{aligned}$$

Since norms are not too hard to compute (they can always be found from symmetric polynomial considerations, often with short-cuts) whereas discriminants involve complicated work with determinants, the following result is sometimes useful:

Proposition 2.18. *Let $K = \mathbb{Q}(\theta)$ be a number field where θ has minimum polynomial p of degree n. The $\mathbb{Q}$-basis $\{1, \theta, \ldots, \theta^{n-1}\}$ has discriminant*

$$\Delta \left[1, \ldots, \theta^{n-1}\right] = (-1)^{n(n-1)/2} N(Dp(\theta))$$

where Dp is the formal derivative of p.

Proof: The proof of Theorem 2.7 yields:

$$\Delta = \Delta[1, \theta, \ldots, \theta^{n-1}] = \prod_{1 \leq i < j \leq n} (\theta_i - \theta_j)^2$$

where $\theta_1, \ldots, \theta_n$ are the conjugates of θ. Now

$$p(t) = \prod_{i=1}^{n} (t - \theta_i)$$

so

$$Dp(t) = \sum_{j=1}^{n} \prod_{\substack{i=1 \\ i \neq j}}^{n} (t - \theta_i)$$

and therefore

$$Dp(\theta_j) = \prod_{\substack{i=1 \\ i \neq j}}^{n} (\theta_j - \theta_i).$$

Multiply all these equations for $j = 1, \ldots, n$:

$$\prod_{j=1}^{n} Dp(\theta_j) = \prod_{\substack{i,j=1 \\ i \neq j}}^{n} (\theta_j - \theta_i).$$

The left-hand side is $N(Dp(\theta))$. On the right, each factor $(\theta_i - \theta_j)$ for $i < j$ appears twice, once as $(\theta_i - \theta_j)$ and once as $(\theta_j - \theta_i)$. The product of these two factors is $-(\theta_i - \theta_j)^2$. Multiplying up, we get Δ multiplied by $(-1)^s$ where s is the number of pairs (i, j) with $1 \leq i < j \leq n$, namely:

$$s = \tfrac{1}{2}n(n-1).$$

$\square$

We close this chapter by noting the following simple identity linking the discriminant and trace:

Proposition 2.19. *If* $\{\alpha_1, \ldots, \alpha_n\}$ *is any* $\mathbb{Q}$-basis of K, then

$$\Delta[\alpha_1, \ldots, \alpha_n] = \det(\mathrm{T}(\alpha_i \alpha_j)).$$

Proof: $\mathrm{T}(\alpha_i \alpha_j) = \sum_{r=1}^{n} \sigma_r(\alpha_i \alpha_j) = \sum_{r=1}^{n} \sigma_r(\alpha_i) \sigma_r(\alpha_j)$. Hence

$$
\begin{aligned}
\Delta[\alpha_1, \ldots, \alpha_n] &= (\det(\sigma_i(\alpha_j)))^2 \\
&= (\det(\sigma_j(\alpha_i)))(\det(\sigma_i(\alpha_j))) \\
&= \det(\sum_{r=1}^{n} \sigma_r(\alpha_i) \sigma_r(\alpha_j)) \\
&= \det(\mathrm{T}(\alpha_i \alpha_j)). \qquad \square
\end{aligned}
$$

2.6 Rings of Integers

We now discuss how to find the ring of integers of a given number field. With the methods available to us, this involves moderately heavy calculation, but by taking advantage of short cuts the technique can be made reasonably efficient. In particular Example 2.3 shows that not every number field has an integral basis of the form $\{1, \theta, \ldots, \theta^{n-1}\}$.

The method is based on the following result:

Theorem 2.20. *Let* G *be an additive subgroup of* $\mathfrak{O}$ *of rank equal to the degree of* K, *with* $\mathbb{Z}$-basis $\{\alpha_1, \ldots, \alpha_n\}$. Then $|\mathfrak{O}/G|^2$ divides $\Delta[\alpha_1, \ldots, \alpha_n]$.

Proof: By Theorem 1.16 there exists a $\mathbb{Z}$-basis for $\mathfrak{O}$ of the form $\{\beta_1, \ldots, \beta_n\}$ such that G has a $\mathbb{Z}$-basis $\{\mu_1 \beta_1, \ldots, \mu_n \beta_n\}$ for suitable $\mu_i \in \mathbb{Z}$. Now

$$\Delta[\alpha_1, \ldots, \alpha_n] = \Delta[\mu_1 \beta_1, \ldots, \mu_n \beta_n]$$

since by Lemma 1.15 a basis-change has a unimodular matrix. The right-hand side is

$$(\mu_1 \ldots \mu_n)^2 \Delta[\beta_1, \ldots, \beta_n] = (\mu_1 \ldots \mu_n)^2 \Delta$$

where Δ is the discriminant of K and so lies in $\mathbb{Z}$. But

$$|\mu_1 \ldots \mu_n| = |\mathfrak{O}/G|.$$

Therefore

$$|\mathfrak{O}/G|^2 \text{ divides } \Delta[\alpha_1, \ldots, \alpha_n].$$

□

In the above situation we use the notation

$$\Delta_G = \Delta[\alpha_1, \ldots, \alpha_n].$$

We then have a generalization of Theorem 2.17:

Proposition 2.21. *Suppose that* $G \neq \mathfrak{O}$. *Then there exists an algebraic integer of the form*

$$\frac{1}{p}(\lambda_1\alpha_1 + \ldots + \lambda_n\alpha_n) \tag{2.7}$$

where $0 \leq \lambda_i \leq p - 1$, $\lambda_i \in \mathbb{Z}$, *and* p *is a prime such that* p^2 *divides* Δ_G.

Proof: If $G \neq \mathfrak{O}$ then $|\mathfrak{O}/G| > 1$. Therefore (by the structure theory for finite abelian groups) there exists a prime p dividing $|\mathfrak{O}/G|$ and an element $u \in \mathfrak{O}/G$ such that $g = pu \in G$. By Theorem 2.20, p^2 divides Δ_G. Further,

$$u = \frac{1}{p}g = \frac{1}{p}(\lambda_1\alpha_1 + \ldots + \lambda_n\alpha_n)$$

since $\{\alpha_i\}$ forms a $\mathbb{Z}$-basis for G.

□

Note that this really *is* a generalization of Theorem 2.17: if Δ_G is squarefree then no such p exists, so that $G = \mathfrak{O}$.

We may use Proposition 2.21 as the basis of a trial-and-error search for algebraic integers in $\mathfrak{O}$ but not in G, because there are only finitely many possibilities (6). The idea is:

(a) Start with an initial guess G for $\mathfrak{O}$.

(b) Compute Δ_G.

(c) For each prime p whose square divides Δ_G, test all numbers of the form (2.6) to see which are algebraic integers.

(d) If any new integers arise, enlarge G to a new G' by adding in the new number (and divide Δ_G by p^2 to get $\Delta_{G'}$).

(e) Repeat until no new algebraic integers are found.

Example 2.22. Find the ring of integers of $\mathbb{Q}(\sqrt[3]{5})$.

Let $\theta \in \mathbb{R}$, $\theta^3 = 5$. The natural first guess is that $\mathfrak{O}$ has $\mathbb{Z}$-basis $\{1, \theta, \theta^2\}$. Let G be the abelian group generated by this set. Let $\omega = e^{2\pi i/3}$ be a cube root of unity. Compute

$$
\Delta_G = \begin{vmatrix} 1 & \theta & \theta^2 \\ 1 & \omega\theta & \omega^2\theta^2 \\ 1 & \omega^2\theta & \omega\theta^2 \end{vmatrix}^2
$$

$$
= \theta^6 \begin{vmatrix} 1 & 1 & 1 \\ 1 & \omega & \omega^2 \\ 1 & \omega^2 & \omega \end{vmatrix}^2
$$

$$
= 5^2 \cdot (\omega^2 + \omega^2 + \omega^2 - \omega - \omega - \omega)^2
$$
$$
= 5^2 \cdot 3^2 \cdot (\omega^2 - \omega)^2
$$
$$
= 3^2 \cdot 5^2 \cdot (-3)
$$
$$
= -3^3 \cdot 5^2.
$$

By Proposition 2.21 we must consider two possibilities.

(a) Can $\alpha = \frac{1}{3}(\lambda_1 + \lambda_2\theta + \lambda_3\theta^2)$ be an algebraic integer, for $0 \le \lambda_i \le 2$?

(b) Can $\alpha = \frac{1}{5}(\lambda_1 + \lambda_2\theta + \lambda_3\theta^2)$ be an algebraic integer, for $0 \le \lambda_i \le 4$?

Consider case (b), which is harder. First use the trace: we have

$$
T(\alpha) = 3\lambda_1/5 \in \mathbb{Z}
$$

so that $\lambda_1 \in 5\mathbb{Z}$. Then

$$
\alpha' = \frac{1}{5}(\lambda_2\theta + \lambda_3\theta^2)
$$

is also an algebraic integer.

Now compute the norm of α'. (It is easier to do this for α' than for α because there are fewer terms, which is why we use the trace first.) We have

$$
\begin{aligned}
N(a\theta + b\theta^2) &= (a\theta + b\theta^2)(a\omega\theta + b\omega^2\theta^2)(a\omega^2\theta + b\omega\theta^2) \\
&= \omega \cdot \omega^2(a\theta + b\theta^2)(a\theta + \omega b\theta^2)(a\theta + \omega^2 b\theta^2) \\
&= (a\theta)^3 + (b\theta^2)^3 \\
&= 5a^3 + 25b^3.
\end{aligned}
$$

Thus in order for α to be an algebraic integer, we must have $N(\alpha') \in \mathbb{Z}$. But $N(\alpha') = (5\lambda_2^3 + 25\lambda_3^3)/125 = (\lambda_2^3 + 5\lambda_3^3)/25$. One way to finish the calculation is just to try all cases:

λ_2	λ_3	$\lambda_2^3 + 5\lambda_3^3$	Divisible by 25?
0	1	5	No
0	2	40	No
0	3	135	No
0	4	320	No
1	0	1	No
1	1	6	No
1	2	41	No
1	3	136	No
1	4	321	No
2	0	8	No
2	1	13	No
2	2	48	No
2	3	143	No
2	4	328	No
3	0	27	No
3	1	32	No
3	2	67	No
3	3	162	No
3	4	347	No
4	0	64	No
4	1	69	No
4	2	104	No
4	3	199	No
4	4	384	No

Whichever argument we use, we have shown that if there are no better ideas, brute force can suffice. But here it is not hard to find a better idea. Suppose $\lambda_2^3 + 5\lambda_3^3 \equiv 0 \pmod{25}$. If $\lambda_3 \equiv 0 \pmod{5}$, then we must also have $\lambda_2 \equiv 0 \pmod 5$. If not, we have $5 \equiv (-\lambda_2/\lambda_3)^3 \pmod{25}$. Therefore 5 is a *cubic residue* (mod 25), that is, is congruent to a cube. The factor 5 shows that we must have $5 \equiv (5k)^3 \pmod{25}$, but then $5 \equiv 0 \pmod{25}$, an impossibility.

Whichever argument we use, we have shown that no new α' occurs in case (b). The analysis in case (a) is similar, and left as Exercise 6 in this chapter.

Note that it is *necessary* for $N(\alpha)$ and $T(\alpha)$ to be rational integers, in order for α to be an algebraic integer; but it may not be *sufficient*. If the use of norms and traces produces a candidate for a new algebraic integer, we still have to check that it is one—for example, by finding its minimum polynomial. However, our main use of $N(\alpha)$ and $T(\alpha)$ is to rule out possible candidates, so this step is not always needed.

Example 2.23. (a) Find the ring of integers of $\mathbb{Q}(\sqrt[3]{175})$.
   (b) Show that it has no $\mathbb{Z}$-basis of the form $\{1, \theta, \theta^2\}$.

(a) Let $t = \sqrt[3]{175} = \sqrt[3]{(5^2 \cdot 7)}$. Consider also $u = \sqrt[3]{5 \cdot 7^2} = \sqrt[3]{245}$. Now

$$
\begin{aligned}
ut &= 35 \\
u^2 &= 7t \\
t^2 &= 5u.
\end{aligned}
$$

Let $\mathfrak{O}$ be the ring of integers of $K = \mathbb{Q}(\sqrt[3]{175})$. We have $u = 35/t \in K$. But $u^3 - 245 = 0$ so $u \in \mathbb{B}$. Therefore $u \in \mathbb{B} \cap K = \mathfrak{O}$. A good initial guess is that $\mathfrak{O} = G$, where G is the abelian group generated by $\{1, t, u\}$.

To see if this is correct, compute Δ_G. The monomorphisms $K \to \mathbb{C}$ are $\sigma_1, \sigma_2, \sigma_3$ where $\sigma_1(t) = t$, $\sigma_2(t) = \omega t$, $\sigma_3(t) = \omega^3 t$. Since $tu = 35$, which must be fixed by each σ_i, we have $\sigma_1(u) = u$, $\sigma_2(u) = \omega^2 u$, $\sigma_3(u) = \omega u$. Therefore

$$
\Delta_G = \begin{vmatrix} 1 & t & u \\ 1 & \omega t & \omega^2 u \\ 1 & \omega^2 t & \omega u \end{vmatrix}^2
$$

which works out as $-3^3 \cdot 5^2 \cdot 7^2$.

There are now three primes to try: $p = 3, 5$, or 7.

If $p = 5$ or 7 then, as in Example 2.22, use of the trace lets us assume that our putative integer is $\frac{1}{p}(at + bu)$ for $a, b, \in \mathbb{Z}$. Now

$$
N(at + bu) = 175a^3 + 245b^3
$$

and we must see whether this can be congruent to $0 \pmod{5^3 \text{ or } 7^3}$ for a, b not congruent to zero.

Suppose $175a^3 + 245b^3 \equiv 0 \pmod{125}$, that is, $35a^3 + 49b^3 \equiv 0 \pmod{25}$. Write this as $10a^3 - b^3 \equiv 0 \pmod{25}$. If $a \equiv 0 \pmod 5$ then also $b \equiv 0 \pmod 5$. If not, $10 \equiv (b/a)^3 \pmod{25}$ is a cubic residue; but then $10 \equiv (5k)^3 \pmod{25}$, hence $10 \equiv 0 \pmod{25}$ which is absurd. The case $p = 7$ is dealt with in the same way.

When $p = 3$ the trace is no help, and we must compute the norm of

$$
\frac{1}{3}(a + bt + cu)
$$

for $a, b, c \in \mathbb{Z}$. The calculation is more complicated, but not too bad since we only have to consider $a, b, c = 0, 1, 2$. No new integers occur.

Therefore $\mathfrak{O} = G$ as we hoped.

(b) Now we have to show that there is no $\mathbb{Z}$-basis of the form $\{1, \theta, \theta^2\}$, where $\theta = a + bt + cu$. Note that $\{1, \theta, \theta^2\}$ is a $\mathbb{Z}$-basis if and only if $\{1, \theta+1, (\theta+1)^2\}$ is a $\mathbb{Z}$-basis, so without loss of generality we may assume that $a = 0$. Now

$$(bt + cu)^2 = b^2 t^2 + 2bctu + c^2 u^2$$
$$= 5b^2 u + 70bc + 7c^2 t.$$

Therefore $\{1, bt + cu, (bt + cu)^2\}$ is a $\mathbb{Z}$-basis if and only if the matrix

$$\begin{vmatrix} 1 & 0 & 0 \\ 0 & b & c \\ 70bc & 7c^2 & 5b^2 \end{vmatrix}$$

is unimodular; that is,

$$5b^3 - 7c^3 = \pm 1.$$

Consider this modulo 7. Cubes are congruent to 0, 1, or -1 (mod 7), so $5(-1, 0, \text{ or } 1) \equiv \pm 1$ (mod 7), a contradiction. Hence no such $\mathbb{Z}$-basis exists.

Example 2.24. Find the ring of integers of $\mathbb{Q}(\sqrt{2}, i)$.

In this example our initial guess turns out not to be good enough, illustrating how to continue the analysis when this unfortunate event occurs.

The obvious guess is $\{1, \sqrt{2}, i, i\sqrt{2}\}$. Let G be the group these generate. We have $\Delta_G = -64$, so $\mathfrak{O}$ may contain elements of the form $\frac{1}{2}g$ (and then possibly $\frac{1}{4}g$ or $\frac{1}{8}g$) for $g \in G$. The norm is

$$N(a + b\sqrt{2} + ci + di\sqrt{2}) = (a^2 - c^2 - 2b^2 + 2d^2)^2 + 4(ac - 2bd)^2.$$

We must find whether this is divisible by 16 for $a, b, c, d = 0$ or 1, and not all zero. By trial and error the only case where this occurs is $b = d = 1$, $a = c = 0$. So

$$\alpha = \tfrac{1}{2}(\theta + \theta i)$$

may be an integer (where $\theta = \sqrt{2}$). In fact

$$\alpha^2 = i$$

so that

$$\alpha^4 + 1 = 0$$

and α *is* an integer.

We therefore revise our initial guess to

$$G' = \{1, \theta, i, \theta i, \tfrac{1}{2}\theta(1+i)\}.$$

Since $2 \cdot \tfrac{1}{2}\theta(1+i) = \theta + \theta i$ this has a $\mathbb{Z}$-basis

$$\{1, \theta, i, \tfrac{1}{2}\theta(1+i)\}.$$

Now

$$\Delta_{G'} = -64/2^2 = -16.$$

A recalculation of the usual kind shows that nothing of the form $\tfrac{1}{2}g$ (where we may now assume that the term in $\tfrac{1}{2}\theta(1+i)$ occurs with nonzero coefficient) has integer norm. So no new integers arise and $\mathfrak{O} = G'$.

2.7   Exercises

1. Which of the following complex numbers are algebraic? Which are algebraic integers?

   (a) $355/113$

   (b) $e^{2\pi i/23}$

   (c) $e^{\pi i/23}$

   (d) $\sqrt{17} + \sqrt{19}$

   (e) $(1 + \sqrt{17})/(2\sqrt{-19})$

   (f) $\sqrt{(1 + \sqrt{2})} + \sqrt{(1 - \sqrt{2})}$

2. Express $\mathbb{Q}(\sqrt{3}, \sqrt[3]{5})$ in the form $\mathbb{Q}(\theta)$.

3. Find all monomorphisms $\mathbb{Q}(\sqrt[3]{7}) \to \mathbb{C}$.

4. Find the discriminant of $\mathbb{Q}(\sqrt{3}, \sqrt{5})$.

5. Let $K = \mathbb{Q}(\sqrt[4]{2})$. Find all monomorphisms $\sigma : K \to \mathbb{C}$ and the minimum polynomials (over $\mathbb{Q}$) and field polynomials (over K) of
   (a) $\sqrt[4]{2}$ (b) $\sqrt{2}$ (c) 2 (d) $\sqrt{2} + 1$. Compare with Theorem 2.6.

6. Complete Example 2.22 by discussing the case $p = 3$.

7. Complete Example 2.23 by discussing the case $p = 3$.

8. Compute integral bases and discriminants of

    (a) $\mathbb{Q}(\sqrt{2}, \sqrt{3})$

    (b) $\mathbb{Q}(\sqrt{2}, i)$

    (c) $\mathbb{Q}(\sqrt[3]{2})$

    (d) $\mathbb{Q}(\sqrt[4]{2})$

9. Let $K = \mathbb{Q}(\theta)$ where $\theta \in \mathfrak{O}_K$. Among the elements

$$\frac{1}{d}(a_0 + \ldots + a_i\theta^i)$$

    $(0 \neq a_i; a_0, \ldots, a_i \in \mathbb{Z})$, where d is the discriminant, pick one with minimal value of $|a_i|$ and call it x_i. Do this for $i = 1, \ldots, n = [K : \mathbb{Q}]$ show that $\{x_1, \ldots, x_n\}$ is an integral basis.

10. If $\alpha_1, \ldots, \alpha_n$ are $\mathbb{Q}$-linearly independent algebraic integers in $\mathbb{Q}(\theta)$, and if

$$\Delta[\alpha_1, \ldots, \alpha_n] = d$$

    where d is the discriminant of $\mathbb{Q}(\theta)$, show that $\{\alpha_i, \ldots, \alpha_n\}$ is an integral basis for $\mathbb{Q}(\theta)$.

11. If $[K : \mathbb{Q}] = n$, $\alpha \in \mathbb{Q}$, show

$$\begin{aligned} \mathrm{N}_K(\alpha) &= \alpha^n, \\ \mathrm{T}_K(\alpha) &= n\alpha. \end{aligned}$$

12. Give examples to show that for fixed α, $\mathrm{N}_K(\alpha)$ and $\mathrm{T}_K(\alpha)$ depend on K. (This is to emphasize that the norm and trace must always be defined in the context of a specific field K; there is no such thing as the norm or trace of α without a specified field.)

13. The norm and trace may be generalized by considering number fields $K \supseteq L$. Suppose $K = L(\theta)$ and $[K : L] = n$. Consider monomorphisms $\sigma : K \to \mathbb{C}$ such that $\sigma(x) = x$ for all $x \in L$. Show that there are precisely n such monomorphisms $\sigma_1, \ldots, \sigma_n$ and describe them. For $\alpha \in K$, define

$$\mathrm{N}_{K/L}(\alpha) = \prod_{i=1}^{n} \sigma_i(\alpha),$$

$$\mathrm{T}_{K/L}(\alpha) = \sum_{i=1}^{n} \sigma_i(\alpha).$$

(Compared with our earlier notation, we have $N_K = N_{K/\mathbb{Q}}$, $T_K = T_{K/\mathbb{Q}}$.) Prove that

$$N_{K/L}(\alpha_1\alpha_2) = N_{K/L}(\alpha_1)N_{K/L}(\alpha_2),$$

$$T_{K/L}(\alpha_1 + \alpha_2) = T_{K/L}(\alpha_1) + T_{K/L}(\alpha_2).$$

Let $K = \mathbb{Q}(\sqrt[4]{3})$, $L = \mathbb{Q}(\sqrt{3})$. Calculate $N_{K/L}(\sqrt{\alpha})$, $T_{K/L}(\alpha)$ for $\alpha = \sqrt[4]{3}$ and $\alpha = \sqrt[4]{3} + \sqrt{3}$.

14. For $K = \mathbb{Q}(\sqrt[4]{3})$, $L = \mathbb{Q}(\sqrt{3})$, calculate $N_{K/L}(\sqrt{3})$ and $N_{K/\mathbb{Q}}(\sqrt{3})$. Deduce that $N_{K/L}(\alpha)$ depends on K and L (provided that $\alpha \in K$). Do the same for $T_{K/L}$.

Quadratic and
Cyclotomic Fields

In this chapter we investigate two special types of number field in the light of our previous work. The quadratic fields are those of degree 2, and are especially important in the study of quadratic forms. The cyclotomic fields are generated by pth roots of unity, and we consider only the case p prime; these fields are central to Kummer's approach to Fermat's Last Theorem and play a substantial role in all subsequent work, including Wiles's proof. We return to both types of field at later stages. For the moment we content ourselves with finding the rings of integers, integral bases, and discriminants.

3.1 Quadratic Fields

A *quadratic field* is a number field K of degree 2 over $\mathbb{Q}$. Then $K = \mathbb{Q}(\theta)$ where θ is an algebraic integer, and θ is a zero of

$$t^2 + at + b \qquad (a, b \in \mathbb{Z}).$$

Thus

$$\theta = \frac{-a \pm \sqrt{(a^2 - 4b)}}{2}.$$

Let $a^2 - 4b = r^2 d$ where $r, d \in \mathbb{Z}$ and d is squarefree. (That this is always possible follows from prime factorization in $\mathbb{Z}$.) Then

$$\theta = \frac{-a \pm r\sqrt{d}}{2}$$

and so $\mathbb{Q}(\theta) = \mathbb{Q}(\sqrt{d})$. This proves:

Proposition 3.1. *The quadratic fields are precisely those of the form $\mathbb{Q}(\sqrt{d})$ for d a squarefree rational integer.* □

Next we determine the ring of integers of $\mathbb{Q}(\sqrt{d})$, for squarefree d. The answer, it turns out, depends on the arithmetic properties of d.

Theorem 3.2. *Let d be a squarefree rational integer. Then the integers of $\mathbb{Q}(\sqrt{d})$ are:*

(a) $\mathbb{Z}[\sqrt{d}]$ *if* $d \not\equiv 1 \pmod 4$,
(b) $\mathbb{Z}[\frac{1}{2} + \frac{1}{2}\sqrt{d}]$ *if* $d \equiv 1 \pmod 4$.

Proof: Every element $\alpha \in \mathbb{Q}(\sqrt{d})$ is of the form $\alpha = r + s\sqrt{d}$ for $r, s \in \mathbb{Q}$. Hence

$$\alpha = \frac{a + b\sqrt{d}}{c}$$

where $a, b, c \in \mathbb{Z}$, $c > 0$, and no prime divides all of a, b, c. Now α is an integer if and only if the coefficients of the minimum polynomial

$$\left(t - \left(\frac{a + b\sqrt{d}}{c} \right) \right) \left(t - \left(\frac{a - b\sqrt{d}}{c} \right) \right)$$

are integers. Thus

$$\frac{a^2 - b^2 d}{c^2} \in \mathbb{Z}, \tag{3.1}$$

$$\frac{2a}{c} \in \mathbb{Z}. \tag{3.2}$$

If c and a have a common prime factor p then (3.1) implies that p divides b (since d is squarefree) which contradicts our previous assumption. Hence from (3.2) $c = 1$ or 2. If $c = 1$ then α is an integer of K in any case, so we may concentrate on the case $c = 2$. Now a and b must both be odd, and $(a^2 - b^2 d)/4 \in \mathbb{Z}$. Hence

$$a^2 - b^2 d \equiv 0 \pmod 4.$$

Now an odd number $2k + 1$ has square $4k^2 + 4k + 1 \equiv 1 \pmod{4}$, hence $a^2 \equiv 1 \equiv b^2 \pmod{4}$, and this implies $d \equiv 1 \pmod{4}$. Conversely, if $d \equiv 1 \pmod{4}$ then for odd a, b we have α an integer because (3.1) and (3.2) hold.

To sum up: if $d \equiv 1 \pmod{4}$ then $c = 1$ and so (a) holds; whereas if $d \equiv 1 \pmod{4}$ we can also have $c = 2$ and a, b odd, whence easily (b) holds. □

The monomorphisms $K \to \mathbb{C}$ are

$$\begin{aligned}
\sigma_1(r + s\sqrt{d}) &= r + s\sqrt{d}, \\
\sigma_2(r + s\sqrt{d}) &= r - s\sqrt{d}.
\end{aligned}$$

We can therefore compute discriminants:

Theorem 3.3. *(a) If $d \not\equiv 1 \pmod{4}$ then $\mathbb{Q}(\sqrt{d})$ has an integral basis of the form $\{1, \sqrt{d}\}$ and discriminant $4d$. (b) If $d \equiv 1 \pmod{4}$ then $\mathbb{Q}(\sqrt{d})$ has an integral basis of the form $\{1, \frac{1}{2} + \frac{1}{2}\sqrt{d}\}$ and discriminant d.*

Proof: The assertions regarding bases are clear from Theorem 3.2. Compute discriminants:

$$\begin{vmatrix} 1 & \sqrt{d} \\ 1 & -\sqrt{d} \end{vmatrix}^2 = (-2\sqrt{d})^2 = 4d,$$

$$\begin{vmatrix} 1 & \frac{1}{2} + \frac{1}{2}\sqrt{d} \\ 1 & \frac{1}{2} - \frac{1}{2}\sqrt{d} \end{vmatrix}^2 = (-\sqrt{d})^2 = d.$$

□

Since the discriminants of isomorphic fields are equal, the fields $\mathbb{Q}(\sqrt{d})$ are not isomorphic for distinct squarefree d. This completes the classification of quadratic fields.

A special case, of historical interest as the first number field to be studied as such, is the *Gaussian field* $\mathbb{Q}(\sqrt{-1})$ or $\mathbb{Q}(i)$. Since $-1 \not\equiv 1 \pmod{4}$ the ring of integers is $\mathbb{Z}[\sqrt{-1}]$ or $\mathbb{Z}(i)$, known as the ring of *Gaussian integers*, and the discriminant is -4.

Incidentally, these results show that Theorem 2.17 is not always applicable: an integral basis *can* have a discriminant that is not squarefree. For instance, the Gaussian integers themselves.

For future reference we note the norms and traces:

$$\begin{aligned}
N(r + s\sqrt{d}) &= r^2 - ds^2, \\
T(r + s\sqrt{d}) &= 2r.
\end{aligned}$$

We also note some useful terminology. A quadratic field $\mathbb{Q}(\sqrt{d})$ is said to be *real* if d is positive, *imaginary* if d is negative. (A real quadratic field contains only real numbers, an imaginary quadratic field contains proper complex numbers as well.)

3.2 Cyclotomic Fields

A *cyclotomic field* is one of the form $\mathbb{Q}(\zeta)$ where $\zeta = e^{2\pi i/m}$ is a primitive complex mth root of unity. (The name means 'circle-cutting' and refers to the equal spacing of powers of ζ around the unit circle in the complex plane.) We consider only the case $m = p$, a prime number. Further, if $p = 2$ then $\zeta = -1$ so that $\mathbb{Q}(\zeta) = \mathbb{Q}$, hence we ignore this case and assume p odd.

Lemma 3.4. *The minimum polynomial of $\zeta = e^{2\pi i/p}$, p an odd prime, over $\mathbb{Q}$ is*

$$f(t) = t^{p-1} + t^{p-2} + \ldots + t + 1.$$

The degree of $\mathbb{Q}(\zeta)$ is $p - 1$.

Proof: We have

$$f(t) = \frac{t^p - 1}{t - 1}.$$

Since $\zeta - 1 \neq 0$ and $\zeta^p = 1$ it follows that $f(\zeta) = 0$, so all we need prove is that f is irreducible. This we do by a standard piece of trickery. We have

$$f(t+1) = \frac{(t+1)^p - 1}{t} = \sum_{r=1}^{p} \binom{p}{r} t^{r-1}.$$

Now the binomial coefficient $\binom{p}{r}$ is divisible by p if $1 \le r \le p - 1$, and $\binom{p}{1} = p$ is not divisible by p^2.

Hence by Eisenstein's criterion (Theorem 1.8) $f(t + 1)$ is irreducible. Therefore $f(t)$ is irreducible, and is the minimum polynomial of ζ. Since $\partial f = p - 1$ we have $[\mathbb{Q}(\zeta) : \mathbb{Q}] = p - 1$ by Theorem 1.11.   □

The powers $\zeta, \zeta^2, \ldots, \zeta^{p-1}$ are also pth roots of unity, not equal to 1, and so by the same argument have $f(t)$ as minimum polynomial. Clearly

$$f(t) = (t - \zeta)(t - \zeta^2) \ldots (t - \zeta^{p-1}) \qquad (3.3)$$

so the conjugates of ζ are $\zeta, \zeta^2, \ldots, \zeta^{p-1}$. Therefore the monomorphisms from $\mathbb{Q}(\zeta)$ to $\mathbb{C}$ are

$$\sigma_i(\zeta) = \zeta^i \qquad (1 \leq i \leq p-1).$$

Because the minimum polynomial $f(t)$ has degree $p-1$, a basis for $\mathbb{Q}(\zeta)$ over $\mathbb{Q}$ is $1, \zeta, \ldots, \zeta^{p-2}$, so for a general element

$$\alpha = a_0 + a_1\zeta + \ldots + a_{p-2}\zeta^{p-2} \qquad (a_i \in \mathbb{Q})$$

we have

$$\sigma_i(a_0 + \zeta + \ldots + a_{p-2}\zeta^{p-2}) = a_0 + \zeta^i + \ldots + a_{p-2}\zeta^{i(p-2)}.$$

From this formula the norm and trace may be calculated using the basic definitions

$$N(\alpha) = \prod_{i=1}^{p-1} \sigma_i(\alpha),$$

$$T(\alpha) = \sum_{i=1}^{p-1} \sigma_i(\alpha).$$

In particular

$$N(\zeta) = \zeta \cdot \zeta^2 \ldots \zeta^{p-1}.$$

Now ζ and $\zeta^i (1 \leq i \leq p-1)$ are conjugates, so have the same norm, which can be calculated by putting $t = 0$ in (3.3) to give

$$N(\zeta) = N(\zeta^i) = (-1)^{p-1}$$

and since p is odd,

$$N(\zeta^i) = 1 \quad (1 \leq i \leq p-1). \qquad (3.4)$$

The trace of ζ^i can be found by a similar argument. We have

$$T(\zeta^i) = T(\zeta) = \zeta + \zeta^2 + \ldots + \zeta^{p-1},$$

and since

$$f(\zeta) = 1 + \zeta + \ldots + \zeta^{p-1} = 0$$

we find

$$T(\zeta^i) = -1 \quad (1 \leq i \leq p-1). \qquad (3.5)$$

For $a \in \mathbb{Q}$ we trivially have

$$\begin{aligned} N(a) &= a^{p-1} \\ T(a) &= (p-1)a. \end{aligned}$$

Since $\zeta^p = 1$, we can use these formulas to extend (3.4) and (3.5) to

$$N(\zeta^s) = 1 \qquad \text{for all } s \in \mathbb{Z} \tag{3.6}$$

and

$$T(\zeta^s) = \begin{cases} -1 & \text{if } s \not\equiv 0 \ (\text{mod } p) \\ p-1 & \text{if } s \equiv 0 \ (\text{mod } p). \end{cases} \tag{3.7}$$

For a general element of $\mathbb{Q}(\zeta)$, the trace is easily calculated:

$$\begin{aligned} T\left(\sum_{i=0}^{p-2} a_i \zeta^i\right) &= \sum_{i=0}^{p-2} T(a_i \zeta^i) \\ &= T(a_0) + \sum_{i=1}^{p-2} T(a_i \zeta^i) \\ &= (p-1)a_0 - \sum_{i=0}^{p-2} a_i \end{aligned}$$

so

$$T\left(\sum_{i=0}^{p-2} a_i \zeta^i\right) = p a_0 - \sum_{i=1}^{p-2} a_i. \tag{3.8}$$

The norm is more complicated in general, but a useful special case is

$$N(1 - \zeta) = \prod_{i=1}^{p-1} (1 - \zeta^i)$$

which can be calculated by putting $t = 1$ in (3.3) to obtain

$$\prod_{i=1}^{p-1} (1 - \zeta^i) = p, \tag{3.9}$$

so

$$N(1 - \zeta) = p. \tag{3.10}$$

We can put these computations to good use, first by showing that the integers of $\mathbb{Q}(\zeta)$ are what one naively might expect:

Theorem 3.5. *The ring $\mathfrak{O}$ of integers of $\mathbb{Q}(\zeta)$ is $\mathbb{Z}[\zeta]$.*

Proof: Suppose $\alpha = a_0 + a_1\zeta + \ldots + a_{p-2}\zeta^{p-2}$ is an integer in $\mathbb{Q}(\zeta)$. We must demonstrate that the rational numbers a_i are actually rational integers.

For $0 \le k \le p-2$ the element

$$\alpha\zeta^{-k} - \alpha\zeta$$

is an integer, so its trace is a rational integer. But

$$
\begin{aligned}
&\mathrm{T}(\alpha\zeta^{-k} - \alpha\zeta) \\
=\ &\mathrm{T}(a_0\zeta^{-k} + \ldots + a_k + \ldots + a_{p-2}\zeta^{p-k-2} - a_0\zeta - \ldots - a_{p-2}\zeta^{p-1}) \\
=\ &pa_k - (a_0 + \ldots + a_{p-2}) - (-a_0 - \ldots - a_{p-2}) \\
=\ &pa_k.
\end{aligned}
$$

Hence $b_k = pa_k$ is a rational integer.

Put $\lambda = 1 - \zeta$. Then

$$
\begin{aligned}
p\alpha &= b_0 + b_1\zeta + \ldots + b_{p-2}\zeta^{p-2} \\
&= c_0 + c_1\lambda + \ldots + c_{p-2}\lambda^{p-2}
\end{aligned}
\tag{3.11}
$$

where, substituting $\zeta = 1 - \lambda$ and expanding,

$$c_i = \sum_{j=i}^{p-2} (-1)^i \binom{j}{i} b_j \in \mathbb{Z}.$$

Since $\lambda = 1 - \zeta$ we also have, symmetrically,

$$b_i = \sum_{j=i}^{p-2} (-1)^i \binom{j}{i} c_j.
\tag{3.12}$$

We claim that all c_i, are divisible by p. By induction we may assume this for all c_i with $i \le k-1$, where $0 \le k \le p-2$. Since $c_0 = b_0 + \ldots + b_{p-2} = p(-\mathrm{T}(\alpha) + b_0)$, we have $p|c_0$, so it is true for $k = 0$. Now by (3.9)

$$
\begin{aligned}
p &= \prod_{i=1}^{p-1}(1 - \zeta^i) \\
&= (1 - \zeta)^{p-1}\prod_{i=1}^{p-1}(1 + \zeta + \ldots + \zeta^{i-1}) \\
&= \lambda^{p-1}\kappa
\end{aligned}
\tag{3.13}
$$

where $\kappa \in \mathbb{Z}[\zeta] \subseteq \mathfrak{O}$. Consider (3.11) as a congruence modulo the ideal $\langle \lambda^{k+1} \rangle$ of $\mathfrak{O}$. By (3.13)

$$p \equiv 0 \qquad (\mathrm{mod}\ \langle \lambda^{k+1} \rangle).$$

so the left-hand side of (3.11) and the terms up to $c_{k-1}\lambda^{k-1}$ vanish. Further, the terms from $c_{k+1}\lambda^{k+1}$ onwards are multiples of λ^{k+1} and also vanish. There remains:

$$c_k\lambda^k \equiv 0 \qquad (\mathrm{mod}\ \langle\lambda^{k+1}\rangle).$$

This is equivalent to

$$c_k\lambda^k = \mu\lambda^{k+1}$$

for some $\mu \in \mathfrak{O}$, from which we obtain

$$c_k = \mu\lambda.$$

Take norms:

$$c_k^{p-1} = \mathrm{N}(c_k) = \mathrm{N}(\mu)\mathrm{N}(\lambda) = p\mathrm{N}(\mu),$$

since $\mathrm{N}(\lambda) = p$ by (3.10). Hence $p|c_k^{p-1}$, so $p|c_k$. By induction $p|c_k$ for all k, and then (3.12) shows that $p|b_k$ for all k. Therefore $a_k \in \mathbb{Z}$ for all k. $\square$

Now we can compute the discriminant.

Theorem 3.6. *Let p be an odd prime and $\zeta = e^{2\pi i/p}$. The discriminant of $\mathbb{Q}(\zeta)$ is*

$$(-1)^{(p-1)/2} \cdot p^{p-2}.$$

Proof: By Theorem 3.5 an integral basis is $\{1, \zeta, \ldots, \zeta^{p-2}\}$. By Proposition 2.18 the discriminant is

$$(-1)^{(p-1)(p-2)/2} \cdot \mathrm{N}(Df(\zeta))$$

with $f(t)$ as above. Since p is odd the first factor reduces to $(-1)^{(p-1)/2}$. To evaluate the second, recall that

$$f(t) = \frac{t^p - 1}{t - 1},$$

so

$$Df(t) = \frac{(t-1)pt^{p-1} - (t^p - 1)}{(t-1)^2}$$

whence

$$Df(\zeta) = \frac{-p\zeta^{p-1}}{\lambda}$$

where $\lambda = 1 - \zeta$ as before. Hence

$$
\begin{aligned}
N(Df(\zeta)) &= \frac{N(p)N(\zeta)^{p-1}}{N(\lambda)} \\
&= \frac{(-p)^{p-1}1^{p-1}}{p} \\
&= p^{p-2}. \qquad \square
\end{aligned}
$$

The case $p = 3$ deserves special mention, for $\mathbb{Q}(\zeta)$ has degree $p - 1 = 2$, so it is a quadratic field. Since

$$
e^{2\pi i/3} = \frac{-1 + \sqrt{-3}}{2}
$$

it is equal to $\mathbb{Q}(\sqrt{-3})$. As a check on our discriminant calculations: Theorem 3.3 gives -3 (since $-3 \equiv 1 \pmod 4$), and Theorem 3.6 gives $(-1)^{2/2}3^1 = -3$ as well.

3.3 Exercises

1. Find integral bases and discriminants for:

   (a) $\mathbb{Q}(\sqrt{3})$

   (b) $\mathbb{Q}(\sqrt{-7})$

   (c) $\mathbb{Q}(\sqrt{11})$

   (d) $\mathbb{Q}(\sqrt{-11})$

   (e) $\mathbb{Q}(\sqrt{6})$

   (f) $\mathbb{Q}(\sqrt{-6})$

2. Let $K = \mathbb{Q}(\zeta)$ where $\zeta = e^{2\pi i/5}$. Calculate $N_K(\alpha)$ and $T_K(\alpha)$ for the following values of α:

   (a) ζ^2 (b) $\zeta + \zeta^2$ (c) $1 + \zeta + \zeta^2 + \zeta^3 + \zeta^4$.

3. Let $K = \mathbb{Q}(\zeta)$ where $\zeta = e^{2\pi i/p}$ for a rational prime p. In the ring of integers $\mathbb{Z}[\zeta]$, show that $\alpha \in \mathbb{Z}[\zeta]$ is a unit if and only if $N_K(\alpha) = \pm 1$.

4. If $\zeta = e^{2\pi i/3}$, $K = \mathbb{Q}(\zeta)$, prove that the norm of $\alpha \in \mathbb{Z}[\zeta]$ is of the form $\frac{1}{4}(a^2 + 3b^2)$ where a, b are rational integers which are either both even or both odd. Using the result of Exercise 3, deduce that there are precisely six units in $\mathbb{Z}[\zeta]$ and find them all.

5. If $\zeta = e^{2\pi i/5}$, $K = \mathbb{Q}(\zeta)$, prove that the norm of $\alpha \in \mathbb{Z}[\zeta]$ is of the form $\frac{1}{4}(a^2 - 5b^2)$ where a, b are rational integers. (*Hint:* in calculating $N(\alpha)$, first calculate $\sigma_1(\alpha)\sigma_4(\alpha)$ where $\sigma_i(\zeta) = \zeta^i$. Show that this is of the form $q + r\theta + s\phi$ where q, r, s are rational integers, $\theta = \zeta + \zeta^4$, $\phi = \zeta^2 + \zeta^3$. In the same way, establish $\sigma_2(\alpha)\sigma_3(\alpha) = q + s\theta + r\phi$.) Using Exercise 3, prove that $\mathbb{Z}[\zeta]$ has an infinite number of units.

6. Let $\zeta = e^{2\pi i/5}$. For $K = \mathbb{Q}(\zeta)$, use the formula

$$N_K(a + b\zeta) = (a^5 + b^5)/(a + b)$$

to calculate the following norms:

(a) $N_K(\zeta + 2)$ (b) $N_K(\zeta - 2)$ (c) $N_K(\zeta + 3)$.

Using the fact that if $\alpha\beta = \gamma$, then $N_K(\alpha)N_K(\beta) = N_K(\gamma)$, deduce that $\zeta + 2$, $\zeta - 2$, $\zeta + 3$ have no proper factors (that is, factors that are not units) in $\mathbb{Z}[\zeta]$.

Factorize 11, 31, 61 in $\mathbb{Z}[\zeta]$.

7. If $\zeta = e^{2\pi i/5}$, as in Exercise 6, calculate

(a) $N_K(\zeta + 4)$ (b) $N_K(\zeta - 3)$.

Deduce that any proper factors of $\zeta + 4$ in $\mathbb{Z}[\zeta]$ have norm 5 or 41. Given $\zeta - 1$ is a factor of $\zeta + 4$, find another factor. Verify that $\zeta - 3$ is a unit times $(\zeta^2 + 2)^2$ in $\mathbb{Z}[\zeta]$.

8. Show that the multiplicative group of non-zero elements of $\mathbb{Z}_7$ is cyclic with generator the residue class of 3. If $\zeta = e^{2\pi i/7}$, define the monomorphism $\sigma : \mathbb{Q}(\zeta) \to \mathbb{C}$ by $\sigma(\zeta) = \zeta^3$. Show that all other monomorphisms from $\mathbb{Q}(\zeta)$ to $\mathbb{C}$ are of the form $\sigma^i(1 \leq i \leq 6)$ where $\sigma^6 = 1$. For any $\alpha \in \mathbb{Q}(\zeta)$, define $c(\alpha) = \alpha\sigma^2(\alpha)\sigma^4(\alpha)$, and show $N(\alpha) = c(\alpha) \cdot \sigma c(\alpha)$. Demonstrate that $c(\alpha) = \sigma^2 c(\alpha) = \sigma^4 c(\alpha)$. Using the relation $1 + \zeta + \ldots + \zeta^6 = 0$, show that every element $\alpha \in \mathbb{Q}(\zeta)$ can be written uniquely as $\sum_{i=1}^{6} a_i\zeta^i (a_i \in \mathbb{Q})$. Deduce that $c(\alpha) = a_1\theta_1 + a_3\theta_2$ where $\theta_1 = \zeta + \zeta^2 + \zeta^4$, $\theta_2 = \zeta^3 + \zeta^5 + \zeta^6$. Show that $\theta_1 + \theta_2 = -1$ and calculate $\theta_1\theta_2$. Verify that $c(\alpha)$ may be written in the form $b_0 + b_1\theta_1$ where b_0, $b_1 \in \mathbb{Q}$, and show that $\sigma c(\alpha) = b_0 + b_1\theta_2$. Deduce that

$$N(\alpha) = b_0^2 - b_0 b_1 + 2b_1^2.$$

Now calculate $N(\zeta + 5\zeta^6)$.

9. Suppose that p is a rational prime and $\zeta = e^{2\pi i/p}$. Given that the group of non-zero elements of $\mathbb{Z}_p$ is cyclic (see Appendix 1, Proposition 6 for a proof) show that there is a monomorphism $\sigma : \mathbb{Q}(\zeta) \to \mathbb{C}$

such that σ^{p-1} is the identity and all monomorphisms from $\mathbb{Q}(\zeta)$ to $\mathbb{C}$ are of the form $\sigma^i(1 \leq i \leq p-1)$. If $p-1 = kr$, define $c_k(\alpha) = \alpha\sigma^r(\alpha)\sigma^{2r}(\alpha)\ldots\sigma^{(k-1)r}(\alpha)$. Show that

$$N(\alpha) = c_k(\alpha) \cdot \sigma c_k(\alpha) \ldots \sigma^{r-1}c_k(\alpha).$$

Prove that every element of $\mathbb{Q}(\zeta)$ is uniquely of the form $\sum_{i=1}^{p-1} a_i\zeta^i$, and by demonstrating that $\sigma^r(c_k(\alpha)) = c_k(\alpha)$, deduce that $c_k(\alpha) = b_1\eta_1 + \ldots + b_k\eta_k$, where

$$\eta_1 = \zeta + \sigma^r(\zeta) + \sigma^{2r}(\zeta) + \ldots + \sigma^{(k-1)r}(\zeta)$$

and $\eta_{i+1} = \sigma^i(\eta_1)$.

Interpret these results in the case $p = 5$, $k = r = 2$, by showing that the residue class of 2 is a generator of the multiplicative group of non-zero elements of $\mathbb{Z}_5$. Demonstrate that $c_2(\alpha)$ is of the form $b_1\eta_1 + b_2\eta_2$ where $\eta_1 = \zeta + \zeta^4$, $\eta_2 = \zeta^2 + \zeta^3$.

Calculate the norms of the following elements in $\mathbb{Q}(\zeta)$:

(a) $\zeta + 2\zeta^2$ (b) $\zeta + \zeta^4$ (c) $15\zeta + 15\zeta^4$ (d) $\zeta + \zeta^2 + \zeta^3 + \zeta^4$.

10. In $\mathbb{Z}[\sqrt{-5}]$, prove that 6 factorizes in two ways:

$$6 = 2 \cdot 3 = (1 + \sqrt{-5})(1 - \sqrt{-5})$$

Verify that $2, 3, 1 + \sqrt{-5}, 1 - \sqrt{-5}$ have no proper factors in $\mathbb{Z}[\sqrt{-5}]$. (*Hint:* Take norms and note that if γ factorizes as $\gamma = \alpha\beta$, then $N(\gamma) = N(\alpha)N(\beta)$ is a factorization of rational integers.) Deduce that it is possible in $\mathbb{Z}[\sqrt{-5}]$ for 2 to have no proper factors, yet 2 divides a product $\alpha\beta$ without dividing either α or β.

4

Factorization into Irreducibles

Now we turn to the vexed but vital question of uniqueness of factorization in the ring of integers of an algebraic number field. Historically, experience with unique factorization of integers and polynomials over a field led to a general intuition that factorization of algebraic integers should also be unique. In the early days of algebraic number theory many experts, including Euler, simply assumed uniqueness without perceiving any need for a proof, and used it implicitly to 'prove' results that were later found to be based on a false assumption.

The reason for this misconception is subtle, and has its origins in the definition of a prime number. Two distinct properties can serve as a definition. The most familiar is that a prime number cannot be factorized into the product of two integers other than itself and 1. For a prime p, this property may be written more generally as:

(a) If $p = ab$ then one of a or b must be a unit.

In $\mathbb{Z}$ the only units are ± 1, so this reduces to the usual definition. However, there is a second property of prime numbers that is also of interest, namely that if a prime number p divides a product of two numbers then it must divide one or the other:

(b) If $p|ab$ then $p|a$ or $p|b$.

What fooled our predecessors was that although these properties are equivalent in the ring of integers—and even in some algebraic number rings— they are not equivalent in all cases. Of deeper psychological significance is that the more familiar property (a) turns out to be less powerful than property (b). Property (b) can be shown in general to imply (a); moreover, it is property (b) that guarantees uniqueness of factorization, not the more comfortable property (a). In contrast, property (a) does not imply (b), and (a) turns out to be inadequate to give uniqueness of factorization.

The way out of this dilemma is to use the less familiar (b) for the definition of a prime. An element p satisfying the weaker assumption (a) is no longer called a prime: it is said to be irreducible. (We have used that word in the same sense earlier in this text.) It can then be proved that if factorization into primes (defined in the new sense) is possible, then it is unique. In contrast, factorization into irreducibles may not be unique even when it is possible. For instance, if we work in $\mathbb{Z}[\sqrt{-6}]$, then there are two factorizations $6 = 2 \cdot 3$ and $6 = \sqrt{-6} \cdot \sqrt{-6}$. Here the elements $2, 3, \sqrt{-6}$ are all irreducible (because they cannot be written as a product of nontrivial factors in $\mathbb{Z}[\sqrt{-6}]$); however, they are not prime. For instance, $\sqrt{-6}$ is a factor of $6 = 2 \cdot 3$, but it is not a factor of either 2 or 3 in the ring $\mathbb{Z}[\sqrt{-6}]$.

We must therefore proceed with care, and an awareness of precisely what we are doing. For instance, if we attempt to factorize an element x in a domain D, it is natural to seek proper factors $x = ab$ (meaning that neither a nor b is a unit). If either of these factors is further reducible, we factorize it, and so on, seeking a factorization

$$x = a_1 a_2 \ldots a_n$$

into factors that cannot be reduced any further. Reflecting on what we are doing, we see that if this search for a factorization terminates, then it naturally leads to elements that are irreducible, definition (a), rather than what we now call primes, definition (b). So initially we concern ourselves with factorization into irreducibles.

In general factorization into irreducibles may not be possible because the procedure may continue indefinitely, but it is possible in the ring $\mathfrak{O}$ of integers of any number field. To prove this we introduce the notion of a noetherian ring, and show that factorization is always possible if the domain D is noetherian; we then demonstrate that $\mathfrak{O}$ is noetherian.

Even though factorization into irreducibles is always possible in $\mathfrak{O}$, we give an extensive list of examples where such a factorization is not unique. In other cases, however, the existence of a generalized version of the division algorithm (which we term a 'Euclidean function') implies that every irreducible is prime. We see that factorization into primes is unique, so some rings $\mathfrak{O}$ possess unique factorization. In particular we characterize such $\mathfrak{O}$ for the fields $\mathbb{Q}(\sqrt{d})$ with d a negative rational integer: there are exactly five of them, corresponding to $d = -1, -2, -3, -7, -11$. We also prove the existence of a Euclidean function for some fields $\mathbb{Q}(\sqrt{d})$ with d positive. In later chapters we see that $\mathfrak{O}$ may have unique factorization without possessing a Euclidean function.

To begin the chapter we consider a little history, and look at an example of the intuitive use of unique factorization, to motivate the ideas.

4.1 Historical Background

In the 18th century and the first part of the 19th there were varying standards of rigour in number theory. Euler, the most prolific mathematician of the 18th century, was primarily interested in obtaining results, and sometimes used intuitive methods of proof which the hindsight of history has shown to be incorrect. For instance, in his famous textbook on algebra, he made several elegant applications of unique factorization to 'prove' number theoretic results in cases where unique factorization is false. Gauss, on the other hand, found it necessary to demonstrate rigorously that the so-called 'Gaussian integers' $\mathbb{Z}[i]$ did factorize uniquely. In 1847, Lamé announced to a meeting of the Paris Academy that he had proved Fermat's Last Theorem, but his proof was seen to depend on uniqueness of factorization and was shown to be inadequate. Kummer had, in fact, published a paper three years earlier that demonstrated the failure of unique factorization for cyclotomic integers, thus destroying Lamé's proof. This result, which formed part of his habilitation thesis and was published as a pamphlet, went unnoticed at the time.

Eisenstein put his finger on the property that characterizes unique factorization in a letter of 1844 which translates

> If one had the theorem which states that the product of two complex numbers can be divisible by a prime number only when one of the factors is—which seems completely obvious—then one would have the whole theory at a single blow; but this theorem is totally false.

By 'the whole theory' he was referring to consequences of unique factorization, which in particular is relevant to Fermat's Last Theorem.

In Eisenstein's letter, 'prime' meant definition (a) of this chapter, and his comment translated into the terminology of this book is 'if every irreducible is prime, then unique factorization holds'. It is also clear from his comment that he knew instances of irreducibles that are not prime, leading to non-unique factorization. All this must have seemed very confusing to average 19th century mathematicians, who were accustomed to using intuitive ideas about factorization.

To give the reader an idea of what this style of mathematical 'proof' was like, before we develop a rigorous theory of unique factorization, we exhibit a concocted, fallacious, but plausible proof of a statement of Fermat

using this intuitive approach. Fermat's proof has not survived, and we are
not suggesting that it resembled our faulty but instructive attempt below.
Indeed it is not hard to reconstruct a *rigorous* proof using ideas that were
known to Fermat: see Weil [80].

A Statement of Fermat. *The equation $y^2 + 2 = x^3$ has only the integer
solutions $y = \pm 5$, $x = 3$.*

Intuitive 'Proof'. Clearly y cannot be even, for then the right-hand side
would be divisible by 8, but the left-hand side only by 2. Factorize in the
ring $\mathbb{Z}[\sqrt{-2}]$, consisting of all $a + b\sqrt{-2}$ for $a, b \in \mathbb{Z}$, to obtain

$$(y + \sqrt{-2})(y - \sqrt{-2}) = x^3.$$

A common factor $c + d\sqrt{-2}$ of $y + \sqrt{-2}$ and $y - \sqrt{-2}$ would also divide
their sum $2y$ and their difference $2\sqrt{-2}$. Taking norms,

$$c^2 + 2d^2 | 4y^2, \qquad c^2 + 2d^2 | 8,$$

hence $c^2 + 2d^2 | 4$. The only solutions of this relation are $c = \pm 1$, $d = 0$,
or $c = 0$, $d = \pm 1$, or $c = \pm 2$, $d = 0$. None of these give proper factors of
$y + \sqrt{-2}$, so $y + \sqrt{-2}$ and $y - \sqrt{-2}$ are coprime. Now the product of two
coprime numbers is a cube only when each is a cube, so

$$y + \sqrt{-2} = (a + b\sqrt{-2})^3,$$

and comparing coefficients of $\sqrt{-2}$,

$$1 = b(3a^2 - 2b^2)$$

for which the only solutions are $b = 1$, $a = \pm 1$. Then $x = 3$, $y = \pm 5$. $\square$

The flaw in this intuitive 'proof', as it stands, is that we are carrying
over the language of factorization of integers to factorization in $\mathbb{Z}[\sqrt{-2}]$
without checking that the usual properties actually hold in $\mathbb{Z}[\sqrt{-2}]$. In
this chapter we develop the appropriate theory and investigate when it
generalizes to a ring of integers in a number field.

4.2   Trivial Factorizations

If u is a unit in a ring R, then any element $x \in R$ can be trivially factorized
as
$$x = uy$$

where $y = u^{-1}x$. An element y is called an *associate* of x if $x = uy$ for a unit u. Recall that a factorization of $x \in R$, $x = yz$ is 'proper' if neither y nor z is a unit. If a factorization is not proper, in which case we call it *trivial*, it is clear that one of the factors is a unit and the other is an associate of x. Before going on to proper factorizations we therefore look at elementary properties of units and associates. We denote the set of units in a ring R by $U(R)$.

Proposition 4.1. *The units $U(R)$ of a ring R form a group under multiplication.* □

Examples 4.2.

(1) $R = \mathbb{Q}$. The units are $U(\mathbb{Q}) = \mathbb{Q}\backslash\{0\}$, which is an infinite group.

(2) $R = \mathbb{Z}$. The units are ± 1, so $U(\mathbb{Z})$ is cyclic of order 2.

(3) $R = \mathbb{Z}[i]$, the Gaussian integers, $a + ib$ $(a, b \in \mathbb{Z})$. The element $a + ib$ is a unit if and only if there exists $c + id$ $(c, d \in \mathbb{Z})$ such that

$$(a + ib)(c + id) = 1.$$

This implies $ac - bd = 1$, $ad + bc = 0$, whence $c = a/(a^2 + b^2)$, $d = -b/(a^2+b^2)$. These have integer solutions only when $a^2 + b^2 = 1$, so $a = \pm 1$, $b = 0$, or $a = 0$, $b = \pm 1$. Hence the units are $\{1, -1, i, -i\}$ and $U(R)$ is cyclic of order 4.

By using norms, we can extend the results of Example 4.2 to the more general case of the units in the ring of integers of $\mathbb{Q}(\sqrt{d})$ for d negative and squarefree:

Proposition 4.3. *The group of units U of the integers in $\mathbb{Q}(\sqrt{d})$ where d is negative and squarefree is as follows:*
(a) For $d = -1$, $U = \{\pm 1, \pm i\}$.
(b) For $d = -3$, $U = \{\pm 1, \pm\omega, \pm\omega^2\}$ where $\omega = e^{2\pi i/3}$.
(c) For all other $d < 0$, $U = \{\pm 1\}$.

Proof: Suppose α is a unit in the ring of integers of $\mathbb{Q}(\sqrt{d})$ with inverse β. Then $\alpha\beta = 1$, so taking norms

$$N(\alpha)N(\beta) = 1.$$

But $N(\alpha)$, $N(\beta)$ are rational integers, so $N(\alpha) = \pm 1$. Writing $\alpha = a + b\sqrt{d}$ $(a, b \in \mathbb{Q})$, then we see that $N(\alpha) = a^2 - db^2$ is positive (for negative d), so

$N(\alpha) = +1$. Hence we are reduced to solving the equation

$$a^2 - db^2 = 1.$$

If $a, b \in \mathbb{Z}$, then for $d = -1$ this reduces to

$$a^2 + b^2 = 1$$

which has the solutions $a = \pm 1$, $b = 0$, or $a = 0$, $b = \pm 1$, already found in Example 4.2. This gives (a). For $d < -3$ we immediately conclude that $b = 0$ (otherwise $a^2 - db^2$ would exceed 1), so the only rational integer solutions are $a = \pm 1$, $b = 0$. If $d \not\equiv 1 \pmod 4$, then $a, b \in \mathbb{Z}$, so the only solutions are those discovered. For $d \equiv 1 \pmod 4$, however, we must also consider the additional possibility $a = A/2$, $b = B/2$ where both A and B are odd rational integers. In this case

$$A^2 - dB^2 = 4.$$

For $d < -3$, we deduce $B = 0$ and there are no additional solutions. This completes (c). For $d = -3$, we find additional solutions $A = \pm 1$, $B = \pm 1$. The case $A = 1$, $B = 1$ gives

$$\alpha = \tfrac{1}{2}(-1 + \sqrt{-3}) = e^{2\pi i/3}$$

which we have denoted by ω. The other three cases give $-\omega, \omega^2, -\omega^2$. These allied with the solutions already found give (b). □

The general case of units in a ring of integers in a number field will be postponed until Appendix B. We now return to simple properties of units and associates.

Proposition 4.1 easily implies that 'being associates' is an equivalence relation on R. The only associate of 0 is 0 itself. Recall that a non-unit $x \in R$ is called an irreducible if it has no proper factors. The zero element $0 = 0.0$ has factors, neither of which is a unit, so in particular an irreducible is non-zero. We now list a few elementary properties of units, associates, and irreducibles. To prove some of these we require the cancellation law, so we must take the ring to be an integral domain.

Proposition 4.4. *For a domain D,*
  (a) *An element x is a unit if and only if $x | 1$.*
  (b) *Any two units are associates and any associate of a unit is a unit.*
  (c) *Elements x, y are associates if and only if $x | y$ and $y | x$.*
  (d) *An element x is irreducible if and only if every divisor of x is an associate of x or a unit.*
  (e) *An associate of an irreducible is irreducible.*

Proof: Most of these follow straight from the definitions. We prove (c) which requires the cancellation law. Suppose $x|y$ and $y|x$; then there exist $a, b \in D$ such that $y = ax$, $x = by$. Substituting,

$$x = bax.$$

Now either $x = 0$, in which case $y = 0$ also and they are associates, or $x \neq 0$ and we cancel x to find

$$1 = ba,$$

so a and b are units. Hence x, y are associates. The converse is trivial. $\square$

Some of these concepts may usefully be expressed in terms of ideals:

Proposition 4.5. *If D is a domain and x, y are non-zero elements of D then*
(a) *$x|y$ if and only if $\langle x \rangle \supseteq \langle y \rangle$.*
(b) *x and y are associates if and only if $\langle x \rangle = \langle y \rangle$.*
(c) *x is a unit if and only if $\langle x \rangle = D$.*
(d) *x is irreducible if and only if $\langle x \rangle$ is maximal among the proper principal ideals of D.*

Proof: (a) If $x|y$ then $y = zx \in \langle x \rangle$ for some $z \in D$, hence $\langle y \rangle \subseteq \langle x \rangle$. Conversely, if $\langle y \rangle \subseteq \langle x \rangle$ then $y \in \langle x \rangle$, so $y = zx$ for some $z \in D$.
(b) is immediate from (a).
(c) If x is a unit then $xv = 1$ for some $v \in D$, hence for any $y \in D$ we have $y = xvy \in \langle x \rangle$ and $D = \langle x \rangle$. If $D = \langle x \rangle$ then since $1 \in D$, $1 = zx$ for some $z \in D$ and x is a unit.
(d) Suppose x is irreducible, with $\langle x \rangle \subsetneq \langle y \rangle \subsetneq D$. Then $y|x$ but is neither a unit, nor an associate of x, contradicting Proposition 4.4(d). Conversely, if no such y exists, then every divisor of x is either a unit or an associate, so x is irreducible. $\square$

4.3 Factorization into Irreducibles

In a domain D, if a non-unit x is reducible, we can write it as

$$x = ab.$$

If either of a or b is reducible, we can express it as a product of proper factors; then carry on the process, seeking to write

$$x = p_1 p_2 \ldots p_m$$

where each p_i is irreducible. We say that *factorization into irreducibles* is *possible* in D if every $x \in D$, not a unit nor zero, is a product of a finite number of irreducibles. In general such a factorization may not be possible; and an example is ready to hand, namely the ring $\mathbb{B}$ of all algebraic integers. For if α is not zero or a unit, neither is $\sqrt{\alpha}$. Since $\alpha = \sqrt{\alpha} \cdot \sqrt{\alpha}$ and $\sqrt{\alpha}$ is an integer, it follows that α is not irreducible. Thus $\mathbb{B}$ has no irreducibles at all, but it does have non-zero non-units, so factorization into irreducibles is not possible.

This trouble does not arise in the ring $\mathfrak{O}$ of integers of a number field— another reason why we concentrate on such rings instead of the whole of $\mathbb{B}$. We prove the possibility of factorization in $\mathfrak{O}$ by introducing a more general notion which makes the arguments involved more transparent. We define a domain D to be *noetherian* if every ideal in D is finitely generated. The adjective commemorates Emmy Noether (1882–1935) who introduced the concept. Having demonstrated the possibility of factorization in any noetherian ring, we will show that $\mathfrak{O}$ is noetherian, so factorization is possible here also.

Two useful properties, which we will see are each equivalent to the noetherian condition are:

The Ascending Chain Condition. Given an ascending chain of ideals:

$$I_0 \subseteq I_1 \subseteq \ldots \subseteq I_n \subseteq \ldots \tag{4.1}$$

then there exists some N for which $I_n = I_N$ for all $n \geq N$. That is, every ascending chain *stops*.

The Maximal Condition. Every non-empty set of ideals has a maximal element, that is an element which is not properly contained in every other element.

This maximal element need not contain *all* the other ideals in the given set: we require only that there is no other element in the set that contains *it*.

Proposition 4.6. *The following conditions are equivalent for an integral domain D:*
  (a) *D is noetherian.*
  (b) *D satisfies the ascending chain condition.*
  (c) *D satisfies the maximal condition.*

Proof: Assume (a). Consider an ascending chain as in (4.1). Let $I = \cup_{n=1}^{\infty} I_n$. Then I is an ideal, so finitely generated: say $I = \langle x_1, \ldots, x_m \rangle$. Each x_i belongs to some $I_{n(i)}$. If we let $N = \max_i n(i)$, then we have $I = I_N$ and $I_n = I_N$ for all $n \geq N$, proving (b).

Now suppose (b) and consider a non-empty set S of ideals. Suppose for a contradiction that S does not have a maximal element. Pick $I_0 \in S$. Since I_0 is not maximal we can pick $I_1 \in S$ with $I_0 \subsetneq I_1$. Inductively, having found I_n, since this is not maximal, we can pick $I_{n+1} \in S$ with $I_n \subsetneq I_{n+1}$. But now we have an ascending chain that does not stop, which is a contradiction. So (b) implies (c). (The reader who wishes to may ponder the use of the axiom of choice in this proof.)

Finally, suppose (c). Let I be any ideal, and let S be the set of all finitely generated ideals contained in I. Then $\{0\} \in S$, so S is non-empty and thus has a maximal element J. If $J \neq I$, pick $x \in I \setminus J$. Then $\langle J, x \rangle$ is finitely generated and strictly larger than J, a contradiction. Hence $J = I$ and I is finitely generated. $\square$

Theorem 4.7. *If a domain D is noetherian, factorization into irreducibles is possible in D.*

Proof: Suppose that D is noetherian, but there exists a non-unit $x \neq 0$ in D which cannot be expressed as a product of a finite number of irreducibles. Choose x so that $\langle x \rangle$ is maximal subject to these conditions on x, which is possible by the maximal condition. By definition, x is not irreducible, so $x = yz$ where y and z are not units. Then $\langle y \rangle \supseteq \langle x \rangle$ by Proposition 4.4(a). If $\langle y \rangle = \langle x \rangle$ then x and y are associates by Proposition 4.4(b), but this is not the case because it implies that z is a unit. So $\langle y \rangle \supsetneq \langle x \rangle$, and similarly $\langle z \rangle \supsetneq \langle x \rangle$. By maximality of $\langle x \rangle$,

$$
\begin{aligned}
y &= p_1 \ldots p_r, \\
z &= q_1 \ldots q_s,
\end{aligned}
$$

where each p_i and q_j is irreducible. Multiply these together to express x as a product of irreducibles, a contradiction. Hence the assumption that there existed a non-unit $\neq 0$ that is not a finite product of irreducibles is false, and factorization into irreducibles is always possible. $\square$

We are now in business:

Theorem 4.8. *The ring of integers $\mathfrak{O}$ in a number field K is noetherian.*

Proof: We prove that every ideal I of $\mathfrak{O}$ is finitely generated. Now $(\mathfrak{O}, +)$ is free abelian of rank n equal to the degree of K by Theorem 2.16. Hence

$(I, +)$ is free abelian of rank $s \leq n$ by Theorem 1.16. If $\{x_1, \ldots, x_s\}$ is a $\mathbb{Z}$-basis for $(I, +)$, then clearly $\langle x_1, \ldots, x_s \rangle = I$, so I is finitely generated and $\mathfrak{O}$ is noetherian.   □

Corollary 4.9. *Factorization into irreducibles is possible in $\mathfrak{O}$.*   □

To get very far in the theory, we need an easy way to detect units and irreducibles in $\mathfrak{O}$. The norm proves to be a convenient tool:

Proposition 4.10. *Let $\mathfrak{O}$ be the ring of integers in a number field K, and let $x, y \in \mathfrak{O}$. Then*
> (a) *x is a unit if and only if $N(x) = \pm 1$.*
> (b) *If x and y are associates, then $N(x) = \pm N(y)$.*
> (c) *If $N(x)$ is a rational prime, then x is irreducible in $\mathfrak{O}$.*

Proof: (a) If $xu = 1$, then $N(x)N(u) = 1$. Since $N(x), N(u) \in \mathbb{Z}$, we have $N(x) = \pm 1$. Conversely, if $N(x) = \pm 1$, then

$$\sigma_1(x)\sigma_2(x)\ldots\sigma_n(x) = \pm 1$$

where the σ_i are the monomorphisms $K \to \mathbb{C}$. One factor, without loss in generality $\sigma_1(x)$, is equal to x; all the other $\sigma_i(x)$ are integers. Put

$$u = \pm \sigma_2(x)\ldots\sigma_n(x).$$

Then $xu = 1$, so $u = x^{-1} \in K$. Hence $u \in K \cap \mathbb{B} = \mathfrak{O}$, and x is a unit.

(b) If x, y are associates, then $x = uy$ for a unit u, so $N(x) = N(uy) = N(u)N(y) = \pm N(y)$ by (a).

(c) Let $x = yz$. Then $N(y)N(z) = N(yz) = N(x) = p$, a rational prime; so one of $N(y)$ and $N(z)$ is $\pm p$ and the other is ± 1. By (a), one of y and z is a unit, so x is irreducible.   □

We have not asserted converses to parts (b) and (c) because these are generally false, as examples in the next section reveal.

4.4   Examples of Non-Unique Factorization into Irreducibles

Factorization in a domain D is *unique* if, whenever

$$p_1 \ldots p_r = q_1 \ldots q_s$$

where every p_i and q_j is irreducible in D, it follows that

(a) $r = s$.

(b) There is a permutation π of $\{1, \ldots, r\}$ such that p_i and $q_{\pi(i)}$ are associates for all $i = 1, \ldots, r$.

In view of our earlier remarks about trivial factorizations, this is the best we can hope for. It says that a factorization into irreducibles (if it exists) is unique except for the order of the factors and the possible presence of units. Variation to this extent is necessary, since even in $\mathbb{Z}$ we have, for instance,

$$3 \cdot 5 = 5 \cdot 3 = (-3)(-5) = (-5)(-3).$$

Unfortunately, factorization into irreducibles need not be unique in a ring of integers of an algebraic number field. Examples are easy to come by, and to drive the point home we give quite a lot of them. They are drawn from quadratic fields, and we state them as positive theorems. The easiest come from imaginary quadratic fields:

Theorem 4.11. *Factorization into irreducibles is not unique in the ring of integers of $\mathbb{Q}(\sqrt{d})$ for (at least) the following values of d: -5, -6, -10, -13, -14, -15, -17, -21, -22, -23, -26, -29, -30.*

Proof: In $\mathbb{Q}(\sqrt{-5})$ we have the factorizations

$$6 = 2 \cdot 3 = (1 + \sqrt{-5})(1 - \sqrt{-5}).$$

We claim that $2, 3, 1 + \sqrt{-5}$ and $1 - \sqrt{-5}$ are irreducible in the ring $\mathfrak{O}$ of integers of $\mathbb{Q}(\sqrt{-5})$. Since the norm is

$$N(a + b\sqrt{-5}) = a^2 + 5b^2$$

their norms are 4, 9, 6, 6, respectively. If $2 = xy$ where $x, y \in \mathfrak{O}$ are non-units, then $4 = N(2) = N(x)N(y)$ so that $N(x) = \pm 2$, $N(y) = \pm 2$. Similarly non-trivial divisors of 3 must, if they exist, have norm ± 3, and non-trivial divisors of $1 \pm \sqrt{-5}$ have norm ± 2 or ± 3. Since $-5 \not\equiv 1 \pmod 4$, the integers in $\mathcal{D}$ are of the form $a + b\sqrt{-5}$ for $a, b \in \mathbb{Z}$ (Theorem 3.2) so

$$a^2 + 5b^2 = \pm 2 \text{ or } \pm 3 \qquad (a, b \in \mathbb{Z}).$$

Now $|b| \geq 1$ implies $|a^2 + 5b^2| \geq 5$, so the only possibility is $|b| = 0$; but then we have $a^2 = \pm 2$ or ± 3, which is impossible in integers. Thus the putative divisors do not exist, and the four factors are all irreducible. Since $N(2) = 4$, $N(1 \pm \sqrt{-5}) = 6$, by Proposition 4.10(b), 2 is not an associate of $1 + \sqrt{-5}$ or $1 - \sqrt{-5}$, so factorization is not unique.

The other stated values of d are dealt with in exactly the same way (with a few slight subtleties noted at the end of the proof) starting from the following factorizations:

$$\mathbb{Q}(\sqrt{-6}): \quad 6 \; = 2 \cdot 3 \qquad = (\sqrt{-6})(-\sqrt{-6})$$
$$\mathbb{Q}(\sqrt{-10}): \quad 14 \; = 2 \cdot 7 \qquad = (2 + \sqrt{-10})(2 - \sqrt{-10})$$
$$\mathbb{Q}(\sqrt{-13}): \quad 14 \; = 2 \cdot 7 \qquad = (1 + \sqrt{-13})(1 - \sqrt{-13})$$
$$\mathbb{Q}(\sqrt{-14}): \quad 15 \; = 3 \cdot 5 \qquad = (1 + \sqrt{-14})(1 - \sqrt{-14})$$
$$\mathbb{Q}(\sqrt{-15}): \quad 4 \; = 2 \cdot 2 \qquad = \left(\frac{1+\sqrt{-15}}{2}\right)\left(\frac{1-\sqrt{-15}}{2}\right)$$
$$\mathbb{Q}(\sqrt{-17}): \quad 18 \; = 2 \cdot 3 \cdot 3 \quad = (1 + \sqrt{-17})(1 - \sqrt{-17})$$
$$\mathbb{Q}(\sqrt{-21}): \quad 22 \; = 2 \cdot 11 \qquad = (1 + \sqrt{-21})(1 - \sqrt{-21})$$
$$\mathbb{Q}(\sqrt{-22}): \quad 26 \; = 2 \cdot 13 \qquad = (2 + \sqrt{-22})(2 - \sqrt{-22})$$
$$\mathbb{Q}(\sqrt{-23}): \quad 6 \; = 2 \cdot 3 \qquad = \left(\frac{1+\sqrt{-23}}{2}\right)\left(\frac{1-\sqrt{-23}}{2}\right)$$
$$\mathbb{Q}(\sqrt{-26}): \quad 27 \; = 3 \cdot 3 \cdot 3 \quad = (1 + \sqrt{-26})(1 - \sqrt{-26})$$
$$\mathbb{Q}(\sqrt{-29}): \quad 30 \; = 2 \cdot 3 \cdot 5 \quad = (1 + \sqrt{-29})(1 - \sqrt{-29})$$
$$\mathbb{Q}(\sqrt{-30}): \quad 34 \; = 2 \cdot 17 \qquad = (2 + \sqrt{-30})(2 - \sqrt{-30}).$$

In cases -15 and -23, note that $d \equiv 1 \pmod 4$ and be careful. For -26 it is easy to prove 3 irreducible. For $1 - \sqrt{-26}$ we are led to the equation $N(x)N(y) = 27$, so $N(x) = \pm 9$, $N(y) = \pm 3$, or the other way round. This leads to $a^2 + 26b^2 = \pm 9$ or ± 3. There *is* a solution for ± 9, but not for ± 3, and the latter is sufficient to show $1 + \sqrt{-26}$ is irreducible.   □

Examining this list, we see that in the ring of integers of $\mathbb{Q}(\sqrt{-17})$ there is an example to show that even the *number* of irreducible factors may differ; the case $\mathbb{Q}(\sqrt{-26})$ shows that the number of distinct factors may differ and that even a (rational) prime power may factorize non-uniquely.

For real quadratic fields there are similar results, but these are harder to find. Also, since the norm is $a^2 - db^2$, it is harder to prove given numbers irreducible. With the same range of values as in Theorem 4.11 we find:

Theorem 4.12. *Factorization into irreducibles is not unique in the ring of integers of $\mathbb{Q}(\sqrt{d})$ for (at least) the following values of d:*

$$10, 15, 26, 30.$$

Proof: In the integers of $\mathbb{Q}(\sqrt{10})$ we have factorizations:

$$6 = 2 \cdot 3 = (4 + \sqrt{10})(4 - \sqrt{10}).$$

We prove $2, 3, 4 \pm \sqrt{10}$ irreducible. Looking at norms this amounts to proving that the equations

$$a^2 - 10b^2 = \pm 2 \text{ or } \pm 3$$

have no solutions in integers a, b. It is no longer helpful to look at the size of $|b|$, because of the minus sign. However, the equation implies

$$a^2 \equiv \pm 2 \text{ or } \pm 3 \qquad (\text{mod } 10)$$

or equivalently

$$a^2 = 2, 3, 7 \text{ or } 8 \qquad (\text{mod } 10).$$

The squares (mod 10) are, in order, 0, 1, 4, 9, 6, 5, 6, 9, 4, 1; by a seemingly remarkable coincidence, the numbers we are looking for are precisely those that do not occur. Hence no solutions exist and the four factors are irreducible. Now 2 and $4 \pm \sqrt{10}$ are not associates, since their norms are 4, 6 respectively.

Similarly:

$$
\begin{array}{lll}
\mathbb{Q}(\sqrt{15}) & : & 10 = 2 \cdot 5 = (5 + \sqrt{15})(5 - \sqrt{15}) \\
\mathbb{Q}(\sqrt{26}) & : & 10 = 2 \cdot 5 = (6 + \sqrt{26})(6 - \sqrt{26}) \\
\mathbb{Q}(\sqrt{30}) & : & 6 = 2 \cdot 3 = (6 + \sqrt{30})(6 - \sqrt{30})
\end{array}
$$

The reader will find it instructive to do their own calculations.   □

The values of d considered in Theorems 4.11 and 4.12 have not, despite appearances, been chosen at random. If we try similar tricks with other d in the range -30 to 30, nothing seems to work. Thus in $\mathbb{Q}(\sqrt{-19})$ we get

$$\left(\frac{1 + \sqrt{-19}}{2}\right)\left(\frac{1 - \sqrt{-19}}{2}\right) = 5$$

but all this shows is that 5 is reducible.

Trying another obvious product in the integers of $\mathbb{Q}(\sqrt{-19})$, we find

$$(2 + \sqrt{-19})(2 - \sqrt{-19}) = 23$$

which just tells us that 23 is also reducible. The case

$$\left(\frac{3 + \sqrt{-19}}{2}\right)\left(\frac{3 - \sqrt{-19}}{2}\right) = 7$$

shows 7 is reducible. After more of these calculations we may alight on

$$35 = 5 \cdot 7 = (4 + \sqrt{-19})(4 - \sqrt{-19}).$$

Will this prove non-uniqueness? No, because neither 5 nor 7 is irreducible, as we have seen; and neither is $4 \pm \sqrt{-19}$.

The complete factorization of 35 is

$$\left(\frac{1 + \sqrt{-19}}{2}\right) \left(\frac{3 + \sqrt{-19}}{2}\right) \left(\frac{1 - \sqrt{-19}}{2}\right) \left(\frac{3 - \sqrt{-19}}{2}\right)$$

and the two apparently distinct factorizations come from different groupings of these in pairs. Eventually we are led to conjecture that the integers of $\mathbb{Q}(\sqrt{-19})$ have unique factorization. This is indeed true, but we shall not be able to prove it until Chapter 10. In fact the ring of integers of $\mathbb{Q}(\sqrt{d})$ for negative squarefree d has unique factorization into irreducibles if and only if d takes one of the values:

$$-1, -2, -3, -7, -11, -19, -43, -67, -163.$$

Numerical evidence available in the time of Gauss pointed to this result. In 1934 Heilbronn and Linfoot [40] showed that at most one further value of d can occur, and that if it does then $|d|$ is very large. In 1952 Heegner [39] offered a proof but it was thought to contain a gap. In 1967 Stark [76] found a proof, as did Baker [3] soon after. Finally Birch [6], Deuring [23] and Siegel [75] filled in the gap in Heegner's proof. The methods of this book are not appropriate to give any of these proofs, but we will prove in Chapter 10 that for these nine values factorization is unique.

The situation for positive d is not at all well understood. Factorization is unique in many more cases, for instance 2, 3, 5, 6, 7, 11, 13, 14, 17, 19, 21, 22, 23, 29, 31, 33, 37, 38, 41, 43, 46, 47, 53, 57, 59, 61, 62, 67, 69, 71, 73, 77, 83, 86, 89, 93, 94, 97, ..., these being all for d less than 100. Gauss conjectured (in the context of his work on quadratic forms) a result equivalent to there being infinitely many real quadratic fields with unique factorization, but this has neither been proved nor disproved. Henri Cohen and Hendrik Lenstra have given a heuristic (that is, plausible but non-rigorous) argument suggesting that about 75.446% of real quadratic fields $\mathbb{Q}(\sqrt{p})$ for prime p have unique factorization. Computational experiments agree with this prediction, see Cohen [17] and te Riele and Williams [83].

4.5   Prime Factorization

So far we have not proved uniqueness of factorization for the ring of integers in any number fields (apart from $\mathbb{Z}$). We now introduce a criterion for factorization to be unique, stated in terms of a special property of the irreducibles. We have already noted that an irreducible p in $\mathbb{Z}$ satisfies the additional property

$$p \mid mn \quad \text{implies} \quad p \mid m \quad \text{or} \quad p \mid n.$$

In this section we show that this property characterizes uniqueness of factorization.

In a domain D an element x is said to be *prime* if it is not zero or a unit and

$$x \mid ab \quad \text{implies} \quad x \mid a \quad \text{or} \quad x \mid b.$$

Note that the zero element satisfies the given property in a domain, but we exclude it to correspond with the definition of prime in $\mathbb{Z}$, where 0 is not usually considered a prime. This convention allows us to state:

Proposition 4.13. *A prime in a domain D is always irreducible.*

Proof: Suppose that D is a domain, $x \in D$ is prime, and $x = ab$. Then $x \mid ab$, so $x \mid a$ or $x \mid b$.

If $x \mid a$, then $a = xc$ $(c \in D)$, so

$$x = xcb$$

and cancelling x (which is non-zero), we see that

$$1 = cb$$

and b is a unit. In the same way, $x \mid b$ implies a is a unit. $\square$

The converse of this result is not true, as Eisenstein lamented in 1844; in many domains there exist irreducibles that are not primes. For example in $\mathbb{Z}[\sqrt{-5}]$

$$6 = 2 \cdot 3 = (1 + \sqrt{-5})(1 - \sqrt{-5}),$$

but 2 does not divide either of $1 + \sqrt{-5}$ or $1 - \sqrt{-5}$, as shown in the proof of Theorem 4.11. So 2 is an irreducible in $\mathbb{Z}[\sqrt{-5}]$, but not prime. The factorizations in the proofs of Theorems 4.11 and 4.12 readily yield other examples. The next theorem tells us that such examples are entirely typical—every domain with non-unique factorization contains irreducibles that are not prime:

Theorem 4.14. *In a domain in which factorization into irreducibles is possible, factorization is unique if and only if every irreducible is prime.*

Proof: Let D be the domain. It is convenient to rephrase the possibility of factorization for all non-zero $x \in D$ as

$$x = up_1 \ldots p_r$$

where u is a unit and $p_1, \ldots, p_r$ are irreducibles. When $r = 0$ this can then be interpreted as $x = u$ is a unit and when $r \geq 1$, then up_1 is an irreducible, so x is a product of the irreducibles $up_1, p_2, \ldots, p_r$.

Now for the proof. Suppose first that factorization is unique and p is an irreducible. We must show p is prime.

$$\text{If } p|ab, \text{ then } pc = ab \ (c \in D).$$

We need consider only the non-trivial case $a \neq 0$, $b \neq 0$ which implies $c \neq 0$ also.

Factorize, a, b, c into irreducibles:

$$
\begin{aligned}
a &= u_1 p_1 \ldots p_n \\
b &= u_2 q_1 \ldots q_m \\
c &= u_3 r_1 \ldots r_s
\end{aligned}
$$

where each u_i is a unit and p_i, q_i and r_i are irreducible. Then

$$p(u_3 r_1 \ldots r_s) = (u_1 p_1 \ldots p_n)(u_2 q_1 \ldots q_m),$$

and unique factorization implies p is an associate (hence divides) one of the p_i or q_j, so divides a or b. Hence p is prime.

Conversely, suppose that every irreducible is prime. We demonstrate that if

$$u_1 p_1 \ldots p_m = u_2 q_1 \ldots q_n \qquad (4.2)$$

where u_1, u_2 are units and the p_i, q_j, are irreducibles, then $m = n$ and there is a permutation π of $\{1, \ldots, m\}$ such that p_i and $q_{\pi(i)}$ are associates $(1 \leq i \leq m)$.

This is trivially true for $m = 0$.

For $m \geq 1$, if (4.2) holds, then $p_m \,|u_2 q_1 \ldots q_n$. But p_m is prime, so by induction on n, $p_m|u_2$ or $p_m|q_j$ for some j. The first of these possibilities implies that p_m is a unit by Proposition 4.4(a), so $p_m|q_j$. Renumber so that $j = n$. Then $p_m|q_n$ and $q_n = p_m u$ where u is a unit, so

$$u_1 p_1 \ldots p_m = u_2 q_1 \ldots q_{n-1} u p_m.$$

Cancel p_m:

$$u_1 p_1 \ldots p_{m-1} = (u_2 u) q_1 \ldots q_{n-1}.$$

By induction we may suppose that $m - 1 = n - 1$ and there is a permutation of $1, \ldots, m - 1$ such that $p_i, q_{\pi(i)}$ are associates $(1 \leq i \leq m - 1)$. We can then extend π to $\{1, \ldots, m\}$ by defining $\pi(m) = m$. $\qquad \square$

A domain D is a *unique factorization domain* if factorization into irreducibles is possible and unique. In a unique factorization domain all irreducibles are primes, so we may speak of a factorization into irreducibles as a 'prime factorization'. Theorem 4.14 tells us that a prime factorization is unique in the usual sense.

We can immediately generalize many ideas on factorization to any unique factorization domain. For instance, if $a, b \in D$, the *highest common factor* h of a, b is defined to be an element that satisfies

(1) $h|a$, $h|b$,

(2) If $h'|a$, $h'|b$, then $h'|h$.

If a is zero, the highest common factor of a, b is b. For $a, b \neq 0$, in a unique factorization domain we can write

$$a = u_1 p_1^{e_1} \ldots p_n^{e_n}$$
$$b = u_2 p_1^{f_1} \ldots p_n^{f_n}$$

where u_1, u_2 are units and the p_i are distinct (that is, non-associate) primes with non-negative integer exponents e_i, f_i. Then it is easy to show that

$$h = u p_1^{m_1} \ldots p_n^{m_n}$$

where u is any unit and m_i is the smaller of e_i, f_i $(1 \leq i \leq n)$. The highest common factor is unique up to multiplication by a unit. We can say that a, b are *coprime* if their highest common factor is 1 (or any other unit).

In the same way we define the *lowest common multiple* l of a, b to satisfy

(3) $a|l$, $b|l$,

(4) If $a|l'$, $b|l'$, then $l|l'$.

For non-zero a, b this is

$$l = u p_1^{k_1} \ldots p_n^{k_n}$$

where k_i is the larger of e_i, f_i in the factorizations noted above.

Without uniqueness of factorization we can no longer guarantee the existence of highest common factors and lowest common multiples. (See Exercise 9 in this chapter.) The language of factorization of integers can be carried over sensibly only to a unique factorization domain.

In the next section we see that if a domain has a property analogous to the division algorithm, then every irreducible is prime and factorization is unique. In later chapters we develop more advanced techniques which prove unique factorization for a wider class of domains that do not have this property.

4.6   Euclidean Domains

The crucial property in the usual proofs of unique factorization in $\mathbb{Z}$ or $K[t]$ for a field K is the existence of a division algorithm. A reasonable generalization of this is the following:

Definition 4.15. Let D be a domain. A *Euclidean function* for D is a function $\phi : D \setminus \{0\} \to \mathbb{N}$ such that
(1) If $a, b \in D \setminus \{0\}$ and $a|b$ then $\phi(a) \leq \phi(b)$.
(2) If $a, b \in D \setminus \{0\}$ then there exist $q, r \in D$ such that $a = bq + r$ where either $r = 0$ or $\phi(r) < \phi(b)$.

Thus for $\mathbb{Z}$ the function $\phi(n) = |n|$ and for $K[t]$, $\phi(p) = \partial p$ (the degree of the polynomial p) are Euclidean functions.

If a domain has a Euclidean function we call it a *Euclidean domain*. We prove that a Euclidean domain has unique factorization by showing that every irreducible is prime. The route is this: first we show that in a Euclidean domain every ideal is principal (a domain with this property is called a *principal ideal domain*). Then we show that the latter property implies all irreducibles are primes.

Theorem 4.16. *Every Euclidean domain is a principal ideal domain.*

Proof: Let D be Euclidean, I an ideal of D. If $I = 0$ it is principal, so we may assume there exists a non-zero element x of I. Further choose x to make $\phi(x)$ as small as possible. If $y \in I$ then by (2) $y = qx + r$ where either $r = 0$ or $\phi(r) < \phi(x)$. Now $r \in I$ so we cannot have $\phi(r) < \phi(x)$ because $\phi(x)$ is minimal. This means that $r = 0$, so y is a multiple of x. Therefore $I = \langle x \rangle$ is principal.   □

Theorem 4.17. *Every principal ideal domain is a unique factorization domain.*

Proof: Let D be a principal ideal domain. Since this implies D is noetherian, factorization into irreducibles is possible by Theorem 4.7. To prove uniqueness we show that every irreducible is prime.

Suppose p is irreducible, then $\langle p \rangle$ is maximal amongst the principal ideals of D by Proposition 4.5(d), but since every ideal is principal, this means that $\langle p \rangle$ is maximal amongst all ideals.

Suppose $p|ab$ but $p \nmid a$. The fact that $p \nmid a$ implies $\langle p, a \rangle \supsetneq \langle p \rangle$, so by maximality, $\langle p, a \rangle = D$. Then $1 \in \langle p, a \rangle$, so

$$1 = cp + da \quad (c, d \in D).$$

Multiply by b:

$$b = cpb + dab.$$

Since $p|ab$, we find that $p|(cpb + dab)$, so $p|b$. This proves p is prime. $\square$

Theorem 4.18. *A Euclidean domain is a unique factorization domain.* $\square$

4.7 Euclidean Quadratic Fields

(*This subsection may be omitted if desired.*)

In order to apply Theorem 4.18 it is necessary to exhibit some number fields for which the ring of integers is Euclidean. We restrict ourselves to the simplest case of quadratic fields $\mathbb{Q}(\sqrt{d})$ for squarefree d, beginning with the easier situation when d is negative.

Theorem 4.19. *The ring of integers $\mathfrak{O}$ of $\mathbb{Q}(\sqrt{d})$ is Euclidean for $d = -1, -2, -3, -7, -11$, with Euclidean function*

$$\phi(\alpha) = |N(\alpha)|.$$

Proof: To begin with we consider the suitability of the function ϕ defined in the theorem. For this to be a Euclidean function, the following two conditions must be satisfied for all $\alpha, \beta \in \mathfrak{O} \setminus 0$:

(a) If $\alpha|\beta$ then $|N(a)| \leq |N(\beta)|$.

(b) There exist $\gamma, \delta \in \mathfrak{O}$ such that $\alpha = \beta\gamma + \delta$ where either $\delta = 0$ or $|N(\delta)| < |N(\beta)|$.

It is clear that (a) holds, for if $\alpha|\beta$ then $\beta = \lambda\alpha$ for $\lambda \in \mathfrak{O}$ and then

$$|N(\beta)| = |N(a\lambda)| = |N(\alpha)N(\lambda)| = |N(\alpha)||N(\lambda)|$$

with rational integer values for the various norms.

To prove (b), we consider the alternative statement:

(c) For any $\epsilon \in \mathbb{Q}(\sqrt{d})$ there exists $\kappa \in \mathfrak{O}$ such that

$$|N(\epsilon - \kappa)| < 1.$$

We prove that (c) is equivalent to (b). First, suppose (b) holds. By Lemma 2.11, $c\epsilon \in \mathfrak{O}$ for some $c \in \mathbb{Z}$. Applying (b) with $\alpha = c\epsilon$, $\beta = c$ we get two possibilities:

(1) $\delta = 0$ and $c\epsilon = c\gamma$ for $\gamma \in \mathfrak{O}$. Then $\epsilon = \gamma \in \mathfrak{O}$ and we may take $\kappa = \epsilon$.

(2) $c\epsilon = c\gamma + \delta$ where $|N(\delta)| < |N(c)|$. Now $c \neq 0$, so this implies

$$|N(\delta/c)| < 1$$

which is the same as

$$|N(\epsilon - \gamma)| < 1$$

so we may take $\kappa = \gamma$. Hence (b) implies (c). To prove that (c) implies (b) we put $\epsilon = \alpha/\beta$ and argue similarly.

This allows us to concentrate on condition (c), which is relatively easy to handle: in spirit it says that everything in $\mathbb{Q}(\sqrt{d})$ is 'near to' an integer.

Suppose $\epsilon = r + s\sqrt{d}$ $(r, s \in \mathbb{Q})$. If $d \not\equiv 1$ (mod 4) we have to find $\kappa = x + y\sqrt{d}$ $(x, y \in \mathbb{Z})$ with

$$|(r - x)^2 - d(s - y)^2| < 1.$$

For $d = -1, -2$ we may do this by taking x and y to be the rational integers nearest to r and s respectively, for then

$$\left|(r - x)^2 - d(s - y)^2\right| \leq \left|\left(\tfrac{1}{2}\right)^2 + 2\left(\tfrac{1}{2}\right)^2\right| = \tfrac{3}{4} < 1.$$

The remaining three values of d to be considered have $d \equiv 1$ (mod 4). In this case we must find

$$\kappa = x + y\left(\frac{1 + \sqrt{d}}{2}\right) \quad (x, y \in \mathbb{Z})$$

such that

$$|(r - x - \tfrac{1}{2}y)^2 - d(s - \tfrac{1}{2}y)^2| < 1.$$

Certainly we can take y to be the rational integer nearest to $2s$, so that $|2s - y| \leq \tfrac{1}{2}$; and then we may find $x \in \mathbb{Z}$ so that $|r - x - \tfrac{1}{2}y| \leq \tfrac{1}{2}$. For $d = -3, -7$, or -11 this means that

$$|(r - x - \tfrac{1}{2}y)^2 - d(s - \tfrac{1}{2}y)^2| \leq \left|\tfrac{1}{4} + \tfrac{11}{16}\right| = \tfrac{15}{16} < 1.$$

The theorem is proved.   □

To complete the picture for negative d we have:

Theorem 4.20. *For squarefree $d < -11$ the ring of integers of $\mathbb{Q}(\sqrt{d})$ is not Euclidean.*

Proof: Let $\mathfrak{O}$ be the ring of integers of $\mathbb{Q}(\sqrt{d})$ and suppose for a contradiction that there exists a Euclidean function ϕ. (We do *not* assume $\phi = |\mathrm{N}|$.) Choose $\alpha \in \mathfrak{O}$ such that $\alpha \neq 0$, α is not a unit, and $\phi(\alpha)$ is minimal subject to this. Let β be any element of $\mathfrak{O}$. Now there exist γ, δ such that $\beta = \alpha\gamma + \delta$ with $\delta = 0$ or $\phi(\delta) < \phi(\alpha)$. By choice of α the latter condition implies that either $\delta = 0$ or δ is a unit.

For $d < -11$, Proposition 4.3 shows that the only units of $\mathbb{Q}(\sqrt{d})$ are ± 1. Hence for every $\beta \in \mathfrak{O}$ we have $\beta \equiv -1, 0$, or $1 \pmod{\langle\alpha\rangle}$ and so $|\mathfrak{O}/\langle\alpha\rangle| \leq 3$.

Now we compute $|\mathfrak{O}/\langle\alpha\rangle|$ using Theorem 1.17. By Theorem 2.16 $(\mathfrak{O}, +)$ is free abelian of rank 2. If $d \not\equiv 1 \pmod 4$ a $\mathbb{Z}$-basis for $\langle\alpha\rangle$ is $\{\alpha, \alpha\sqrt{d}\}$ since a $\mathbb{Z}$-basis for $\mathfrak{O}$ is $\{1, \sqrt{d}\}$. If $\alpha = a + b\sqrt{d}$ $(a, b \in \mathbb{Z})$ the $\mathbb{Z}$-basis for $\langle\alpha\rangle$ is

$$\{a + b\sqrt{d}, db + a\sqrt{d}\}.$$

Hence by Theorem 1.17

$$|\mathfrak{O}/\langle\alpha\rangle| = \left\| \begin{matrix} a & b \\ db & a \end{matrix} \right\| = \left| a^2 - db^2 \right| = |\mathrm{N}(\alpha)|.$$

Similar calculations apply for $d \equiv 1 \pmod 4$ with the same end result. (These calculations are a special case of Corollary 5.10). It follows that $|\mathrm{N}(\alpha)| \leq 3$. Thus if $d \not\equiv 1 \pmod 4$ we have $|a^2 - db^2| \leq 3$ with $a, b \in \mathbb{Z}$. If $d \equiv 1 \pmod 4$ then $a = A/2$, $b = B/2$, for $A, B \in \mathbb{Z}$; and then $|A^2 - dB^2| \leq 12$. Since $d < -11$ the only solutions are $a = \pm 1$, $b = 0$; so $|\mathrm{N}(\alpha)| = 1$ and hence α is a unit. This contradicts the choice of α. $\quad\square$

These two theorems together show that for negative d the ring of integers of $\mathbb{Q}(\sqrt{d})$ is Euclidean if and only if $d = -1, -2, -3, -7, -11$. Further, when it is Euclidean it has as Euclidean function the absolute value of the norm. For brevity call such fields *norm-Euclidean*.

The determination of the norm-Euclidean quadratic fields with d positive has been a long process involving many mathematicians. Leonard Dickson proved $\mathbb{Q}(\sqrt{d})$ Euclidean for $d = 2, 3, 5, 13$, mistakenly asserting there are no others. Oskar Perron added 6, 7, 11, 17, 21, 29 to the list. Oppenheimer, Robert Remak, and László Rédei added 19, 33, 37, 41, 55, 73. Rédei claimed 97 as well but this was disproved by Eric Barnes and Peter Swinnerton-Dyer. Hans Heilbronn proved the list finite in 1934, and the problem was finished off by Harold Chatland and Harold Davenport [15] in 1950 (and independently Kustaa Inkeri [43] in 1949) who proved:

Theorem 4.21. *The ring of integers of $\mathbb{Q}(\sqrt{d})$, for positive d, is norm-Euclidean if and only if $d = 2, 3, 5, 6, 7, 11, 13, 17, 19, 21, 29, 33, 37, 41, 55, 73$.*

We cannot prove this theorem here. A good survey of the problem and related questions, with references, is given by Narkiewicz [59].

Unlike the case $d < 0$, a real quadratic field $\mathbb{Q}(d)$ can be Euclidean but not norm-Euclidean, as Clark [16] proved in 1994 for $d = 69$. Samuel [71] suggested that $\mathbb{Z}(\sqrt{14})$ might also have such properties. Malcolm Harper proved this in his PhD thesis in 2000, and published a proof [38] in 2004. Using similar methods he also proved that for positive d, whenever the discriminant of $\mathbb{Q}(\sqrt{d})$ is less than or equal to 500, the ring of integers is Euclidean if and only if it is a principal ideal domain, that is, has unique prime factorization.

4.8   Consequences of Unique Factorization

When the integers in a number field have unique factorization, we can carry over many arguments of the type used in the factorization of integers (taking a little care at first). For example, the proof of the statement of Fermat in Section 4.1 makes it clear that, since $\mathbb{Z}[\sqrt{-2}]$ (the ring of integers in $\mathbb{Q}(\sqrt{-2})$) has unique factorization, the intuitive 'proof' given there is, in fact, valid. We now prove another example of the same sort of thing, again a statement of Fermat:

Theorem 4.22. *The only integer solutions of the equation*

$$y^2 + 4 = z^3 \tag{4.3}$$

are $y = \pm 11$, $z = 5$ and $y = \pm 2$, $z = 2$.

Proof: First suppose y odd, and work in the ring $\mathbb{Z}[i]$, which is a unique factorization domain by Theorem 4.17. Then (4.3) factorizes as

$$(2 + iy)(2 - iy) = z^3.$$

A common factor $a + ib$ of $2 + iy$, $2 - iy$ is also a factor of their sum, 4, and difference, $2y$, so taking norms

$$a^2 + b^2 | 16, \qquad a^2 + b^2 | 4y^2,$$

implying
$$a^2 + b^2 | 4.$$

The only solutions of this relation are $a = \pm 1$, $b = 0$, or $a = 0$, $b = \pm 1$, or $a = \pm 1$, $b = \pm 1$, none of which turn out to give a proper factor $a + ib$ of

$2 + iy$. Hence $2 + iy$, $2 - iy$ are coprime. By unique factorization in $\mathbb{Z}[i]$, if their product is a cube then one is $\epsilon\alpha^3$ and the other is $\epsilon^{-1}\beta^3$ where ϵ is a unit, and $\alpha, \beta \in \mathbb{Z}[i]$. By Proposition 4.3 the units in $\mathbb{Z}[i]$ are $\pm i$, ± 1, which are all cubes, so

$$2 + iy = (a + ib)^3$$

for some $a, b \in \mathbb{Z}$. Taking complex conjugates,

$$2 - iy = (a - ib)^3.$$

Adding the two equations,

$$4 = 2a(a^2 - 3b^2)$$

so

$$a(a^2 - 3b^2) = 2.$$

Now a divides 2, so $a = \pm 1$ or ± 2; and the choice of a determines b. It is easy to see that the only solutions are $a = -1$, $b = \pm 1$, or $a = 2$, $b = \pm 1$. Then

$$z^3 = ((a + ib)(a - ib))^3 = (a^2 + b^2)^3,$$

so $z = a^2 + b^2 = 2$, 5 respectively. Then $y^2 + 4 = 8$, 125, so $y = \pm 2$, ± 11. This gives the solutions with $y = \pm 11$ as the only ones for y odd.

Now suppose y even, so that $y = 2Y$. Then z is even as well, say $z = 2Z$, and

$$Y^2 + 1 = 2Z^3.$$

Then Y must be odd, say $Y = 2k + 1$. The highest common factor of $Y + i$ and $Y - i$ divides the difference $2i = (1 + i)^2$. Now $1 + i$ divides $Y + i$ and $Y - i$ but $(1 + i)^2$ does not, so the highest common factor of $Y + i$ and $Y - i$ is $1 + i$. But

$$(1 + iY)(1 - iY) = 2Z^3$$

and the common factor $1 + i$ occurs twice on the left (bearing in mind that $1 + iY = i(Y - i)$, $1 - iY = -i(Y + i)$). Hence there must be a factorization

$$1 + iY = (1 + i)(a + ib)^3$$

whence as before

$$1 = (a + b)(a^2 - 4ab + b^2)$$

so $a = \pm 1$, $b = 0$, or $a = 0$, $b = \pm 1$. These imply $y = \pm 2$, which correspond to the other two solutions stated. $\qquad\square$

4.9   The Ramanujan-Nagell Theorem

We now give a more intricate and impressive example of how unique factorization properties of algebraic number fields are used to prove theorems on Diophantine equations. Using uniqueness of factorization in $\mathbb{Q}(\sqrt{-7})$ Nagell verified a conjecture of Ramanujan:

Theorem 4.23. *The only solutions of the equation*

$$x^2 + 7 = 2^n \tag{4.4}$$

in integers x, n are:

$$\pm x = 1 \quad 3 \quad 5 \quad 11 \quad 181$$
$$n = 3 \quad 4 \quad 5 \quad 7 \quad 15.$$

Proof: Work in $\mathbb{Q}(\sqrt{-7})$, whose ring of integers has unique factorization by Theorem 4.17. Clearly a solution for x is odd, and we suppose x is positive.

Assume first that n is even. Then we have a factorization of integers:

$$(2^{n/2} + x)(2^{n/2} - x) = 7$$

so that

$$2^{n/2} + x = 7, \quad 2^{n/2} - x = 1.$$

Now

$$2^{1+n/2} = 8$$

and $n = 4$, $x = 3$.

Let $n > 3$ be odd. We have the factorization into primes:

$$2 = \left(\frac{1 + \sqrt{-7}}{2}\right)\left(\frac{1 - \sqrt{-7}}{2}\right).$$

Now x is odd, $x = 2k + 1$, so $x^2 + 7 = 4k^2 + 4k + 8$ is divisible by 4. Put $m = n - 2$ and rewrite (4.4) as

$$\frac{x^2 + 7}{4} = 2^m$$

so that

$$\left(\frac{x + \sqrt{-7}}{2}\right)\left(\frac{x - \sqrt{-7}}{2}\right) = \left(\frac{1 + \sqrt{-7}}{2}\right)^m \left(\frac{1 - \sqrt{-7}}{2}\right)^m.$$

where the right-hand side is a prime factorization. Neither $(1 + \sqrt{-7})/2$ nor $(1 - \sqrt{-7})/2$ is a common factor of the terms on the left because such a factor would divide their difference, $\sqrt{-7}$, which is impossible by taking norms. Comparing the two factorizations, since the only units in the integers of $\mathbb{Q}(\sqrt{-7})$ are ± 1, (Proposition 4.2), we must have

$$\frac{x \pm \sqrt{-7}}{2} = \pm \left(\frac{1 \pm \sqrt{-7}}{2} \right)^m$$

from which we derive

$$\pm\sqrt{-7} = \left(\frac{1 + \sqrt{-7}}{2} \right)^m - \left(\frac{1 - \sqrt{-7}}{2} \right)^m.$$

We claim that the positive sign cannot occur. For, putting $\left(\frac{1+\sqrt{-7}}{2} \right) = a$, $\left(\frac{1-\sqrt{-7}}{2} \right) = b$, we have

$$a^m - b^m = a - b.$$

Then

$$a^2 \equiv (1 - b)^2 \equiv 1 \qquad (\bmod\ b^2)$$

since $ab = 2$, so

$$a^m \equiv a(a^2)^{(m-1)/2} \equiv a \qquad (\bmod\ b^2)$$

whence

$$a \equiv a - b \qquad (\bmod\ b^2),$$

a contradiction.

Hence the sign must be negative, and expanding by the binomial theorem

$$-2^{m-1} = \binom{m}{1} - \binom{m}{3}7 + \binom{m}{5}7^2 \ldots \pm \binom{m}{m}7^{(m-1)/2}$$

so

$$-2^{m-1} \equiv m \qquad (\bmod\ 7). \tag{4.5}$$

Now $2^6 \equiv 1 \pmod 7$ and it follows easily that the only solutions of (4.5) are

$$m \equiv 3, 5, \text{ or } 13 \qquad (\bmod\ 42).$$

We prove that only $m \equiv 3, 5, 13$ can occur. It suffices to show that we cannot have two solutions of the original equation that are congruent modulo

42. So let m, m_1 be two such solutions, and let 7^l be the largest power of 7 dividing $m - m_1$. Then

$$a^{m_1} = a^m a^{m_1-m} = a^m \left(\tfrac{1}{2}\right)^{m_1-m} (1 + \sqrt{-7})^{m_1-m}. \qquad (4.6)$$

Now

$$\left(\tfrac{1}{2}\right)^{m_1-m} = \left[\left(\tfrac{1}{2}\right)^6\right]^{(m_1-m)/6} \equiv 1 \qquad (\text{mod } 7^{l+1}),$$

(since 2 is prime to 7^{l+1} the fraction $\tfrac{1}{2}$ has a unique meaning here). Moreover,

$$\left(1 + \sqrt{-7}\right)^{m_1-m} \equiv 1 + (m_1 - m)\sqrt{-7} \qquad (\text{mod } 7^{l+1})$$

(first raise to powers $7, 7^2, \ldots, 7^l$, then $(m - m_1)/7^l$). Since

$$a^m \equiv \frac{1 + m\sqrt{-7}}{2m} \qquad (\text{mod } 7)$$

substituting in (4.6) gives

$$a^{m_1} \equiv a^m + \frac{m_1 - m}{2m}\sqrt{-7} \qquad (\text{mod } 7^{l+1})$$

and

$$b^{m_1} = b^m - \frac{m_1 - m}{2m}\sqrt{-7} \qquad (\text{mod } 7^{l+1}).$$

But

$$a^m - b^m = a^{m_1} - b^{m_1}$$

so

$$(m - m_1)\sqrt{-7} \equiv 0 \qquad (\text{mod } 7^{l+1}).$$

Since m and m_1 are rational integers,

$$m \equiv m_1 \qquad (\text{mod } 7^{l+1})$$

which contradicts the definition of l.   □

4.10   Exercises

1. Which of the following elements of $\mathbb{Z}[i]$ are irreducible ($i = \sqrt{-1}$): $1 + i$, $3 - 7i$, 5, 7, $12i$, $-4 + 5i$?

2. Write down the group of units of the ring of integers of: $\mathbb{Q}(\sqrt{-1})$, $\mathbb{Q}(\sqrt{-2})$, $\mathbb{Q}(\sqrt{-3})$, $\mathbb{Q}(\sqrt{-5})$, $\mathbb{Q}(\sqrt{-6})$.

3. Is the group of units of the integers in $\mathbb{Q}(\sqrt{3})$ finite?

4. Show that a homomorphic image of a noetherian ring is noetherian.

5. Find all ideals of $\mathbb{Z}$ which contain $\langle 120 \rangle$.
   Show that every ascending chain of ideals of $\mathbb{Z}$ starting with $\langle 120 \rangle$ stops, by direct examination of the possibilities.

6. Find a ring that is not noetherian.

7. Check the calculations required to complete Theorems 4.10, 4.11.

8. Is $10 = (3 + i)(3 - i) = 2 \cdot 5$ an example of non-unique factorization in $\mathbb{Z}[i]$? Give reasons for your answer.

9. Show that 6 and $2(1 + \sqrt{-5})$ both have 2 and $1 + \sqrt{-5}$ as factors, but do not have a highest common factor in $\mathbb{Z}[\sqrt{-5}]$. Do they have a least common multiple? (*Hint:* Consider norms.)

10. Let D be any integral domain. Suppose an element $x \in D$ has a factorization
$$x = up_1 \ldots p_n$$
where u is a unit and $p_1, \ldots, p_n$ are *primes*. Show that given any factorization
$$x = vq_1 \ldots q_m$$
where v is a unit and $q_1, \ldots, q_m$ are irreducibles, then $m = n$ and there exists a permutation π of $\{1, \ldots, n\}$ such that $p_i, q_{\pi(i)}$ are associates $(1 \leq i \leq n)$.

11. Show in $\mathbb{Z}[\sqrt{-5}]$ that $\sqrt{-5}|(a + b\sqrt{-5})$ if and only if $5|a$. Deduce that $\sqrt{-5}$ is prime in $\mathbb{Z}[\sqrt{-5}]$. Hence conclude that the element 5 factorizes uniquely into irreducibles in $\mathbb{Z}[\sqrt{-5}]$, although $\mathbb{Z}[\sqrt{-5}]$ does not have unique factorization.

12. Suppose D is a unique factorization domain, and a, b are coprime non-units. Deduce that if
$$ab = c^n$$
for $c \in D$, there exists a unit $e \in D$ such that ea and $e^{-1}b$ are nth powers in D.

13. Let p be an odd rational prime and $\zeta = e^{2\pi i/p}$. If α is a prime element in $\mathbb{Z}[\zeta]$, prove that the rational integers which are divisible by α are precisely the rational integer multiples of some prime rational integer q. (*Hint:* $\alpha|N(\alpha)$, so α divides some rational prime factor q of $N(\alpha)$. Now show α is not a factor of any $m \in \mathbb{Z}$ prime to q.)

14. Prove that the ring of integers of $\mathbb{Q}(e^{2\pi i/5})$ is Euclidean.

15. Prove that the ring of integers of $\mathbb{Q}(\sqrt{2}, i)$ is Euclidean.

16. Let $\mathbb{Q}_2$ be the set of all rational numbers a/b, where $a, b \in \mathbb{Z}$ and b is odd. Prove that $\mathbb{Q}_2$ is a domain, and that the only irreducibles in $\mathbb{Q}_2$ are 2 and its associates.

17. Generalize Exercise 16 to the ring $\mathbb{Q}_\pi$, where π is a finite set of ordinary primes, this being defined as the set of all rationals a/b with b prime to the elements of π.

18. The following purports to be a proof that in any number field K the ring of integers contains infinitely many irreducibles. Find the error.

    Assume $\mathfrak{O}$ has only finitely many irreducibles $p_1, \ldots, p_n$. The number $1 + p_1 \ldots p_n$ must be divisible by some irreducible q, and this cannot be any of $p_1, \ldots, p_n$. This is a contradiction. Of course the argument breaks down unless we can find at least one irreducible in $\mathfrak{O}$; but since not every element of $\mathfrak{O}$ is a unit this is easy: let x be any non-unit and let p be some irreducible factor of x.

    (*Hint*: The 'proof' does not use any properties of $\mathfrak{O}$ beyond the existence of irreducible factorization and the fact that not every element is a unit. Now $\mathbb{Q}_2$ has these properties ...)

19. Give a correct proof of the statement in Exercise 18.

5

Ideals

After the somewhat traumatic realization that factorization into irreducibles is unique in some rings of integers but not in others, we seek some way to minimize the damage. Kummer, and then Dedekind, took steps to develop more insightful theories. Kummer had the bright idea that if he could not factorize a number uniquely in a given ring of integers, then perhaps he could extend the ring to a bigger one in which further factorization might be not only possible, but unique. For example, we pointed out in Chapter 4 that there are two factorizations $6 = 2 \cdot 3 = \sqrt{-6} \cdot \sqrt{-6}$ in $\mathbb{Z}[\sqrt{-6}]$, but $\sqrt{-6}$ does not divide 2 or 3 in this ring. In fact, $2/\sqrt{-6} = \sqrt{-2/3}$, and $3/\sqrt{-6} = \sqrt{-3/2}$, neither of which belongs to $\mathbb{Z}[\sqrt{-6}]$. Kummer's idea: throw them into the pot to create a larger ring. He called the new elements introduced in this way 'ideal numbers'.

Dedekind looked at the same ideas from a different direction, introducing the notion of an 'ideal' in ring theory: a special kind of subring. The word referred to the reformulation of Kummer's notion of ideal numbers. Dedekind showed that although unique factorization may fail for numbers, a simple and elegant theory of unique factorization can be developed for ideals. In this theory, the essential building blocks are 'prime ideals', which are defined by adapting the definition of a prime element from the previous chapter.

Just as it is often easier to work with both a ring of integers and its corresponding field of quotients, we generalize the concept of ideal to 'fractional ideal'; this generalization has the advantage that the non-zero fractional ideals form a group under multiplication. From this the uniqueness of factorization into prime ideals follows easily. Several standard consequences of unique factorization are easily deduced.

We define the norm of an ideal as a generalization of the norm of an element and prove that the new norm has the corresponding multiplicative property. We use this to show that every ideal can be generated by at most two elements. This tightens the noose around the neck of non-uniqueness of factorization. In the previous chapter we saw that factorization of elements is unique in a principal ideal domain (where every ideal is generated by a single element). We refine this result to show that factorization of elements into irreducibles is unique in a ring of integers if and only if every ideal is principal.

5.1   Historical Background

To motivate Kummer's introduction of ideal numbers and Dedekind's reformulation of this concept in terms of ideals, we look more closely at some examples where unique factorization fails, in the hope that some pattern may emerge.

Many previous examples exhibit no obvious pattern, but others seem to have significant features. For instance, consider:

$$\mathbb{Q}\left(\sqrt{15}\right): \quad 2 \cdot 5 = (5 + \sqrt{15})(5 - \sqrt{15})$$
$$\mathbb{Q}\left(\sqrt{30}\right): \quad 2 \cdot 3 = (6 + \sqrt{30})(6 - \sqrt{30})$$
$$\mathbb{Q}\left(\sqrt{-10}\right): \quad 2 \cdot 7 = (2 + \sqrt{-10})(2 - \sqrt{-10}).$$

In these we see a curious phenomenon: there is a prime p occurring on the left, and on the right a factor $a + b\sqrt{d}$ where a and d are multiples of p. It looks as though $\sqrt{p}$ is somehow a common factor of both sides—but $\sqrt{p}$ does not lie in the given number field. As a specific case, consider the first example; here $\sqrt{5}$ looks a likely candidate for a common factor but $\sqrt{5}$ is not an element of $\mathbb{Q}(\sqrt{15})$. Leaving aside the niceties for the moment, introduce $\sqrt{5}$ into the factorization to get

$$5 + \sqrt{15} = \sqrt{5}(\sqrt{5} + \sqrt{3})$$
$$5 - \sqrt{15} = \sqrt{5}(\sqrt{5} - \sqrt{3}).$$

Multiply and cancel the 5 to get

$$2 = (\sqrt{5} + \sqrt{3})(\sqrt{5} - \sqrt{3}).$$

We now see that the two given factorizations of 10 are obtained by grouping the factors in

$$(\sqrt{5})(\sqrt{5})(\sqrt{5} + \sqrt{3})(\sqrt{5} - \sqrt{3})$$

in two different ways.

Perhaps by introducing new numbers, such as $\sqrt{5}$, we can restore unique factorization. Can our problem be that we are not factorizing in the right context? In other words, if factorization of some element in the ring of integers of the given number field K is not unique, can we extend K to a field L where it *is*? In our example to factorize the element 10 we extended $\mathbb{Q}(\sqrt{15})$ to $\mathbb{Q}(\sqrt{3}, \sqrt{5})$. What about the others? The factorizations of 14 in $\mathbb{Q}(\sqrt{-10})$ can be found by extending to $\mathbb{Q}(\sqrt{2}, \sqrt{-5})$ to get two possible groupings of the factors in

$$14 = (\sqrt{2})(\sqrt{2})(\sqrt{2} + \sqrt{-5})(\sqrt{2} - \sqrt{-5}).$$

The case of 6 in $\mathbb{Q}(\sqrt{30})$ is even more interesting: we have

$$6 = (\sqrt{2})(\sqrt{2})(\sqrt{3})(\sqrt{3})(\sqrt{6} + \sqrt{5})(\sqrt{6} - \sqrt{5}),$$

and the last two factors are *units* because $(\sqrt{6} + \sqrt{5})(\sqrt{6} - \sqrt{5}) = 6 - 5 = 1$.

This is one way to view Kummer's theory. Start with a number field K and extend to a field L. Then $\mathfrak{O}_K \subseteq \mathfrak{O}_L$. Neither of these rings of integers need have unique factorization, but an element in $\mathfrak{O}_K$ may factorize uniquely into elements in $\mathfrak{O}_L$.

At the outset Kummer did not describe the theory in this way. His method involved detailed computations, described in Edwards [24, 25], and a radically new notion of 'ideal' prime factors for elements that have no prime factors in $\mathfrak{O}_K$. These extra 'ideal' numbers may be interpreted as elements introduced from $\mathfrak{O}_L$ for factorization purposes, but Kummer's description was more mysterious: things that behaved like numbers but were not.

A simpler and more natural formulation of the theory by Dedekind in terms of ideals clarified matters. To motivate this approach, consider a factorization

$$x = ab$$

in a ring R. Recall from Chapter 1 that the product of ideals IJ is just the set of finite sums $\sum x_i y_i$ ($x_i \in I, y_i \in J$). Therefore the ideal generated by x is the product of the ideals generated by a and by b:

$$\langle x \rangle = \langle a \rangle \langle b \rangle .$$

More generally a product

$$x = p_1 \ldots p_n$$

of elements in R corresponds to a product of principal ideals

$$\langle x \rangle = \langle p_1 \rangle \ldots \langle p_n \rangle .$$

When considering unique factorization, the formulation in terms of ideals is marginally better, for if we replace p_1 by up_1 where u is a unit, the ideals $\langle p_1 \rangle$ and $\langle up_1 \rangle$ are the same, see Proposition 4.4(b). Thus, when the factors are unique up to multiplication by units and order, the ideals $\langle p_1 \rangle, \ldots, \langle p_n \rangle$ are *unique* (up to order). So passing to ideals eliminates the problems introduced by units.

How does this tie in with the earlier discussion? First consider the example

$$10 = (\sqrt{5})(\sqrt{5})(\sqrt{5} + \sqrt{3})(\sqrt{5} - \sqrt{3})$$

in the integers of $\mathbb{Q}(\sqrt{3}, \sqrt{5})$. Let $K = \mathbb{Q}(\sqrt{15})$, $L = \mathbb{Q}(\sqrt{3}, \sqrt{5})$; then this factorization holds in the ring of integers $\mathfrak{O}_L$. In this ring we also have the corresponding factorization of principal ideals:

$$\langle 10 \rangle = \langle \sqrt{5} \rangle \langle \sqrt{5} \rangle \langle \sqrt{5} + \sqrt{3} \rangle \langle \sqrt{5} - \sqrt{3} \rangle.$$

We may intersect the ideals in this factorization with $\mathfrak{O}_K$, and once more we get ideals in $\mathfrak{O}_K$, but these ideals *need not be principal*, and in this case they are not. For instance, let $I = \langle \sqrt{5} + \sqrt{3} \rangle \cap \mathfrak{O}_K$. Then $\sqrt{3}(\sqrt{5}+\sqrt{3}) = \sqrt{15} + 3 \in I$, and $\sqrt{5}(\sqrt{5} + \sqrt{3}) = 5 + \sqrt{15} \in I$, so their difference

$$(5 + \sqrt{15}) - (3 + \sqrt{15}) = 2 \in I.$$

If I were principal, say $I = \langle a + b\sqrt{15} \rangle$, then 2 would be a multiple of $a + b\sqrt{15}$, and taking norms,

$$a^2 - 15b^2 | 4.$$

Suppose that I is principal, say $I = \langle k \rangle$. Now $N(5 + \sqrt{15}) = 10$ and $N(3 + \sqrt{15}) = -6$, so $N(k)|2$. We know that $N(k) \neq \pm 1$ since I is proper. If $N(k) = \pm 2$ then there exist $a, b \in \mathbb{Z}$ with $a^2 - 15b^2 = \pm 2$. But, taken modulo 5, this leads to a contradiction. So I is not principal.

The moral is now clear. If we wish to factorize the principal ideal $\langle x \rangle$ in a ring of integers $\mathfrak{O}_K$, we might get a unique factorization of ideals,

$$\langle x \rangle = I_1 \ldots I_n,$$

but the ideals $I_1, \ldots, I_n$ might not be principal.

Factorization into ideals proves to be most useful, however; for the ideals in $\mathfrak{O}_K$ are not far off being principal, having (as we shall see) at most two generators.

5.2   Prime Factorization of Ideals

Throughout this chapter $\mathfrak{O}$ is the ring of integers of a number field K of degree n. We use small Gothic letters $\mathfrak{a}, \mathfrak{b}, \mathfrak{c}, \ldots$ to denote ideals (and later 'fractional ideals') of $\mathfrak{O}$. We are interested in two special types of ideal, which we define in a general context as follows. Let R be a ring. Then an ideal $\mathfrak{a}$ of R is *maximal* if $\mathfrak{a}$ is a proper ideal of R and there are no ideals of R strictly between $\mathfrak{a}$ and R. The ideal $\mathfrak{a} \neq R$ of R is *prime* if, whenever $\mathfrak{b}$ and $\mathfrak{c}$ are ideals of R with $\mathfrak{bc} \subseteq \mathfrak{a}$, then either $\mathfrak{b} \subseteq \mathfrak{a}$ or $\mathfrak{c} \subseteq \mathfrak{a}$.

We can see where the latter definition comes from by considering the special case where all three ideals concerned are principal, say $\mathfrak{a} = \langle a \rangle$, $\mathfrak{b} = \langle b \rangle$, $\mathfrak{c} = \langle c \rangle$. Since $x|y$ is equivalent to $\langle y \rangle \subseteq \langle x \rangle$ by Proposition 4.5(a), the statement

$$\mathfrak{bc} \subseteq \mathfrak{a} \text{ implies either } \mathfrak{b} \subseteq \mathfrak{a} \text{ or } \mathfrak{c} \subseteq \mathfrak{a}$$

translates into

$$a|bc \text{ implies either } a|b \text{ or } a|c.$$

If R is an integral domain, then the zero ideal is prime, and here we find $\langle p \rangle$ is prime if and only if p is a prime or zero. (See Exercise 5 in this chapter.) Excluding 0 from the list of prime elements but including $\langle 0 \rangle$ as a prime ideal is a quirk of historical development. Elements came first and 0 was excluded from the list of primes of $\mathbb{Z}$. On the other hand, the definition we have given for a prime ideal implies the following simple characterizations:

Lemma 5.1. *Let R be a ring and $\mathfrak{a}$ an ideal of R. Then*
   (a) $\mathfrak{a}$ *is maximal if and only if $R/\mathfrak{a}$ is a field.*
   (b) $\mathfrak{a}$ *is prime if and only if $R/\mathfrak{a}$ is a domain.*

Proof: The ideals of $R/\mathfrak{a}$ are in bijective correspondence with the ideals of R lying between $\mathfrak{a}$ and R. Hence $\mathfrak{a}$ is maximal if and only if $R/\mathfrak{a}$ has no non-zero proper ideals. Now it is easy to show that a ring S has no non-zero proper ideals if and only if S is a field. Taking $S = R/\mathfrak{a}$ proves (a).

To prove (b), first suppose $\mathfrak{a}$ is prime. If $x, y \in R$ are such that in $R/\mathfrak{a}$ we have

$$(\mathfrak{a} + x)(\mathfrak{a} + y) = 0$$

and then $xy \in \mathfrak{a}$, so $\langle x \rangle \langle y \rangle \subseteq \mathfrak{a}$. Hence either $\langle x \rangle \subseteq \mathfrak{a}$ or $\langle y \rangle \subseteq \mathfrak{a}$, so either $x \in \mathfrak{a}$ or $y \in \mathfrak{a}$. Hence one of $(\mathfrak{a}+x)$ or $(\mathfrak{a}+y)$ is zero in $R/\mathfrak{a}$, and therefore $R/\mathfrak{a}$ has no zero-divisors so it is a domain. Conversely suppose $R/\mathfrak{a}$ is a domain. Then $|R/\mathfrak{a}| \neq 1$ so $\mathfrak{a} \neq R$. Suppose if possible that $\mathfrak{bc} \subseteq \mathfrak{a}$ but $\mathfrak{b} \not\subseteq \mathfrak{a}$, $\mathfrak{c} \not\subseteq \mathfrak{a}$. Then we can find elements $b \in \mathfrak{b}$, $c \in \mathfrak{c}$, with $b, c \notin \mathfrak{a}$ but $bc \in \mathfrak{a}$. This means that $(\mathfrak{a} + b)$ and $(\mathfrak{a} + c)$ are zero-divisors in $R/\mathfrak{a}$, which is a contradiction. $\square$

Corollary 5.2. *Every maximal ideal is prime.*   □

Next we list some important properties of the ring of integers of a number field:

Theorem 5.3. *The ring of integers $\mathfrak{O}$ of a number field K has the following properties:*
  (a) *It is a domain, with field of fractions K.*
  (b) *It is noetherian.*
  (c) *If $\alpha \in K$ satisfies a monic polynomial equation with coefficients in $\mathfrak{O}$ then $\alpha \in \mathfrak{O}$.*
  (d) *Every non-zero prime ideal of $\mathfrak{O}$ is maximal.*

Proof: Part (a) is obvious. For part (b) note that by Theorem 2.16 the group $(\mathfrak{O}, +)$ is free abelian of rank n. It follows by Theorem 1.16 that if $\mathfrak{a}$ is an ideal of $\mathfrak{O}$ then $(\mathfrak{a}, +)$ is free abelian of rank $\leq n$. Now any $\mathbb{Z}$-basis for $(\mathfrak{a}, +)$ generates $\mathfrak{a}$ as an ideal, so every ideal of $\mathfrak{O}$ is finitely generated and $\mathfrak{O}$ is noetherian. Part (c) is immediate from Theorem 2.10. To prove part (d) let $\mathfrak{p}$ be a prime ideal of $\mathfrak{O}$. Let $0 \neq \alpha \in \mathfrak{p}$. Then

$$N = \mathrm{N}(\alpha) = \alpha_1, \ldots \alpha_n \in \mathfrak{p}$$

(the α_i being the conjugates of α) since $\alpha_1 = \alpha$. Therefore $\langle N \rangle \subseteq \mathfrak{p}$, so $\mathfrak{O}/\mathfrak{p}$ is a quotient ring of $\mathfrak{O}/N\mathfrak{O}$ which, being a finitely generated abelian group with every element of finite order, is finite. Since $\mathfrak{O}/\mathfrak{p}$ is a domain by Lemma 5.1(b) and is finite, it is a field by Theorem 1.5. Hence $\mathfrak{p}$ is a maximal ideal by Lemma 5.1 (a).   □

Part (d) of Theorem 5.3 is by no means typical of general rings. For example if $R = \mathbb{R}[x, y]$, the ring of polynomials in indeterminates x, y with real coefficients, the ideal $\langle x \rangle$ is prime but not maximal because $R/\langle x \rangle \cong \mathbb{R}[y]$ is a domain but not a field. A ring that satisfies conditions 5.3(a)–(d) is called a *Dedekind ring*. The proof of unique factorization of ideals, which we give shortly, is valid for all Dedekind rings—although in applications we require only the special case when the ring is a ring $\mathfrak{O}$ of integers in a number field.

To prove uniqueness we need to study the 'arithmetic' of non-zero ideals of $\mathfrak{O}$, especially their behaviour under multiplication. Clearly this multiplication is commutative and associative with $\mathfrak{O}$ itself as an identity. However, inverses need not exist, so we do not have a group structure. It turns out that we can capture a group if we spread our net wider. Note that an ideal may be described as an $\mathfrak{O}$-submodule of $\mathfrak{O}$, so we look at $\mathfrak{O}$-submodules of the field K. The particular submodules of interest to give the group structure we desire will turn out to be characterized by the following property:

an $\mathfrak{O}$-submodule $\mathfrak{a}$ of K is called a *fractional ideal* of $\mathfrak{O}$ if there exists some non-zero $c \in \mathfrak{O}$ such that $c\mathfrak{a} \subseteq \mathfrak{O}$. In other words, the set $\mathfrak{b} = c\mathfrak{a}$ is an ideal of $\mathfrak{O}$, and $\mathfrak{a} = c^{-1}\mathfrak{b}$; thus the fractional ideals of $\mathfrak{O}$ are subsets of K of the form $c^{-1}\mathfrak{b}$ where $\mathfrak{b}$ is an ideal of $\mathfrak{O}$ and c is a non-zero element of $\mathfrak{O}$. (This explains the name.)

Example 5.4. The fractional ideals of $\mathbb{Z}$ are of the form $r\mathbb{Z}$ where $r \in \mathbb{Q}$.

Of course if every ideal of $\mathfrak{O}$ is principal, then the fractional ideals are of the form $c^{-1} \langle d \rangle = c^{-1}d\mathfrak{O}$ where d is a generator. By Theorem 5.3(a) this means the fractional ideals in a principal ideal domain $\mathfrak{O}$ are just $\alpha\mathfrak{O}$ where $\alpha \in K$. The interest in fractional ideals is greater because $\mathfrak{O}$ need not be a principal ideal domain.

In general, an ideal is clearly a fractional ideal and, conversely, a fractional ideal $\mathfrak{a}$ is an ideal if and only if $\mathfrak{a} \subseteq \mathfrak{O}$. The product of fractional ideals is once more a fractional ideal. In fact, if $\mathfrak{a}_1 = c_1^{-1}\mathfrak{b}_1$, $\mathfrak{a}_2 = c_2^{-1}\mathfrak{b}_2$ where $\mathfrak{b}_1$, $\mathfrak{b}_2$, are ideals and c_1, c_2 are non-zero elements of $\mathfrak{O}$, then $\mathfrak{a}_1\mathfrak{a}_2 = (c_1c_2)^{-1}\mathfrak{b}_1\mathfrak{b}_2$. The multiplication of fractional ideals is commutative and associative with $\mathfrak{O}$ acting as an identity.

Theorem 5.5. *The non-zero fractional ideals of $\mathfrak{O}$ form an abelian group under multiplication.*

It is convenient to prove this result along with the main theorem of the chapter:

Theorem 5.6. *Every non-zero ideal of $\mathfrak{O}$ can be written as a product of prime ideals, uniquely up to the order of the factors.*

Proof: We prove Theorems 5.5 and 5.6 together in a series of nine steps.

(1)*Let $\mathfrak{a} \neq 0$ be an ideal of $\mathfrak{O}$. Then there exist prime ideals $\mathfrak{p}_1, \ldots, \mathfrak{p}_r$ such that $\mathfrak{p}_1 \ldots \mathfrak{p}_r \subseteq \mathfrak{a}$.*

For a contradiction, suppose not. By Theorem 5.3(b) $\mathfrak{O}$ is noetherian, so we may choose $\mathfrak{a}$ to be maximal subject to the non-existence of such $\mathfrak{p}$'s. Then $\mathfrak{a}$ is not prime (since we could then take $\mathfrak{p}_1 = \mathfrak{a}$), so there exist ideals $\mathfrak{b}$, $\mathfrak{c}$ of $\mathfrak{O}$ with $\mathfrak{b}\mathfrak{c} \subseteq \mathfrak{a}$, $\mathfrak{b} \not\subseteq \mathfrak{a}$, $\mathfrak{c} \not\subseteq \mathfrak{a}$. Let

$$\mathfrak{a}_1 = \mathfrak{a} + \mathfrak{b}, \qquad \mathfrak{a}_2 = \mathfrak{a} + \mathfrak{c}.$$

Then $\mathfrak{a}_1\mathfrak{a}_2 \subseteq \mathfrak{a}$, $\mathfrak{a}_1 \supsetneq \mathfrak{a}$, $\mathfrak{a}_2 \supsetneq \mathfrak{a}$. By maximality of $\mathfrak{a}$ there exist prime ideals $\mathfrak{p}_1, \ldots, \mathfrak{p}_s, \mathfrak{p}_{s+1}, \ldots, \mathfrak{p}_r$ such that

$$\mathfrak{p}_1 \ldots \mathfrak{p}_s \subseteq \mathfrak{a}_1,$$
$$\mathfrak{p}_{s+1} \ldots \mathfrak{p}_r \subseteq \mathfrak{a}_2.$$

Hence

$$\mathfrak{p}_1 \ldots \mathfrak{p}_r \subseteq \mathfrak{a}_1 \mathfrak{a}_2 \subseteq \mathfrak{a}$$

contrary to the choice of $\mathfrak{a}$.

(2) *Definition of what will turn out to be the inverse of an ideal:*

For each ideal $\mathfrak{a}$ of $\mathfrak{O}$, define

$$\mathfrak{a}^{-1} = \{x \in K \mid x\mathfrak{a} \subseteq \mathfrak{O}\}.$$

It is clear that $\mathfrak{a}^{-1}$ is an $\mathfrak{O}$-submodule. If $\mathfrak{a} \neq 0$ then for any $c \in \mathfrak{a}$, $c \neq 0$, we have $c\mathfrak{a}^{-1} \subseteq \mathfrak{O}$, so $\mathfrak{a}^{-1}$ is a fractional ideal. Clearly $\mathfrak{O} \subseteq \mathfrak{a}^{-1}$, so $\mathfrak{a} = \mathfrak{a}\mathfrak{O} \subseteq \mathfrak{a}\mathfrak{a}^{-1}$. From the definition,

$$\mathfrak{a}\mathfrak{a}^{-1} = \mathfrak{a}^{-1}\mathfrak{a} \subseteq \mathfrak{O}.$$

Therefore the fractional ideal $\mathfrak{a}\mathfrak{a}^{-1}$ is actually an *ideal*. Our aim will be to prove $\mathfrak{a}\mathfrak{a}^{-1} = \mathfrak{O}$.

A further useful fact for ideals $\mathfrak{p}$, $\mathfrak{a}$ is that $\mathfrak{a} \subseteq \mathfrak{p}$ implies $\mathfrak{O} \subseteq \mathfrak{p}^{-1} \subseteq \mathfrak{a}^{-1}$.

(3) *If $\mathfrak{a}$ is a proper ideal, then $\mathfrak{a}^{-1} \supsetneq \mathfrak{O}$.*

Since $\mathfrak{a} \subseteq \mathfrak{p}$ for some maximal ideal $\mathfrak{p}$, whence $\mathfrak{p}^{-1} \subseteq \mathfrak{a}^{-1}$, it is sufficient to prove $\mathfrak{p}^{-1} \neq \mathfrak{O}$ for $\mathfrak{p}$ maximal. We must therefore find a non-integer in $\mathfrak{p}^{-1}$. We start with any $a \in \mathfrak{p}$, $a \neq 0$. Using step (1) we choose the smallest r such that

$$\mathfrak{p}_1 \ldots \mathfrak{p}_r \subseteq \langle a \rangle$$

for $\mathfrak{p}_1, \ldots, \mathfrak{p}_r$ prime. Since $\langle a \rangle \subseteq \mathfrak{p}$ and $\mathfrak{p}$ is prime (remember maximal implies prime), some $\mathfrak{p}_i \subseteq \mathfrak{p}$. Without loss of generality $\mathfrak{p}_1 \subseteq \mathfrak{p}$. Hence $\mathfrak{p}_1 = \mathfrak{p}$ since prime ideals in $\mathfrak{O}$ are maximal by Theorem 5.3 (d). Moreover,

$$\mathfrak{p}_2 \ldots \mathfrak{p}_r \nsubseteq \langle a \rangle$$

by minimality of r. Hence we can find $b \in \mathfrak{p}_2 \ldots \mathfrak{p}_r \setminus \langle a \rangle$. But $b\mathfrak{p} \subseteq \langle a \rangle$ so $ba^{-1}\mathfrak{p} \subseteq \mathfrak{O}$ and $ba^{-1} \in \mathfrak{p}^{-1}$. But $b \notin a\mathfrak{O}$ and so $ba^{-1} \notin \mathfrak{O}$, whence $\mathfrak{p}^{-1} \neq \mathfrak{O}$.

(4) *If $\mathfrak{a}$ is a non-zero ideal and $\mathfrak{a}S \subseteq \mathfrak{a}$ for any subset $S \subseteq K$, then $S \subseteq \mathfrak{O}$.*

We must show that if $\mathfrak{a}\theta \subseteq \mathfrak{a}$ for $\theta \in S$, then $\theta \in \mathfrak{O}$. Because $\mathfrak{O}$ is noetherian, $\mathfrak{a} = \langle a_1, \ldots, a_m \rangle$, where not all the a_i are zero. Then $\mathfrak{a}\theta \subseteq \mathfrak{a}$ implies

$$a_1\theta \;=\; b_{11}a_1 + \ldots + b_{1m}a_m$$
$$\vdots \quad \ldots \qquad\qquad (b_{ij} \in \mathcal{D})$$
$$a_m\theta \;=\; b_{m1}a_1 + \ldots + b_{mm}a_m.$$

The equations

$$(b_{11} - \theta)\,x_1 + \ldots + b_{1m}x_m \;=\; 0$$
$$\vdots$$
$$b_{m1}x_1 + \ldots + (b_{mm} - \theta)\,x_m \;=\; 0$$

have a non-zero solution $x_1 = a_1, \ldots, x_m = a_m$, so, as in Lemma 2.8, the determinant of the array of coefficients is zero. This gives a monic polynomial equation in θ with coefficients in $\mathfrak{O}$, so $\theta \in \mathfrak{O}$ by Theorem 5.3(c). (We could short-cut part of this proof by noting, as in the proof of Lemma 2.8, that the b_{ij} may be taken to be *rational* integers which gives $\theta \in \mathfrak{O}$ directly.)

We are now in a position to take an important step in the proof of Theorem 5.5:

(5) *If $\mathfrak{p}$ is a maximal ideal, then $\mathfrak{p}\mathfrak{p}^{-1} = \mathfrak{O}$.*

From (2), $\mathfrak{p}\mathfrak{p}^{-1}$ is an ideal where $\mathfrak{p} \subseteq \mathfrak{p}\mathfrak{p}^{-1} \subseteq \mathfrak{O}$. Since $\mathfrak{p}$ is maximal, $\mathfrak{p}\mathfrak{p}^{-1}$ is equal to $\mathfrak{p}$ or $\mathfrak{O}$. But if $\mathfrak{p}\mathfrak{p}^{-1} = \mathfrak{O}$, then (4) would imply $\mathfrak{p}^{-1} \subseteq \mathfrak{O}$, contradicting (3). So $\mathfrak{p}\mathfrak{p}^{-1} = \mathfrak{O}$.

We can now extend (5) to any ideal $\mathfrak{a}$:

(6) *For every ideal $\mathfrak{a} \neq 0$, $\mathfrak{a}\mathfrak{a}^{-1} = \mathfrak{O}$.*

If not, choose $\mathfrak{a}$ maximal subject to $\mathfrak{a}\mathfrak{a}^{-1} \neq \mathfrak{O}$. Then $\mathfrak{a} \subseteq \mathfrak{p}$ where $\mathfrak{p}$ is maximal. From (2), $\mathfrak{O} \subseteq \mathfrak{p}^{-1} \subseteq \mathfrak{a}^{-1}$, so

$$\mathfrak{a} \subseteq \mathfrak{a}\mathfrak{p}^{-1} \subseteq \mathfrak{a}\mathfrak{a}^{-1} \subseteq \mathfrak{O}.$$

In particular, $\mathfrak{a}\mathfrak{p}^{-1} \subseteq \mathfrak{O}$ implies $\mathfrak{a}\mathfrak{p}^{-1}$ is an ideal. Now we cannot have $\mathfrak{a} = \mathfrak{a}\mathfrak{p}^{-1}$, for that would imply $\mathfrak{p}^{-1} \subseteq \mathfrak{O}$ by (4), contradicting (3) once more. So $\mathfrak{a} \subsetneq \mathfrak{a}\mathfrak{p}^{-1}$ and the maximality condition on $\mathfrak{a}$ implies that $\mathfrak{a}\mathfrak{p}^{-1}$ satisfies

$$\mathfrak{a}\mathfrak{p}^{-1}\left(\mathfrak{a}\mathfrak{p}^{-1}\right)^{-1} = \mathfrak{O}.$$

By the definition of $\mathfrak{a}^{-1}$ this implies:

$$\mathfrak{p}^{-1}\left(\mathfrak{a}\mathfrak{p}^{-1}\right)^{-1} \subseteq \mathfrak{a}^{-1}.$$

Thus

$$\mathfrak{O} = \mathfrak{a}\mathfrak{p}^{-1}\left(\mathfrak{a}\mathfrak{p}^{-1}\right)^{-1} \subseteq \mathfrak{a}\mathfrak{a}^{-1} \subseteq \mathfrak{O}$$

from which the result follows.

(7) *Every fractional ideal $\mathfrak{a}$ has an inverse $\mathfrak{a}^{-1}$ such that $\mathfrak{a}\mathfrak{a}^{-1} = \mathfrak{O}$.*

The set $\mathfrak{F}$ of fractional ideals is already known to be a commutative semigroup, so given a fractional ideal $\mathfrak{a}$, we only need to find another fractional ideal $\mathfrak{a}'$ such that $\mathfrak{a}\mathfrak{a}' = \mathfrak{O}$, then $\mathfrak{a}'$ will be the required inverse. But there exists an ideal $\mathfrak{b}$ and a non-zero element $c \in \mathfrak{O}$ such that $\mathfrak{a} = c^{-1}\mathfrak{b}$. Let $\mathfrak{a}' = c\mathfrak{b}^{-1}$, then $\mathfrak{a}\mathfrak{a}' = \mathfrak{O}$ as required.

This, of course, proves Theorem 5.5.

(8) *Every non-zero ideal $\mathfrak{a}$ is a product of prime ideals.*

If not, let $\mathfrak{a}$ be maximal subject to the condition of not being a product of prime ideals. Then $\mathfrak{a}$ is not prime, but we will have $\mathfrak{a} \subseteq \mathfrak{p}$ for some maximal (hence prime) ideal, and as in (6),

$$\mathfrak{a} \subsetneq \mathfrak{a}\mathfrak{p}^{-1} \subseteq \mathfrak{O}.$$

By the maximality condition on $\mathfrak{a}$,

$$\mathfrak{a}\mathfrak{p}^{-1} = \mathfrak{p}_2 \ldots \mathfrak{p}_r$$

for prime ideals $\mathfrak{p}_2, \ldots, \mathfrak{p}_r$, whence

$$\mathfrak{a} = \mathfrak{p}\mathfrak{p}_2 \ldots \mathfrak{p}_r.$$

(9) *Prime factorization is unique.*

By analogy with factorization of elements, for ideals $\mathfrak{a}$, $\mathfrak{b}$ we say that $\mathfrak{a}$ divides $\mathfrak{b}$ (written $\mathfrak{a}|\mathfrak{b}$) if there is an ideal $\mathfrak{c}$ such that $\mathfrak{b} = \mathfrak{a}\mathfrak{c}$. This condition is equivalent to $\mathfrak{a} \supseteq \mathfrak{b}$ since we may then take $\mathfrak{c} = \mathfrak{a}^{-1}\mathfrak{b}$. The definition of prime ideal $\mathfrak{p}$ shows that if $\mathfrak{p}|\mathfrak{a}\mathfrak{b}$ then either $\mathfrak{p}|\mathfrak{a}$ or $\mathfrak{p}|\mathfrak{b}$. If we now have prime ideals $\mathfrak{p}_1, \ldots, \mathfrak{p}_r, \mathfrak{q}_1, \ldots, \mathfrak{q}_s$ with

$$\mathfrak{p}_1 \ldots \mathfrak{p}_r = \mathfrak{q}_1 \ldots \mathfrak{q}_s,$$

then $\mathfrak{p}_1$ divides some $\mathfrak{q}_i$, so by maximality $\mathfrak{p}_1 = \mathfrak{q}_i$. Multiplying by $\mathfrak{p}_1^{-1}$ and using induction we obtain uniqueness of prime factorization up to the order of the factors.

This proves Theorem 5.6.   □

In fact, the fractional ideals also factorize uniquely if we allow negative powers of prime ideals. Namely, if $\mathfrak{a}$ is a fractional ideal with $0 \neq c \in \mathfrak{O}$ such that $c\mathfrak{a}$ is an ideal, we have

$$\langle c \rangle = \mathfrak{p}_1 \ldots \mathfrak{p}_r, \qquad c\mathfrak{a} = \mathfrak{q}_1 \ldots \mathfrak{q}_s,$$

so that

$$\mathfrak{a} = \mathfrak{p}_1^{-1} \ldots \mathfrak{p}_r^{-1} \mathfrak{q}_1 \ldots \mathfrak{q}_s.$$

One result in the proofs of Theorems 5.5 and 5.6, which is worth isolating, occurs in step (9):

Proposition 5.7. *For ideals* $\mathfrak{a}$, $\mathfrak{b}$ *of* $\mathfrak{O}$,

$$\mathfrak{a} \mid \mathfrak{b} \quad \text{if and only if } \mathfrak{a} \supseteq \mathfrak{b}. \qquad \qquad \square$$

This tells us that in $\mathfrak{O}$ the factors of an ideal $\mathfrak{b}$ are precisely the ideals containing $\mathfrak{b}$. The definition of a prime ideal $\mathfrak{p}$ also translates into a notation directly analogous to that of a prime element:

$$\mathfrak{p} \mid \mathfrak{ab} \quad \text{implies} \quad \mathfrak{p} \mid \mathfrak{a} \quad \text{or} \quad \mathfrak{p} \mid \mathfrak{b}.$$

Extended Worked Example. Factorization of the ideal $\langle 18 \rangle$ in $\mathbb{Z}[\sqrt{-17}]$.
Theorem 4.11 displays the factorization

$$18 = 2 \cdot 3 \cdot 3 = (1 + \sqrt{-17})(1 - \sqrt{-17}).$$

Consider the ideal $\mathfrak{p}_1 = \langle 2, 1 + \sqrt{-17} \rangle$ whose generators are both factors of 18. Clearly $18 \in \mathfrak{p}_1$, so $\langle 18 \rangle \subseteq \mathfrak{p}_1$, which means that $\mathfrak{p}_1$ is a factor of $\langle 18 \rangle$. In fact we also have

$$1 - \sqrt{-17} = 2 - (1 + \sqrt{-17}) \in \mathfrak{p}_1$$

so

$$18 = (1 + \sqrt{-17})(1 - \sqrt{-17}) \in \mathfrak{p}_1^2$$

which means that $\langle 18 \rangle \subseteq \mathfrak{p}_1^2$ and $\mathfrak{p}_1^2$ is a factor of $\langle 18 \rangle$. Now the elements of $\mathfrak{p}_1$ are of the form

$$2(a + b\sqrt{-17}) + (1 + \sqrt{-17})(c + d\sqrt{-17})$$

$$= (2a + c - 17d) + (2b + c + d)\sqrt{-17}$$

$$= r + s\sqrt{-17}$$

where

$$r - s = 2a - 2b - 18d,$$

which is always even. Clearly r may be taken to be any integer and then s may be any integer of the same parity (odd or even). This implies $\mathfrak{p}_1$ is not the whole ring $\mathbb{Z}[\sqrt{-17}]$. On the other hand, $\mathfrak{p}_1$ is maximal, for if $m+n\sqrt{-17}$ is any element not in $\mathfrak{p}_1$ then one of m, n is even and the other odd, so

$$\langle \mathfrak{p}_1, m+n\sqrt{-17} \rangle = \mathbb{Z}\left[\sqrt{-17}\right].$$

Similarly, considering

$$\mathfrak{p}_2 = \langle 3, 1+\sqrt{-17} \rangle,$$

we find that an element of $\mathfrak{p}_2$ is of the form

$$r+s\sqrt{-17} = (3a+c-17d)+(3b+c+d)\sqrt{-17}$$

where $r-s = 3(a+b-6d)$. Thus r, s can be any integers subject to the constraint

$$r \equiv s \quad (\mathrm{mod}\ 3).$$

Once more we find $\mathfrak{p}_2$ maximal and $18 = 2 \cdot 3 \cdot 3 \in \mathfrak{p}_2^2$, so $\mathfrak{p}_2^2$ is a factor of $\langle 18 \rangle$.

Finally, considering

$$\mathfrak{p}_3 = \langle 3, 1-\sqrt{-17} \rangle,$$

we get another prime ideal such that $\mathfrak{p}_3^2$ is a factor of $\langle 18 \rangle$, and a calculation similar to the previous ones shows that $r+s\sqrt{-17} \in \mathfrak{p}_3$ if and only if

$$r+s \equiv 0 \quad (\mathrm{mod}\ 3).$$

Using the factorization theory of Theorem 5.6, we find that

$$\mathfrak{p}_1^2\mathfrak{p}_2^2\mathfrak{p}_3^2 \supseteq \langle 18 \rangle.$$

The final step, to show that $\langle 18 \rangle = \mathfrak{p}_1^2\mathfrak{p}_2^2\mathfrak{p}_3^2$, is best performed using a counting argument. Since every element in $\mathbb{Z}[\sqrt{-17}]$ is either in $\mathfrak{p}_1$ or of the form $1+x$ for $x \in \mathfrak{p}_1$, the number of elements in the quotient ring $\mathbb{Z}[-17]/\mathfrak{p}_1$ is

$$|\mathbb{Z}[\sqrt{-17}]/\mathfrak{p}_1| = 2.$$

Similarly

$$|\mathbb{Z}[\sqrt{-17}]/\mathfrak{p}_r| = 3 \quad (r = 2, 3).$$

In the next section we call

$$|\mathfrak{O}/\mathfrak{p}|$$

the *norm* of the ideal $\mathfrak{p}$ and write it as $N(\mathfrak{p})$. The crucial property of this new type of norm is that it is multiplicative:

$$N(\mathfrak{ab}) = N(\mathfrak{a})N(\mathfrak{b}).$$

Granted this fact, we deduce that

$$N(\mathfrak{p}_1^2\mathfrak{p}_2^2\mathfrak{p}_3^2) = 2^2 \cdot 3^2 \cdot 3^2 = 18^2.$$

Now the norm of the ideal $\langle 18 \rangle$ is

$$N(\langle 18 \rangle) = |\mathbb{Z}[\sqrt{-17}]/\langle 18 \rangle|$$

and since every element of $\mathbb{Z}[\sqrt{-17}]$ is uniquely of the form

$$a + b\sqrt{-17} + x$$

where a, b are integers in the range 0 to 17 and $x \in \langle 18 \rangle$, we find 18 choices each for a, b so

$$N(\langle 18 \rangle) = 18^2.$$

Suppose $\langle 18 \rangle$ factorizes as

$$\langle 18 \rangle = \mathfrak{p}_1^2\mathfrak{p}_2^2\mathfrak{p}_3^2\mathfrak{a}$$

for some ideal $\mathfrak{a}$. Then taking norms and using the multiplicative property, we find $N(\mathfrak{a}) = 1$, so $\mathfrak{a}$ is the whole ring and

$$\langle 18 \rangle = \mathfrak{p}_1^2\mathfrak{p}_2^2\mathfrak{p}_3^2. \tag{5.1}$$

If we consider the factorization of elements $18 = 2 \cdot 3 \cdot 3$, we obtain

$$\langle 2 \rangle \langle 3 \rangle^2 = \mathfrak{p}_1^2\mathfrak{p}_2^2\mathfrak{p}_3^2. \tag{5.2}$$

By unique factorization for ideals, both $\langle 2 \rangle$, $\langle 3 \rangle$ are products of prime ideals from the set $\{\mathfrak{p}_1, \mathfrak{p}_2, \mathfrak{p}_3\}$. Now $2 \in \mathfrak{p}_1$ but $2 \notin \mathfrak{p}_2$, $2 \notin \mathfrak{p}_3$, so $\mathfrak{p}_1 | \langle 2 \rangle$, $\mathfrak{p}_2 \nmid \langle 2 \rangle$, $\mathfrak{p}_3 \nmid \langle 2 \rangle$, thus

$$\langle 2 \rangle = \mathfrak{p}_1^q.$$

Similarly $3 \notin \mathfrak{p}_1$, $3 \in \mathfrak{p}_2$, $3 \in \mathfrak{p}_3$ implies

$$\langle 3 \rangle = \mathfrak{p}_2^r\mathfrak{p}_3^s.$$

Substitute in Equation (5.2) to get

$$\mathfrak{p}_1^q\mathfrak{p}_2^{2r}\mathfrak{p}_3^{2s} = \mathfrak{p}_1^2\mathfrak{p}_2^2\mathfrak{p}_3^2.$$

Unique factorization of ideals implies that

$$q = 2, \quad r = s = 1,$$

$$\langle 2 \rangle = \mathfrak{p}_1^2, \quad \langle 3 \rangle = \mathfrak{p}_2 \mathfrak{p}_3. \tag{5.3}$$

(The reader may find it instructive to check these by direct calculation.)

A similar argument using

$$\langle 18 \rangle = \langle 1 + \sqrt{-17} \rangle \langle 1 - \sqrt{-17} \rangle = \mathfrak{p}_1^2 \mathfrak{p}_2^2 \mathfrak{p}_3^2 \tag{5.4}$$

where $1 + \sqrt{-17} \in \mathfrak{p}_1, \mathfrak{p}_2; 1 + \sqrt{-17} \notin \mathfrak{p}_3; 1 - \sqrt{-17} \in \mathfrak{p}_1, \mathfrak{p}_3; 1 - \sqrt{-17} \notin \mathfrak{p}_2$
gives

$$\langle 1 + \sqrt{-17} \rangle = \mathfrak{p}_1^m \mathfrak{p}_2^n, \quad \langle 1 - \sqrt{-17} \rangle = \mathfrak{p}_1^r \mathfrak{p}_3^s.$$

Substitute in (5.4) to get $m = r = 1$, $n = s = 2$, so

$$\langle 1 + \sqrt{-17} \rangle = \mathfrak{p}_1 \mathfrak{p}_2^2, \quad \langle 1 - \sqrt{-17} \rangle = \mathfrak{p}_1 \mathfrak{p}_3^2. \tag{5.5}$$

By (5.3) and (5.5) the two alternative factorizations of the element 18 come from alternative groupings of the ideals:

$$\begin{aligned}
\langle 18 \rangle &= \left(\mathfrak{p}_1^2 \right) \left(\mathfrak{p}_2 \mathfrak{p}_3 \right)^2 = \langle 2 \rangle \langle 3 \rangle^2 \\
&= \left(\mathfrak{p}_1 \mathfrak{p}_2^2 \right) \left(\mathfrak{p}_1 \mathfrak{p}_3^2 \right) = \langle 1 + \sqrt{-17} \rangle \langle 1 - \sqrt{-17} \rangle.
\end{aligned}$$

We define the norm of an ideal and prove its multiplicative property in the next section, once we have dealt with some simple consequences of unique factorization. Later on we develop other properties of the norm which will help streamline the calculations in the above example.

5.3   The Norm of an Ideal

Once unique factorization is proved, several useful consequences follow in the usual way. In particular, any two non-zero ideals $\mathfrak{a}$ and $\mathfrak{b}$ have a *greatest common divisor* $\mathfrak{g}$ and a *least common multiple* $\mathfrak{l}$ with the following properties:

$$\mathfrak{g}|\mathfrak{a}, \mathfrak{g}|\mathfrak{b} \text{ and if } \mathfrak{g}' \text{ has the same properties then } \mathfrak{g}'|\mathfrak{g},$$

$$\mathfrak{a}|\mathfrak{l}, \mathfrak{b}|\mathfrak{l} \text{ and if } \mathfrak{l}' \text{ has the same properties then } \mathfrak{l}|\mathfrak{l}'.$$

In fact, suppose we factorize $\mathfrak{a}$ and $\mathfrak{b}$ into primes as:

$$\mathfrak{a} = \prod \mathfrak{p}_i^{e_i}, \quad \mathfrak{b} = \prod \mathfrak{p}_i^{f_i}$$

with distinct prime ideals $\mathfrak{p}_i$. Then we clearly have

$$\mathfrak{g} = \prod \mathfrak{p}_i^{\min(e_i, f_i)}$$

$$\mathfrak{l} = \prod \mathfrak{p}_i^{\max(e_i, f_i)}.$$

There are useful alternative expressions:

Lemma 5.8. *If $\mathfrak{a}$ and $\mathfrak{b}$ are ideals of $\mathfrak{D}$ and $\mathfrak{g}$, $\mathfrak{l}$ are their greatest common divisor and least common multiple, respectively, then*

$$\mathfrak{g} = \mathfrak{a} + \mathfrak{b}, \quad \mathfrak{l} = \mathfrak{a} \cap \mathfrak{b}.$$

Proof: By Proposition 5.7 $\mathfrak{r} | \mathfrak{a}$ if and only if $\mathfrak{r} \supseteq \mathfrak{a}$. Therefore $\mathfrak{g}$ is the smallest ideal containing $\mathfrak{a}$ and $\mathfrak{b}$, and $\mathfrak{c}$ is the largest ideal contained in $\mathfrak{a}$ and $\mathfrak{b}$. The rest is obvious.   □

The proof of Theorem 5.3 shows that if $\mathfrak{a}$ is a non-zero ideal of $\mathfrak{D}$ then the quotient ring $\mathfrak{D}/\mathfrak{a}$ is finite. Define the *norm* of $\mathfrak{a}$ to be

$$N(\mathfrak{a}) = |\mathfrak{D}/\mathfrak{a}|.$$

Then $N(\mathfrak{a})$ is a positive integer. There is no reason to confuse this norm with the old norm of an element $N(a)$ since it applies only to ideals. But in fact there is a connection between the two norms, as we see in a moment.

Theorem 5.9. (a) *Every ideal $\mathfrak{a}$ of $\mathfrak{D}$ with $\mathfrak{a} \neq 0$ has a $\mathbb{Z}$-basis $\{\alpha_1, \ldots, \alpha_n\}$ where n is the degree of K.*
  (b) *We have*

$$N(\mathfrak{a}) = \left| \frac{\Delta[\alpha_1, \ldots, \alpha_n]}{\Delta} \right|^{1/2}$$

where Δ is the discriminant of K.

Proof: By Theorem 2.16 $(\mathfrak{D}, +)$ is free abelian of rank n. Since $\mathfrak{D}/\mathfrak{a}$ is finite, Theorem 1.17 implies that $(\mathfrak{a}, +)$ is free abelian of rank n, hence has a $\mathbb{Z}$-basis $\{\alpha_1, \ldots, \alpha_n\}$. This proves (a).
  For part (b) let $\{\omega_1, \ldots, \omega_n\}$ be a $\mathbb{Z}$-basis for $\mathfrak{D}$, and suppose that $\alpha_i = \sum c_{ij} \omega_j$. By Theorem 1.17,

$$N(a) = |\mathfrak{D}/\mathfrak{a}| = |\det c_{ij}|.$$

By equation (2.2) before Theorem 2.7

$$\Delta[\alpha_1, \ldots, \alpha_n] = (\det c_{ij})^2 \Delta[\omega_1, \ldots, \omega_n]$$
$$= (N(\mathfrak{a}))^2 \Delta.$$

Now take square roots and remember that $N(\mathfrak{a})$ is positive.   □

Corollary 5.10. *If* $\mathfrak{a} = \langle a \rangle$ *is a principal ideal then* $N(\mathfrak{a}) = |N(a)|$.

Proof: A $\mathbb{Z}$-basis for $\mathfrak{a}$ is given by $\{a\omega_1, \dots, a\omega_n\}$. The result follows from the definition of $\Delta[\alpha_1, \dots, \alpha_n]$ and Theorem 5.9.   □

This corollary helps us to perform a straightforward calculation of the norm of a principal ideal.

Example 5.11. If $\mathfrak{O}$ is the ring of integers of $\mathbb{Q}(\sqrt{d})$ for a squarefree rational integer d, then
$$N(\langle a + b\sqrt{d} \rangle) = |a^2 - bd^2| \,.$$
In particular in $\mathfrak{O} = \mathbb{Z}[\sqrt{-17}]$
$$N(\langle 18 \rangle) = 18^2.$$

The new norm, like the old, is *multiplicative*:

Theorem 5.12. *If* $\mathfrak{a}$ *and* $\mathfrak{b}$ *are non-zero ideals of* $\mathfrak{O}$, *then*
$$N(\mathfrak{ab}) = N(\mathfrak{a})N(\mathfrak{b}).$$

Proof: By uniqueness of factorization and induction on the number of factors, it is sufficient to prove
$$N(\mathfrak{ap}) = N(\mathfrak{a})N(\mathfrak{p}) \tag{5.6}$$
where $\mathfrak{p}$ is prime.

We establish
$$|\mathfrak{O}/\mathfrak{ap}| = |\mathfrak{O}/\mathfrak{a}| \, |\mathfrak{a}/\mathfrak{ap}| \tag{5.7}$$
and
$$|\mathfrak{a}/\mathfrak{ap}| = |\mathfrak{O}/\mathfrak{p}| \,. \tag{5.8}$$
Then (5.6) follows immediately from (5.7, 5.8), and the definition of the norm.

Equation (5.7) is a consequence of the isomorphism theorem for rings. Define $\phi : \mathfrak{O}/\mathfrak{ap} \to \mathfrak{O}/\mathfrak{a}$ by
$$\phi(\mathfrak{ap} + x) = \mathfrak{a} + x$$

then ϕ is a surjective ring homomorphism with kernel $\mathfrak{a}/\mathfrak{ap}$; Lagrange's theorem (applied to the additive groups) gives Equation (5.7).

To establish Equation (5.8), first note that unique factorization implies $\mathfrak{a} \neq \mathfrak{ap}$, so $\mathfrak{a} \supsetneq \mathfrak{ap}$. Now we show that there is no ideal $\mathfrak{b}$ strictly between $\mathfrak{a}$ and $\mathfrak{ap}$, for if

$$\mathfrak{a} \supseteq \mathfrak{b} \supseteq \mathfrak{ap},$$

then, as fractional ideals,

$$\mathfrak{a}^{-1}\mathfrak{a} \supseteq \mathfrak{a}^{-1}\mathfrak{b} \supseteq \mathfrak{a}^{-1}\mathfrak{ap},$$

so

$$\mathfrak{O} \supseteq \mathfrak{a}^{-1}\mathfrak{b} \supseteq \mathfrak{p}.$$

Since $\mathfrak{a}^{-1}\mathfrak{b} \subseteq \mathfrak{O}$, we see that it is actually an ideal, and since $\mathfrak{p}$ is maximal by Theorem 5.3 (d),

$$\mathfrak{a}^{-1}\mathfrak{b} = \mathfrak{O} \quad \text{or} \quad \mathfrak{a}^{-1}\mathfrak{b} = \mathfrak{p}$$

so

$$\mathfrak{b} = \mathfrak{a} \quad \text{or} \quad \mathfrak{b} = \mathfrak{ap}.$$

This means that for any element $a \in \mathfrak{a} \backslash \mathfrak{ap}$,

$$\mathfrak{ap} + \langle a \rangle = \mathfrak{a}. \tag{5.9}$$

Fix such an a and define $\theta : \mathfrak{O} \to \mathfrak{a}/\mathfrak{ap}$ by

$$\theta(x) = \mathfrak{ap} + ax.$$

Then θ is an $\mathfrak{O}$-module homomorphism, surjective by (5.9), whose kernel is an ideal satisfying

$$\mathfrak{p} \subseteq \ker \theta.$$

Now $\ker \theta \neq \mathfrak{O}$ (for that would mean $\mathfrak{a}/\mathfrak{ap} \cong \mathfrak{O}/\ker \theta = 0$, which would contradict $\mathfrak{a} \neq \mathfrak{ap}$), and $\mathfrak{p}$ is maximal, so

$$\ker \theta = \mathfrak{p}.$$

Hence $\mathfrak{O}/\mathfrak{p} \cong \mathfrak{a}/\mathfrak{ap}$ as $\mathfrak{O}$-modules, leading to (5.8). $\square$

Example 5.13. If $\mathfrak{O} = \mathbb{Z}[\sqrt{-17}]$, $\mathfrak{p}_1 = \langle 2, 1 + \sqrt{-17} \rangle$, $\mathfrak{p}_2 = \langle 3, 1 + \sqrt{-17} \rangle$, $\mathfrak{p}_3 = \langle 3, 1 - \sqrt{-17} \rangle$, then

$$N(\mathfrak{p}_1^2\mathfrak{p}_2^2\mathfrak{p}_3^2) = 2^2 \cdot 3^2 \cdot 3^2 = 18^2.$$

This particular calculation completes the details of the extended example in the previous section.

It is convenient to introduce yet another usage for the word 'divides'. If $\mathfrak{a}$ is an ideal of $\mathfrak{O}$ and b an element of $\mathfrak{O}$ such that $\mathfrak{a}|\langle b \rangle$, then we also write $\mathfrak{a}|b$ and say that $\mathfrak{a}$ divides b. It is clear that $\mathfrak{a}|b$ if and only if $b \in \mathfrak{a}$; however, the new notation has certain distinct advantages. For example, if $\mathfrak{p}$ is a prime ideal and $\mathfrak{p}|\langle a \rangle \langle b \rangle$, then we must have $\mathfrak{p}|\langle a \rangle$ or $\mathfrak{p}|\langle b \rangle$. Thus for $\mathfrak{p}$ prime,

$$\mathfrak{p}|ab \quad \text{implies} \quad \mathfrak{p}|a \text{ or } \mathfrak{p}|b.$$

This new notation allows us to emphasize the correspondence between factorization of elements and principal ideals which would otherwise be less evident.

Theorem 5.14. *Let* $\mathfrak{a}$ *be an ideal of* $\mathfrak{O}, \mathfrak{a} \neq 0$.

(a) *If* $N(\mathfrak{a})$ *is prime, then so is* $\mathfrak{a}$.
(b) $N(\mathfrak{a})$ *is an element of* $\mathfrak{a}$, *or equivalently* $\mathfrak{a}|N(\mathfrak{a})$.
(c) *If* $\mathfrak{a}$ *is prime it divides exactly one rational prime* p, *and then*

$$N(\mathfrak{a}) = p^m$$

where $m \leq n$, *the degree of* K.

Proof: For part (a) write $\mathfrak{a}$ as a product of prime ideals and equate norms. For part (b) note that since $N(\mathfrak{a}) = |\mathfrak{O}/\mathfrak{a}|$ it follows that for any $x \in \mathfrak{O}$ we have $N(\mathfrak{a})x \in \mathfrak{a}$. Now put $x = 1$. For part (c) we note that by part (b)

$$\mathfrak{a}|N(\mathfrak{a}) = p_1^{m_1} \ldots p_r^{m_r}$$

so, considering principal ideals in place of the p_i, we have $\mathfrak{a}|p_i$ for some rational prime p_i. If p and q are distinct rational primes, both divisible by $\mathfrak{a}$, we can find integers u, v such that $up + vq = 1$, so $\mathfrak{a}|1$ and $\mathfrak{a} = \mathfrak{O}$, a contradiction. Then

$$N(\mathfrak{a})|N(\langle p \rangle) = p^n$$

so $N(\mathfrak{a}) = p^m$ for some $m \leq n$. $\qquad\qquad\qquad\qquad\qquad\qquad\square$

Example 5.15. If $\mathfrak{O} = \mathbb{Z}[\sqrt{-17}]$, $\mathfrak{p}_1 = \langle 2, 1 + \sqrt{-17} \rangle$, then because $N(\mathfrak{p}_1) = 2$ we immediately deduce that $\mathfrak{p}_1$ is prime. Note that $N(\mathfrak{p}_1) = 2 \in \mathfrak{p}_1$, as asserted by Theorem 5.14(b).

Example 5.16. A prime ideal $\mathfrak{a}$ can satisfy

$$N(\mathfrak{a}) = p^m$$

where $m > 1$, which means that the norm of a prime ideal need not be prime. For instance $\mathfrak{O} = \mathbb{Z}[i]$, $\mathfrak{a} = \langle 3 \rangle$. Here 3 is irreducible in $\mathbb{Z}[i]$, hence prime because $\mathbb{Z}[i]$ has unique factorization. It is an easy deduction (Exercise 1 at the end of this chapter) that if an element is prime, so is the ideal it generates. Hence $\langle 3 \rangle$ is prime in $\mathbb{Z}[i]$, but

$$N\left(\langle 3 \rangle \right) = 3^2.$$

The next theorem collects together several useful finiteness assertions:

Theorem 5.17. (a) *Every non-zero ideal of $\mathfrak{O}$ has a finite number of divisors.*

(b) *A non-zero rational integer belongs to only a finite number of ideals of $\mathfrak{O}$.*

(c) *Only finitely many ideals of $\mathfrak{O}$ have given norm.*

Proof: (a) is an immediate consequence of prime factorization, (b) is a special case of (a), and (c) follows from (b) using Theorem 5.14 (b).   □

Example 5.18. Consider our earlier calculation

$$\langle 18 \rangle = \mathfrak{p}_1^2 \mathfrak{p}_2^2 \mathfrak{p}_3^2$$

in $\mathbb{Z}[\sqrt{-17}]$ where $\mathfrak{p}_1 = \langle 2, 1 + \sqrt{-17} \rangle$, $\mathfrak{p}_2 = \langle 3, 1 + \sqrt{-17} \rangle$, and $\mathfrak{p}_3 = \langle 3, 1 - \sqrt{-17} \rangle$. We find that the only prime divisors of $\langle 18 \rangle$ are $\mathfrak{p}_1, \mathfrak{p}_2, \mathfrak{p}_3$. If 18 belongs to some ideal $\mathfrak{a}$, then $\langle 18 \rangle \subseteq \mathfrak{a}$, whence $\mathfrak{a} | \langle 18 \rangle$, so $\mathfrak{a} | \mathfrak{p}_1^2 \mathfrak{p}_2^2 \mathfrak{p}_3^2$ and $\mathfrak{a} = \mathfrak{p}_1^q \mathfrak{p}_2^r \mathfrak{p}_3^s$ where q, r, s are 0, 1, or 2. Thus 18 belongs only to a finite number of ideals.

How many ideals $\mathfrak{a}$ have norm 18? By Theorem 5.14 (b) this can happen only when $\mathfrak{a} | 18$, so

$$\mathfrak{a} = \mathfrak{p}_1^q \mathfrak{p}_2^r \mathfrak{p}_3^s$$

which implies

$$N(\mathfrak{a}) = 2^q \cdot 3^r \cdot 3^s.$$

This norm is 18 only when $q = 1$ and $r + s = 2$, which means that $\mathfrak{a}$ is $\mathfrak{p}_1 \mathfrak{p}_2^2$, $\mathfrak{p}_1 \mathfrak{p}_2 \mathfrak{p}_3$ or $\mathfrak{p}_1 \mathfrak{p}_3^2$.

We know that every ideal of $\mathfrak{O}$ is finitely generated. In fact, two generators suffice. First, we prove:

Lemma 5.19. *If $\mathfrak{a}$, $\mathfrak{b}$ are non-zero ideals of $\mathfrak{O}$, there exists $\alpha \in \mathfrak{a}$ such that*

$$\alpha \mathfrak{a}^{-1} + \mathfrak{b} = \mathfrak{O}.$$

Proof: First note that if $\alpha \in \mathfrak{a}$ we have $\mathfrak{a}|\alpha$ so that $\alpha\mathfrak{a}^{-1}$ is an ideal and not just a fractional ideal. Now $\alpha\mathfrak{a}^{-1} + \mathfrak{b}$ is the greatest common divisor of $\alpha\mathfrak{a}^{-1}$ and $\mathfrak{b}$, so it is sufficient to choose $\alpha \in \mathfrak{a}$ so that

$$\alpha\mathfrak{a}^{-1} + \mathfrak{p}_i = \mathfrak{O} \quad (i = 1, \ldots, r)$$

where $\mathfrak{p}_1, \ldots, \mathfrak{p}_r$ are the distinct prime ideals dividing $\mathfrak{b}$. This follows if

$$\mathfrak{p}_i \nmid \alpha\mathfrak{a}^{-1}$$

since $\mathfrak{p}_i$ is a maximal ideal. So it is sufficient to choose $\alpha \in \mathfrak{a} \setminus \mathfrak{a}\mathfrak{p}_i$ for all $i = 1, \ldots, r$.

If $r = 1$ this is easy, for unique factorization of ideals implies $\mathfrak{a} \neq \mathfrak{a}\mathfrak{p}_i$. For $r > 1$ let

$$\mathfrak{a}_i = \mathfrak{a}\mathfrak{p}_1 \ldots \mathfrak{p}_{i-1}\mathfrak{p}_{i+1} \ldots \mathfrak{p}_r.$$

By the case $r = 1$ we can choose

$$\alpha_i \in \mathfrak{a}_i \setminus \mathfrak{a}_i\mathfrak{p}_i.$$

Define

$$\alpha = \alpha_1 + \ldots + \alpha_r.$$

Then each $\alpha_i \in \mathfrak{a}_i \subseteq \mathfrak{a}$, so $\alpha \in \mathfrak{a}$. Suppose if possible that $\alpha \in \mathfrak{a}\mathfrak{p}_i$. If $j \neq i$ then $\alpha_j \in \mathfrak{a}_j \subseteq \mathfrak{a}\mathfrak{p}_i$, so

$$\alpha_i = \alpha - \alpha_1 - \ldots - \alpha_{i-1} - \alpha_{i+1} - \ldots - \alpha_r \in \mathfrak{a}\mathfrak{p}_i.$$

Hence $\mathfrak{a}\mathfrak{p}_i | \langle \alpha_i \rangle$. On the other hand $\mathfrak{a}_i | \langle \alpha_i \rangle$. We have $\mathfrak{a}_i\mathfrak{p}_i | \langle \alpha_i \rangle$. This contradicts the choice of α_i.

Theorem 5.20. *Let $\mathfrak{a} \neq 0$ be an ideal of $\mathfrak{O}$, and $0 \neq \beta \in \mathfrak{a}$. Then there exists $\alpha \in \mathfrak{a}$ such that $\mathfrak{a} = \langle \alpha, \beta \rangle$.*

Proof: Let $\mathfrak{b} = \beta\mathfrak{a}^{-1}$. By Lemma 5.19 there exists $\alpha \in \mathfrak{a}$ such that

$$\alpha\mathfrak{a}^{-1} + \mathfrak{b} = \alpha\mathfrak{a}^{-1} + \beta\mathfrak{a}^{-1} = \mathfrak{O},$$

hence

$$(\langle \alpha \rangle + \langle \beta \rangle)\,\mathfrak{a}^{-1} = \mathfrak{O},$$

so that

$$\mathfrak{a} = \langle \alpha \rangle + \langle \beta \rangle = \langle \alpha, \beta \rangle. \qquad \square$$

This theorem demonstrates that the earlier extended example, where each ideal considered has at most two generators, is typical.

We are now in a position to characterize those $\mathfrak{O}$ for which factorization of elements into irreducibles is unique:

Theorem 5.21. *Factorization of elements of $\mathfrak{O}$ into irreducibles is unique if and only if every ideal of $\mathfrak{O}$ is principal.*

Proof: If every ideal is principal, then unique factorization of elements follows by Theorem 4.17. To prove the converse, if factorization of elements is unique, it is enough to prove that every *prime* ideal is principal, since every other ideal, being a product of prime ideals, is then principal.

Let $\mathfrak{p} \neq 0$ be a prime ideal of $\mathfrak{O}$. By Theorem 5.14(b) there exists a rational integer $N = N(\mathfrak{p})$ such that $\mathfrak{p}|N$. We can factorize N as a product of irreducible elements in $\mathfrak{O}$, say

$$N = \pi_1 \ldots \pi_s.$$

Since $\mathfrak{p}|N$ and $\mathfrak{p}$ is a prime ideal, it follows that $\mathfrak{p}|\pi_i$, or equivalently, $\mathfrak{p}|\langle\pi_i\rangle$. But factorization being unique in $\mathfrak{O}$, the irreducible π_i is actually *prime* by Theorem 4.14, and then the principal ideal $\langle\pi_i\rangle$ is prime (Exercise 1 at the end of the chapter). Thus $\mathfrak{p}|\langle\pi_i\rangle$, where both $\mathfrak{p}, \langle\pi_i\rangle$ are prime, and by uniqueness of factorization,

$$\mathfrak{p} = \langle\pi_i\rangle$$

so $\mathfrak{p}$ is principal. $\qquad\qquad\square$

Using this theorem we can nicely round off the relationship between factorization of elements and ideals. To do this, consider an element π that is irreducible but not prime. Then the ideal $\langle\pi\rangle$ is not prime, so it has a proper factorization into prime ideals:

$$\langle\pi\rangle = \mathfrak{p}_1 \ldots \mathfrak{p}_r.$$

None of these $\mathfrak{p}_i$ can be principal, for if $\mathfrak{p}_i = \langle a \rangle$ then $\langle a \rangle | \langle \pi \rangle$, implying $a|\pi$. Since π is irreducible, a is either a unit, contradicting $\langle a \rangle$ being prime, or an associate of π, whence $\langle\pi\rangle = \mathfrak{p}_i$, contradicting $\langle\pi\rangle$ having a proper factorization.

Tying up the loose ends, we see that if $\mathfrak{O}$ has unique factorization of elements into irreducibles then these irreducibles are all primes, and factorization of elements corresponds precisely to factorization of the corresponding principal ideals. On the other hand, if $\mathfrak{O}$ does not have unique factorization of elements, then not all irreducibles are prime, and any non-prime irreducible generates a principal ideal which has a proper factorization into non-principal ideals. We may add in the latter case that such non-principal ideals have precisely two generators.

Example 5.22. In $\mathbb{Z}[\sqrt{-17}]$, the elements 2, 3 are irreducible (proved by considering norms) and not prime, with

$$\langle 2 \rangle = \langle 2, 1 + \sqrt{-17} \rangle^2$$
$$\langle 3 \rangle = \langle 3, 1 + \sqrt{-17} \rangle \langle 3, 1 - \sqrt{-17} \rangle.$$

5.4   Non-Unique Factorization in Cyclotomic Fields

We mentioned in the introductory section that unique prime factorization *fails* in the cyclotomic field of 23rd roots of unity. (The failure, rather than the precise value $n = 23$, is the crucial point!) In this (optional) section we use the tools developed in this chapter to demonstrate this result. The calculations are somewhat tedious and have been abbreviated where feasible: the energetic reader may care to check the details. A few tricks, inspired by the structure of the group of units of the ring $\mathbb{Z}_{23}$, are used, but we lack the space to motivate them. For further details see the admirable book by Edwards [25].

Let $\zeta = e^{2\pi i/23}$, and let $K = \mathbb{Q}(\zeta)$. By Theorem 3.5 the ring of integers $\mathfrak{O}_K$ is $\mathbb{Z}[\zeta]$. The group of units of $\mathbb{Z}_{23}$ is generated by -2, whose powers in order are

$$1, -2, 4, -8, -7, 14, -5, \ldots. \tag{5.10}$$

For reasons that will emerge later we introduce two elements

$$\theta_0 = \zeta + \zeta^4 + \zeta^{-7} + \zeta^{-5} + \ldots$$
$$\theta_1 = \zeta^{-2} + \zeta^{-8} + \zeta^{14} + \ldots.$$

The powers that occur are alternate elements in the sequence (5.10). We have

$$\theta_0 + \theta_1 = \zeta + \zeta^2 + \ldots + \zeta^{22} = -1,$$

$$\theta_0\theta_1 = 6.$$

The norm of a general element $f(\zeta)$, with f a polynomial over $\mathbb{Z}$ of degree

≤ 22, can be broken up as

$$\begin{aligned} \mathrm{N}(f(\zeta)) &= \prod_{j=1}^{22} \mathrm{N}f(\zeta^j) \\ &= \prod_{j \text{ even}} \mathrm{N}f(\zeta^j) \cdot \prod_{j \text{ odd}} \mathrm{N}f(\zeta^j) \\ &= G(\zeta^2)G(\zeta^{-2}) \end{aligned} \tag{5.11}$$

where

$$G(\zeta) = f(\zeta)f(\zeta^4)f(\zeta^{-7})f(\zeta^{-5})f(\zeta^3)f(\zeta^{-11})f(\zeta^2)f(\zeta^8)f(\zeta^9)f(\zeta^{-10})f(\zeta^6).$$

By definition, $G(\zeta)$ is invariant under the linear mapping α sending ζ^i to ζ^{4j}. But it is easy to check that an element fixed by α must be of the form $a + b\theta_0$ for $a, b \in \mathbb{Z}$. (Either use Galois theory, or a direct argument based on the linear independence over $\mathbb{Q}$ of $\{1, \zeta, \ldots, \zeta^{22}\}$.)

We pull out of a hat the element

$$\mu = 1 - \zeta + \zeta^{21} = 1 - \zeta + \zeta^{-2},$$

which Kummer found by a great deal of (fairly systematic) experimentation. Using (5.10) and a lot of paper and ink we eventually find that

$$\mathrm{N}(\mu) = (-31 + 28\theta_0)(-31 + 28\theta_1) = 6533 = 47 \cdot 139.$$

By Theorem 5.14 the principal ideal $\mathfrak{m} = \langle \mu \rangle$ cannot be prime, hence it must be nontrivial product of prime ideals, say

$$\mathfrak{m} = \mathfrak{p}\mathfrak{q}.$$

Taking norms we must (without loss of generality) have $\mathrm{N}(\mathfrak{p}) = 47$, $\mathrm{N}(\mathfrak{q}) = 139$. If K has unique factorization then every ideal, in particular $\mathfrak{p}$, is principal by Theorem 5.21. Hence $\mathfrak{p} = \langle v \rangle$ for some $v \in \mathbb{Z}[\zeta]$. Clearly $\mathrm{N}(v) = \pm 47$ by Corollary 5.10.

We claim this is impossible. We have already observed that $G(\zeta)$ can always be expressed in the form $a + b\theta_0$ $(a, b \in \mathbb{Z})$; and then $G(\zeta^{-2})$ must be equal to $a + b\theta_1$. Set $f(\zeta) = v$:

$$\pm 47 = (a + b\theta_0)(a + b\theta_1) = a^2 - ab + 6b^2.$$

Multiplying by 4 and regrouping,

$$(2a - b)^2 + 23b^2 = \pm 188.$$

The sign must be positive. A simple trial-and-error analysis (involving only two cases) shows that 188 cannot be written in the form $P^2 + 23Q^2$. This contradiction establishes that prime factorization of elements cannot be unique in $\mathfrak{O}_K$.

5.5   Exercises

1. In a domain D, show that a principal ideal $\langle p \rangle$ is prime if and only if p is prime or zero.

2. In $\mathbb{Z}[\sqrt{-5}]$, define the ideals

$$\begin{aligned}
\mathfrak{p} &= \langle 2, 1 + \sqrt{-5} \rangle, \\
\mathfrak{q} &= \langle 3, 1 + \sqrt{-5} \rangle, \\
\mathfrak{r} &= \langle 3, 1 - \sqrt{-5} \rangle.
\end{aligned}$$

Prove that these are maximal ideals, hence prime. Show that

$$\begin{aligned}
\mathfrak{p}^2 &= \langle 2 \rangle, & \mathfrak{q}\mathfrak{r} &= \langle 3 \rangle, \\
\mathfrak{p}\mathfrak{q} &= \langle 1 + \sqrt{-5} \rangle, & \mathfrak{p}\mathfrak{r} &= \langle 1 - \sqrt{-5} \rangle.
\end{aligned}$$

Show that the factorizations of 6 given in the proof of Theorem 4.11 come from two different groupings of the factorization into prime ideals $\langle 6 \rangle = \mathfrak{p}^2 \mathfrak{q}\mathfrak{r}$.

3. Calculate the norms of the ideals mentioned in Exercise 2 and check multiplicativity.

4. Prove that the ideals $\mathfrak{p}$, $\mathfrak{q}$, $\mathfrak{r}$ of Exercise 2 cannot be principal.

5. Show the principal ideals $\langle 2 \rangle$, $\langle 3 \rangle$ in Exercise 2 are generated by irreducible elements but the ideals are not prime.

6. In $\mathbb{Z}[\sqrt{-6}]$ we have

$$6 = 2 \cdot 3 = (\sqrt{-6})(-\sqrt{-6}).$$

Factorize these elements further in the extension ring $\mathbb{Z}[\sqrt{2}, \sqrt{-3}]$ as

$$6 = (-1)\sqrt{2}\sqrt{2}\sqrt{-3}\sqrt{-3}.$$

Show that if $\mathfrak{J}_1$ is the principal ideal in $\mathbb{Z}[\sqrt{2}, \sqrt{-3}]$ generated by $\sqrt{2}$, then

$$\mathfrak{p}_1 = \mathfrak{J}_1 \cap \mathbb{Z}[\sqrt{-6}] = \langle 2, \sqrt{-6} \rangle.$$

Demonstrate that $\mathfrak{p}_1$ is maximal in $\mathbb{Z}[\sqrt{-6}]$, hence prime, and find another prime ideal $\mathfrak{p}_2$ in $\mathbb{Z}[\sqrt{-6}]$ such that

$$\langle 6 \rangle = \mathfrak{p}_1^2 \mathfrak{p}_2^2.$$

7. Factorize $14 = 2 \cdot 7 = (2 + \sqrt{-10})(2 - \sqrt{-10})$ further in $\mathbb{Z}[\sqrt{-5}, \sqrt{2}]$ and by intersecting appropriate ideals with $\mathbb{Z}[\sqrt{-10}]$, factorize the ideal $\langle 14 \rangle$ into prime (maximal) ideals in $\mathbb{Z}[\sqrt{-10}]$.

8. Suppose that $\mathfrak{p}$, $\mathfrak{q}$ are distinct prime ideals in $\mathfrak{O}$. Show that $\mathfrak{p} + \mathfrak{q} = \mathfrak{O}$ and $\mathfrak{p} \cap \mathfrak{q} = \mathfrak{p}\mathfrak{q}$.

9. If $\mathfrak{O}$ is a principal ideal domain, prove that every fractional ideal is of the form $\{a\phi | a \in \mathfrak{O}\}$ for some $\phi \in K$. Does the converse hold?

10. Find all fractional ideals of $\mathbb{Z}$ and of $\mathbb{Z}[\sqrt{-1}]$.

11. In $\mathbb{Z}[\sqrt{-5}]$, find a $\mathbb{Z}$-basis $\{\alpha_1, \alpha_2\}$ for the ideal $\langle 2, 1 + \sqrt{-5} \rangle$. Check the formula

$$N\left(\langle 2, 1 + \sqrt{-5} \rangle\right) = \left| \frac{\Delta[\alpha_1, \alpha_2]}{d} \right|^{1/2}$$

of Theorem 5.9.

12. Find all the ideals in $\mathbb{Z}[\sqrt{-5}]$ that contain the element 6.

13. Find all the ideals in $\mathbb{Z}[\sqrt{2}]$ with norm 18.

14. If K is a number field of degree n with integers $\mathfrak{O}$, show that if $m \in \mathbb{Z}$ and $\langle m \rangle$ is the ideal in $\mathfrak{O}$ generated by m, then

$$N\left(\langle m \rangle\right) = |m|^n.$$

15. In $\mathbb{Z}[\sqrt{-29}]$

$$30 = 2 \cdot 3 \cdot 5 = (1 + \sqrt{-29})(1 - \sqrt{-29})$$

Show that $\langle 30 \rangle \subseteq \langle 2, 1 + \sqrt{-29} \rangle$ and verify that $\mathfrak{p}_1 = \langle 2, 1 + \sqrt{-29} \rangle$ has norm 2 and is thus prime. Check that $1 - \sqrt{-29} \in \mathfrak{p}_1$ and deduce $\langle 30 \rangle \subseteq \mathfrak{p}_1^2$. Find prime ideals $\mathfrak{p}_2, \mathfrak{p}_2', \mathfrak{p}_3, \mathfrak{p}_3'$ with norms 3 or 5 such that $\langle 30 \rangle \subseteq \mathfrak{p}_i \mathfrak{p}_i'$ $(i = 2, 3)$. Deduce that $\mathfrak{p}_1^2 \mathfrak{p}_2 \mathfrak{p}_2' \mathfrak{p}_3 \mathfrak{p}_3' | \langle 30 \rangle$ and by calculating norms, or otherwise, show that

$$\langle 30 \rangle = \mathfrak{p}_1^2 \mathfrak{p}_2 \mathfrak{p}_2' \mathfrak{p}_3 \mathfrak{p}_3'.$$

Comment on how this relates to the two factorizations:

$$\langle 30 \rangle = \langle 2 \rangle \langle 3 \rangle \langle 5 \rangle,$$
$$\langle 30 \rangle = \langle 1 + \sqrt{-29} \rangle \langle 1 - \sqrt{-29} \rangle.$$

16. Find all ideals in $\mathbb{Z}[\sqrt{-29}]$ containing the element 30.

II

Geometric Methods

6

Lattices

At this stage we take a radical new view of the theory, turning from purely algebraic methods to techniques inspired by geometry. This approach requires a different attitude of mind from the reader, in which formal ideas are built on a visual foundation. We begin with basic properties of lattices: subsets of $\mathbb{R}^n$ which in some sense generalize the way $\mathbb{Z}$ is embedded in $\mathbb{R}$. We characterize lattices topologically as the discrete subgroups of $\mathbb{R}^n$. We introduce the fundamental domain and quotient torus corresponding to a lattice and relate the two concepts. Finally we define a concept of volume for subsets of the quotient torus.

6.1 Lattices

Let $e_1, \ldots, e_m$ be a linearly independent set of vectors in $\mathbb{R}^n$ (so that $m \leq n$). The additive subgroup of $(\mathbb{R}^n, +)$ generated by $e_1, \ldots, e_m$ is called a *lattice* of *dimension m*, *generated* by $e_1, \ldots, e_m$. Figure 6.1 shows a lattice of dimension 2 in $\mathbb{R}^2$, generated by $(1, 2)$ and $(2, -1)$. (Do not confuse this with any other uses of the word 'lattice' in algebra.) Obviously, as regards the group-theoretic structure, a lattice of dimension m is a free abelian group of rank m, so we can apply the terminology and theory of free abelian groups to lattices.

We now give a topological characterization of lattices. Let $\mathbb{R}^n$ be equipped with the usual metric (à la Pythagoras), where $||x - y||$ denotes the distance between x and y, and denote the (closed) ball centre x radius

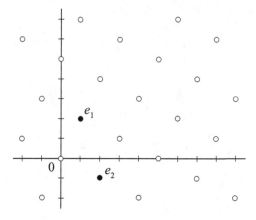

Figure 6.1. The lattice in $\mathbb{R}^2$ generated by $e_1 = (1, 2)$ and $e_2 = (2, -1)$.

r by $B_r[x]$. Recall that a subset $X \subseteq \mathbb{R}^n$ is *bounded* if $X \subseteq B_r[0]$ for some r. We say that a subset of $\mathbb{R}^n$ is *discrete* if and only if it intersects every $B_r[0]$ in a finite set.

Theorem 6.1. *An additive subgroup of $\mathbb{R}^n$ is a lattice if and only if it is discrete.*

Proof: Suppose L is a lattice. By passing to the subspace spanned by L we may assume L has dimension n. Let L be generated by $e_1, \ldots, e_n$; then these vectors form a basis for the space $\mathbb{R}^n$. Every $v \in \mathbb{R}^n$ has a unique representation

$$v = \lambda_1 e_1 + \ldots + \lambda_n e_n \quad (\lambda_i \in \mathbb{R}).$$

Define $f : \mathbb{R}^n \to \mathbb{R}^n$ by

$$f(\lambda_1 e_1 + \ldots + \lambda_n e_n) = (\lambda_1, \ldots, \lambda_n).$$

Then $f(B_r[0])$ is bounded, say

$$\|f(v)\| \le k \text{ for } v \in B_r[0].$$

If $\sum a_i e_i \in B_r[0]$ $(a_i \in \mathbb{Z})$, then certainly $\|(a_1, \ldots, a_n)\| \le k$. This implies

$$|a_i| \le \|(a_1, \ldots, a_n)\| \le k. \tag{6.1}$$

The number of integer solutions of (6.1) is finite and so $L \cap B_r[0]$, being a subset of the solutions of (6.1), is also finite, and L is discrete.

Conversely, let G be a discrete subgroup of $\mathbb{R}^n$. We prove by induction on n that G is a lattice. Let $\{g_1, \ldots, g_m\}$ be a maximal linearly independent subset of G, let V be the subspace spanned by $\{g_1, \ldots, g_{m-1}\}$, and let $G_0 = G \cap V$. Then G_0 is discrete so, by induction, is a lattice. Hence there exist linearly independent elements $h_1, \ldots, h_{m'}$ generating G_0. Since the elements $g_1, \ldots, g_{m-1} \in G_0$ we have $m' = m - 1$, and we can replace $\{g_1, \ldots, g_{m-1}\}$ by $\{h_1, \ldots, h_{m-1}\}$, or equivalently assume that every element of G_0 is a $\mathbb{Z}$-linear combination of $g_1, \ldots, g_{m-1}$. Let T be the subset of all $x \in G$ of the form

$$x = a_1 g_1 + \ldots + a_m g_m$$

with $a_i \in \mathbb{R}$, such that

$$0 \le a_i < 1 \quad (i = 1, \ldots, m-1)$$
$$0 \le a_m \le 1.$$

Then T is bounded, hence finite since G is discrete, and we may therefore choose $x' \in T$ with smallest non-zero coefficient a_m, say

$$x' = b_1 g_1 + \ldots + b_m g_m.$$

Certainly $\{g_1, \ldots, g_{m-1}, x'\}$ is linearly independent. Now starting with any vector $g \in G$ we can select integer coefficients c_i so that

$$g' = g - c_m x' - c_1 g_1 - \ldots - c_{m-1} g_{m-1}$$

lies in T, and the coefficient of g_m in g' is less than b_m, but non-negative. By choice of x' this coefficient must be zero, so $g' \in G_0$. Hence $\{x', g_1, \ldots, g_{m-1}\}$ generates G, and G is a lattice. $\qquad\square$

If L is a lattice generated by $\{e_1, \ldots, e_n\}$ we define the *fundamental domain* T to consist of all elements $\sum a_i e_i$ $(a_i \in \mathbb{R})$ for which

$$0 \le a_i < 1.$$

Note that this depends on the choice of generators.

Lemma 6.2. *Each element of $\mathbb{R}^n$ lies in exactly one of the sets $T + l$ for $l \in L$.*

Proof: Chop off the integer parts of the coefficients. $\qquad\square$

Figure 6.2 illustrates the concept of a fundamental domain, and the result of Lemma 6.2, for the lattice of Figure 6.1.

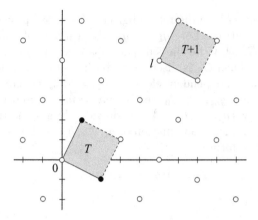

Figure 6.2. A fundamental domain T for the lattice of Figure 6.1, and a translate $T + l$. Dotted lines indicate omission of boundaries.

6.2 The Quotient Torus

Let L be a lattice in $\mathbb{R}^n$, and assume to start that L has dimension n. We study the quotient group $\mathbb{R}^n / L$.

Let $\mathbb{S}$ denote the set of all complex numbers of modulus 1. Under multiplication $\mathbb{S}$ is a group, called for obvious reasons the *circle group*.

Lemma 6.3. *The quotient group $\mathbb{R}/\mathbb{Z}$ is isomorphic to the circle group $\mathbb{S}$.*

Proof: Define a map $\phi : \mathbb{R} \to \mathbb{S}$ by

$$\phi(x) = e^{2\pi i x}.$$

Then ϕ is a surjective homomorphism with kernel $\mathbb{Z}$. $\square$

Next let $\mathbb{T}^n$ denote the direct product of n copies of $\mathbb{S}$, and call this the *n-dimensional torus*. For instance, $\mathbb{T}^2 = \mathbb{S} \times \mathbb{S}$ is the usual torus (with a group structure) as sketched in Figure 6.3.

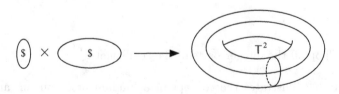

Figure 6.3. The Cartesian product of two circles is a torus.

Theorem 6.4. *If L is an n-dimensional lattice in $\mathbb{R}^n$ then $\mathbb{R}^n/L$ is isomorphic to the n-dimensional torus $\mathbb{T}^n$.*

Proof: Let $\{e_1, \ldots, e_n\}$ be generators for L. Then $\{e_1, \ldots, e_n\}$ is a basis for $\mathbb{R}^n$. Define $\phi : \mathbb{R}^n \to \mathbb{T}^n$ by

$$\phi(a_1 e_1 + \ldots + a_n e_n) = (e^{2\pi i a_1}, \ldots, e^{2\pi i a_n}).$$

Then ϕ is a surjective homorphism, and the kernel of ϕ is L.   □

Lemma 6.5. *The map ϕ defined above, when restricted to the fundamental domain T, yields a bijection $T \to \mathbb{T}^n$.*   □

Geometrically, $\mathbb{T}^n$ is obtained by 'gluing' (that is, identifying) opposite faces of the closure of the fundamental domain, as in Figure 6.4.

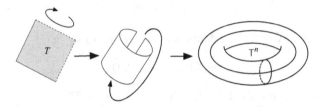

Figure 6.4. The quotient of Euclidean space by a lattice of the same dimension is a torus, obtained by identifying opposite edges of a fundamental domain.

If the dimension of L is less than n, we have a similar result:

Theorem 6.6. *Let L be an m-dimensional lattice in $\mathbb{R}^n$. Then $\mathbb{R}^n/L$ is isomorphic to $\mathbb{T}^m \times \mathbb{R}^{n-m}$.*

Proof: Let V be the subspace spanned by L, and choose a complement W so that $\mathbb{R}^n = V \oplus W$. Then $L \subseteq V$, $V/L \cong \mathbb{T}^m$ by Theorem 6.4, $W \cong \mathbb{R}^{n-m}$.   □

For example, $\mathbb{R}^2/\mathbb{Z} \cong \mathbb{T}^1 \times \mathbb{R}$, which geometrically is a cylinder as in Figure 6.5.

The *volume* $v(X)$ of a subset $X \subseteq \mathbb{R}^n$ is defined in the usual way: for precision we take it to be the value of the multiple integral

$$\int_X \mathrm{d}x_1 \ldots \mathrm{d}x_n$$

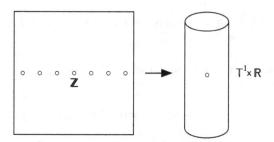

Figure 6.5. The quotient of Euclidean space by a lattice of smaller dimension is a cylinder.

where $(x_1, \ldots, x_n)$ are coordinates. Of course the volume exists only when the integral does.

Let $L \subseteq \mathbb{R}^n$ be a lattice of dimension n, so that $\mathbb{R}^n/L \cong \mathbb{T}^n$. Let T be a fundamental domain of L. We have noted the existence of a bijection

$$\phi : T \to \mathbb{T}^n.$$

For any subset X of $\mathbb{T}^n$ we define the *volume* $v(X)$ by

$$v(X) = v(\phi^{-1}(X))$$

which exists if and only if $\phi^{-1}(X)$ has a volume in $\mathbb{R}^n$.

Let $\nu : \mathbb{R}^n \to \mathbb{T}^n$ be the natural homomorphism with kernel L. It is intuitively clear that ν is 'locally volume-preserving', that is, for each $x \in \mathbb{R}^n$ there exists a ball $B_\epsilon[x]$ such that for all subsets $X \subseteq B_\epsilon[x]$ for which $v(X)$ exists we have

$$v(X) = v(\nu(X))$$

It is also intuitively clear that if an injective map is locally volume-preserving then it is volume-preserving. We prove a result that combines these two intuitive ideas:

Theorem 6.7. *If X is a bounded subset of $\mathbb{R}^n$ and $v(X)$ exists, and if $v(\nu(X)) \neq v(X)$, then $\nu|_X$ is not injective.*

Proof: Assume $\nu|_X$ is injective. Now X, being bounded, intersects only a finite number of the sets $T + l$, for T a fundamental domain and $l \in L$. Put

$$X_l = X \cap (T + l).$$

Then

$$X = X_{l_1} \cup \ldots \cup X_{l_n}.$$

For each l_i define

$$Y_{l_i} = X_{l_i} - l_i,$$

so that $Y_{l_i} \subseteq T$. We claim that the Y_{l_i} are disjoint. Since $\nu(x - l_i) = \nu(x)$ for all $x \in \mathbb{R}^n$ this follows from the assumed injectivity of ν. Now

$$v(X_{l_i}) = v(Y_{l_i})$$

for all i. Also

$$\nu(X_{l_i}) = \phi(Y_{l_i})$$

where ϕ is the bijection $T \to \mathbb{T}^n$. Now we compute:

$$
\begin{aligned}
v(\nu(X)) &= v\left(\nu\left(\bigcup X_{l_i}\right)\right) \\
&= v\left(\bigcup Y_{l_i}\right) \\
&= \sum v(Y_{l_i}) \text{ by disjointness} \\
&= \sum v(X_{l_i}) \\
&= v(X),
\end{aligned}
$$

a contradiction. $\qquad\square$

The idea of the proof is illustrated in Figure 6.6.

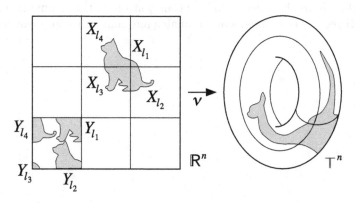

Figure 6.6. Proof of Theorem 6.7: if a locally volume-preserving map does not preserve volume globally, then it cannot be injective. The volume-preserving case is illustrated here; the parts of the cat do not overlap.

6.3   Exercises

1. Let L be a lattice in $\mathbb{R}^2$ with $L \subseteq \mathbb{Z}^2$. Prove that the volume of a fundamental domain T is equal to the number of points of $\mathbb{Z}^2$ lying in T.

2. Generalize the previous exercise to $\mathbb{R}^n$ and link this to Lemma 9.3 by using Theorem 1.17.

3. Sketch the lattices in $\mathbb{R}^2$ generated by:

   (a) $(0,1)$ and $(1,0)$.

   (b) $(-1,2)$ and $(2,2)$.

   (c) $(1,1)$ and $(2,3)$.

   (d) $(-2,-7)$ and $(4,-3)$.

   (e) $(1,20)$ and $(1,-20)$.

   (f) $(1,\pi)$ and $(\pi,1)$.

4. Sketch fundamental domains for these lattices.

5. Hence show that the fundamental domain of a lattice is not uniquely determined until we specify a set of generators.

6. Verify that nonetheless the volume of a fundamental domain of a given lattice is independent of the set of generators chosen.

7. Find two different fundamental domains for the lattice in $\mathbb{R}^3$ generated by $(0,0,1)$, $(0,2,0)$, $(1,1,1)$. Show by direct calculation that they have the same volume. Can you prove this geometrically by dissecting the fundamental domains into mutually congruent pieces?

7

Minkowski's Theorem

The aim of this chapter is to prove a marvellous theorem due to Hermann Minkowski in 1896. This asserts the existence, within a suitable set X, of a non-zero point of a lattice L, provided the volume of X is sufficiently large relative to that of a fundamental domain of L. The idea behind the proof is deceptive in its simplicity: X cannot be squashed into a space whose volume is less than that of X, unless X is allowed to overlap itself. Minkowski discovered that this essentially trivial observation has many non-trivial and important consequences, and used it as a foundation for an extensive theory of the 'geometry of numbers'. As immediate and accessible instances of its application we prove the two- and four-squares theorems of classical number theory.

7.1   Minkowski's Theorem

A subset $X \subseteq \mathbf{R}^n$ is *convex* if whenever x, $y \in X$ then all points on the straight line segment joining x to y also lie in X. In algebraic terms, X is convex if, whenever x, $y \in X$, the point

$$\lambda x + (1 - \lambda)y$$

belongs to X for all real λ, $0 \leq \lambda \leq 1$.

For example a circle, a square, an ellipse, or a triangle is convex in $\mathbf{R}^2$, but an annulus or crescent is not (Figure 7.1). A subset $X \subseteq \mathbf{R}^n$ is *(centrally) symmetric* if $x \in X$ implies $-x \in X$. Geometrically this

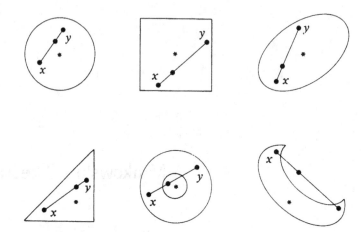

Figure 7.1. Convex and non-convex sets. The circular disc, square, ellipse, and triangle are convex; the annulus and crescent are not. The circle, square, ellipse, and annulus are centrally symmetric about *; the triangle and crescent are not.

means that X is invariant under reflection in the origin. Among the sets in Figure 7.1, assuming the origin to be at the positions marked with an asterisk, the circle, square, ellipse, and annulus are symmetric, but the triangle and crescent are not.

We may now state Minkowski's theorem.

Theorem 7.1. (Minkowski's Theorem.) *Let L be an n-dimensional lattice in $\mathbf{R}^n$ with fundamental domain T, and let X be a bounded symmetric convex subset of $\mathbf{R}^n$. If*

$$v(X) > 2^n v(T)$$

then X contains a non-zero point of L.

Proof: Double the size of L to obtain a lattice $2L$ with fundamental domain $2T$ of volume $2^n v(T)$. Consider the torus

$$\mathbf{T}^n = \mathbf{R}^n / 2L.$$

By definition,

$$v(\mathbf{T}^n) = v(2T) = 2^n v(T).$$

Now the natural map $\nu : \mathbf{R}^n \to \mathbf{T}^n$ cannot preserve the volume of X, since this is strictly larger than $v(\mathbf{T}^n)$: since $\nu(X) \subseteq \mathbf{T}^n$ we have

$$v(\nu(X)) \leq v(\mathbf{T}^n) = 2^n v(T) < v(X).$$

By Theorem 6.7 $\nu|_X$ is not injective, so there exist $x_1 \neq x_2$, $x_1, x_2 \in X$ such that $\nu(x_1) = \nu(x_2)$, or equivalently

$$x_1 - x_2 \in 2L. \tag{7.1}$$

But $x_2 \in X$, so $-x_2 \in X$ by symmetry; and now by convexity

$$\tfrac{1}{2}(x_1) + \tfrac{1}{2}(-x_2) \in X,$$

that is,

$$\tfrac{1}{2}(x_1 - x_2) \in X.$$

But by (7.1)

$$\tfrac{1}{2}(x_1 - x_2) \in L.$$

Therefore

$$0 \neq \tfrac{1}{2}(x_1 - x_2) \in X \cap L$$

as required. $\square$

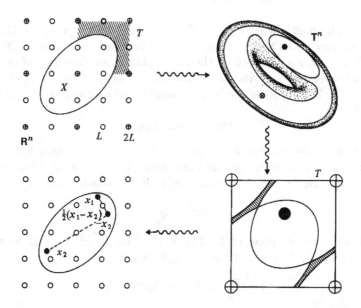

Figure 7.2. Proof of Minkowski's theorem. Expand the original lattice ($\circ$) to double the size ($\oplus$) and form the quotient torus. By computing volumes, the natural quotient map is not injective when restricted to the given convex set. From point x_1 and x_2 with the same image we may construct a non-zero lattice point $\tfrac{1}{2}(x_1 - x_2)$.

The geometrical reasoning is illustrated in Figure 7.2. The decisive step in the proof is that since $\mathbf{T}^n$ has smaller volume than X it is impossible to squash X into $\mathbf{T}^n$ without overlap: the ancient platitude of quarts and pint pots. That such olde-worlde wisdom becomes, in the hands of Minkowski, a weapon of devastating power, was the wonder of the 19th century and a lesson for the 21st. We unleash this power at several crucial stages in the forthcoming battle. (Our original Thespian metaphor has been abandoned in favour of a military one, reinforcing the change of viewpoint from that of the algebraic *voyeur* to that of the geometric participant.) As a more immediate affirmation, we now give two traditional applications to number theory: the 'two-squares' and 'four-squares' theorems.

7.2   The Two-Squares Theorem

We use Minkowski's method to prove a wonderful theorem of Fermat:

Theorem 7.2. (Two-Squares Theorem.) *If p is prime of the form $4k+1$ then p is a sum of two integer squares.*

Proof: The multiplicative group G of the field $\mathbf{Z}_p$ is cyclic; see Garling [31] Corollary 1 to Theorem 12.3, p. 105 or Stewart [78] Corollary 20.9, p. 230. It has order $p-1 = 4k$. It therefore contains an element u of order 4. Then $u^2 \equiv -1 \pmod{p}$ since -1 is the only element of order 2 in G.

Let $L \subseteq \mathbf{Z}^2$ be the lattice in $\mathbf{R}^2$ consisting of all pairs (a,b) $(a,b \in \mathbf{Z})$ such that

$$b \equiv ua \pmod{p}.$$

This is a subgroup of $\mathbf{Z}^2$ of index p (an easy verification left to the reader) so the volume of a fundamental domain for L is p. By Minkowski's theorem any circle, centre the origin, of radius r, which has area

$$\pi r^2 > 4p$$

contains a non-zero point of L. This is the case for $r^2 = 3p/2$. So there exists a point $(a,b) \in L$, not the origin, for which

$$0 \neq a^2 + b^2 \leq r^2 = 3p/2 < 2p.$$

But modulo p

$$a^2 + b^2 \equiv a^2 + u^2 a^2 \equiv 0.$$

Therefore $a^2 + b^2$ is a multiple of p lying strictly between 0 and $2p$, so it must equal p. □

The reader should draw the lattice L and the appropriate circle in a few cases ($p = 5, 13, 17$) and check that the relevant lattice point exists and provides suitable a, b.

Theorem 7.2 goes back to Fermat, who stated it in a letter to Marin Mersenne in 1640. He sent a sketch proof, quite different from the geometric one given here, to Pierre de Carcavi in 1659. Euler gave a complete proof in 1754.

7.3 The Four-Squares Theorem

Refining this argument leads to another famous theorem, first proved by Lagrange:

Theorem 7.3. (Four-Squares Theorem.) *Every positive integer is a sum of four integer squares.*

Proof: We prove the theorem for primes p, and then extend the result to all integers. Now

$$2 = 1^2 + 1^2 + 0^2 + 0^2$$

so we may suppose p is odd. We claim that the congruence

$$u^2 + v^2 + 1 \equiv 0 \pmod{p}$$

has a solution $u, v \in \mathbf{Z}$. This is because u^2 takes exactly $(p+1)/2$ distinct values as u runs through $0, \ldots, p-1$; and $-1 - v^2$ also takes on $(p+1)/2$ values. For the congruence to have no solution, all these values, $p+1$ in total, are distinct, but then $p + 1 \leq p$ which is absurd.

For such a choice of u, v consider the lattice $L \subseteq \mathbf{Z}^4$ consisting of (a, b, c, d) such that

$$c \equiv ua + vb, \quad d \equiv ub - va \pmod{p}.$$

Then L has index p^2 in $\mathbf{Z}^4$ (another easy computation) so the volume of a fundamental domain is p^2. Now a 4-dimensional sphere, centre the origin, radius r, has volume

$$\pi^2 r^4 / 2$$

and we choose r to make this greater than $16p^2$; say $r^2 = 1.9p$.

There exists a lattice point $0 \neq (a, b, c, d)$ in this 4-sphere, so

$$0 \neq a^2 + b^2 + c^2 + d^2 \leq r^2 = 1.9p < 2p.$$

Modulo p, it is easy to verify that $a^2 + b^2 + c^2 + d^2 \equiv 0$, hence as before it must equal p.

To deal with an arbitrary integer n, it suffices to factorize n into primes and then use the identity

$$(a^2 + b^2 + c^2 + d^2)(A^2 + B^2 + C^2 + D^2)$$

$$= (aA - bB - cC - dD)^2 + (aB + bA + cD - dC)^2$$

$$+(aC - bD + cA + dB)^2 + (aD + bC - cB + dA)^2.$$

$\square$

Theorem 7.3 also goes back to Fermat. Euler spent 40 years trying to prove it, and Lagrange succeeded in 1770.

7.4 Exercises

1. Which of the following solids are convex? Sphere, pyramid, icosahedron, cube, torus, ellipsoid, parallelepiped.

2. How many different convex solids can be made by joining n unit cubes face to face, so that their vertices coincide, for $n = 1, 2, 3, 4, 5, 6$; counting two solids as different if and only if they cannot be mapped to each other by rigid motions? What is the result for general n?

3. Verify the two-squares theorem on all primes less than 200.

4. Verify the four-squares theorem on all integers less than 100.

5. Prove that not every integer is a sum of three squares.

6. Prove that the number $\mu(n)$ of pairs of integers (x, y) with $x^2 + y^2 < n$ satisfies $\mu(n)/n \to \pi$ as $n \to \infty$.

8

Geometric Representation of Algebraic Numbers

The purpose of this chapter is to develop a method for embedding a number field K in a real vector space of dimension equal to the degree of K, in such a way that ideals in K map to lattices in this vector space. This clever idea opens the way to applications of Minkowski's theorem. The embedding is defined in terms of the monomorphisms $K \to \mathbb{C}$, and we have to distinguish between those that map K into $\mathbb{R}$ and those that do not.

8.1 The Space $\mathbb{L}^{st}$

Let $K = \mathbb{Q}(\theta)$ be a number field of degree n, where θ is an algebraic integer. Let $\sigma_1, \ldots, \sigma_n$ be the set of all monomorphisms $K \to \mathbb{C}$ (see Theorem 2.4). If $\sigma_i(K) \subseteq \mathbb{R}$, which happens if and only if $\sigma_i(\theta) \in \mathbb{R}$, we say that σ_i, is *real*; otherwise σ_i is *complex*. As usual denote complex conjugation by bars and define

$$\bar{\sigma}_i(\alpha) = \overline{\sigma_i(\alpha)}.$$

Since complex conjugation is an automorphism of $\mathbb{C}$ it follows that $\bar{\sigma}_i$ is a monomorphism $K \to \mathbb{C}$, so equals σ_j for some j. Now $\sigma_i = \bar{\sigma}_i$ if and only if σ_i is real, and $\bar{\bar{\sigma}}_i = \sigma_i$, so the complex monomorphisms come in conjugate pairs. Hence

$$n = s + 2t$$

where s is the number of real monomorphisms and $2t$ is the number of complex ones. We standardize the numeration in such a way that the system of all monomorphisms $K \to \mathbb{C}$ is

$$\sigma_1, \ldots, \sigma_s; \sigma_{s+1}, \bar{\sigma}_{s+1}, \ldots, \sigma_{s+t}, \bar{\sigma}_{s+t},$$

where $\sigma_1, \ldots, \sigma_s$ are real and the rest complex.

Further define

$$\mathbb{L}^{st} = \mathbb{R}^s \times \mathbb{C}^t,$$

the set of all $(s+t)$-tuples

$$x = (x_1, \ldots, x_s; x_{s+1}, \ldots, x_{s+t})$$

where $x_1, \ldots, x_s \in \mathbb{R}$ and $x_{s+1}, \ldots, x_{s+t} \in \mathbb{C}$. Then $\mathbb{L}^{st}$ is a vector space over $\mathbb{R}$, and a ring (with coordinatewise operations): in fact it is an $\mathbb{R}$-algebra. As vector space over $\mathbb{R}$ it has dimension $s + 2t = n$.

For $x \in \mathbb{L}^{st}$, define the *norm*

$$N(x) = x_1 \ldots x_s |x_{s+1}|^2 \ldots |x_{s+t}|^2. \tag{8.1}$$

(There is no confusion with other uses of the word 'norm', and we will see why it is desirable to use this apparently overworked word in a moment.) The norm has two obvious properties:

(a) $N(x)$ is real for all x,

(b) $N(xy) = N(x)N(y)$.

Define a map

$$\sigma : K \to \mathbb{L}^{st}$$

by

$$\sigma(\alpha) = (\sigma_1(\alpha), \ldots, \sigma_s(\alpha); \sigma_{s+1}(\alpha), \ldots, \sigma_{s+t}(\alpha))$$

for $\alpha \in K$. Clearly

$$\begin{aligned} \sigma(\alpha + \beta) &= \sigma(\alpha) + \sigma(\beta) \\ \sigma(\alpha\beta) &= \sigma(\alpha)\sigma(\beta) \end{aligned} \tag{8.2}$$

for all $\alpha, \beta \in K$, so σ is a ring homomorphism. If r is a rational number then

$$\sigma(r\alpha) = r\sigma(\alpha)$$

so σ is a $\mathbb{Q}$-algebra homomorphism. Furthermore,

$$N(\sigma(\alpha)) = N(\alpha) \tag{8.3}$$

since the latter is defined to be

$$\sigma_1(\alpha) \ldots \sigma_s(\alpha)\sigma_{s+1}(\alpha)\bar{\sigma}_{s+1}(\alpha) \ldots \sigma_{s+t}(\alpha)\bar{\sigma}_{s+t}(\alpha)$$

which equals the former.

For example, let $K = \mathbb{Q}(\theta)$ where $\theta \in \mathbb{R}$ satisfies

$$\theta^3 - 2 = 0.$$

Then the conjugates of θ are $\theta, \omega\theta, \omega^2\theta$ where ω is a complex cube root of unity. The monomorphisms $K \to \mathbb{C}$ are given by

$$\sigma_1(\theta) = \theta, \quad \sigma_2(\theta) = \omega\theta, \quad \bar{\sigma}_2(\theta) = \omega^2\theta.$$

Hence $s = t = 1$.

An element of K, say

$$x = q + r\theta + s\theta^2$$

where $q, r, s \in \mathbb{Q}$, maps into $\mathbb{L}^{1,1}$ according to

$$\sigma(x) = (q + r\theta + s\theta^2, q + r\omega\theta + s\omega^2\theta^2).$$

The kernel of σ is an ideal of K since σ is a ring homomorphism. Since K is a field, σ is either identically zero or injective. But

$$\sigma(1) = (1, 1, \ldots, 1) \neq 0$$

so σ must be injective. Much stronger is the following result:

Theorem 8.1. *If $\alpha_1, \ldots, \alpha_n$ is a basis for K over $\mathbb{Q}$ then $\sigma(\alpha_1), \ldots, \sigma(\alpha_n)$ are linearly independent over $\mathbb{R}$.*

Proof: Linear independence over $\mathbb{Q}$ is immediate since σ is injective, but we need more than this. Let

$$\sigma_k(\alpha_l) = x_k^{(l)} \qquad (k = 1, \ldots, s)$$
$$\sigma_{s+j}(\alpha_l) = y_j^{(l)} + iz_j^{(l)} \qquad (j = 1, \ldots, t)$$

where $x_k^{(l)}, y_k^{(l)}, z_k^{(l)}$ are real. Then

$$\sigma(\alpha_l) = (x_1^{(l)}, \ldots, x_s^{(l)}; y_1^{(l)} + iz_1^{(l)}, \ldots, y_t^{(l)} + iz_t^{(l)}),$$

and it is sufficient to prove that the determinant

$$D = \begin{vmatrix} x_1^{(1)} & \cdots & x_s^{(1)} & y_1^{(1)} & z_1^{(1)} & \cdots & y_t^{(1)} & z_t^{(1)} \\ \vdots & & \vdots & \vdots & \vdots & & \vdots & \vdots \\ x_1^{(n)} & \cdots & x_s^{(n)} & y_1^{(n)} & z_1^{(n)} & \cdots & y_t^{(n)} & z_t^{(n)} \end{vmatrix}$$

is non-zero. Put

$$
E = \begin{vmatrix}
x_1^{(1)} & \cdots & x_s^{(1)} & y_1^{(1)} + iz_1^{(1)} & y_1^{(1)} - iz_1^{(1)} & \cdots \\
\vdots & & \vdots & \vdots & \vdots & \vdots \\
x_1^{(n)} & \cdots & x_s^{(n)} & y_1^{(n)} + iz_1^{(n)} & y_1^{(n)} - iz_1^{(n)} & \cdots
\end{vmatrix}
$$

$$
= \begin{vmatrix}
\sigma_1(\alpha_1) & \cdots & \sigma_s(\alpha_1) & \sigma_{s+1}(\alpha_1)\bar{\sigma}_{s+1}(\alpha_1) & \cdots \\
\vdots & & \vdots & \vdots & \vdots \\
\sigma_1(\alpha_n) & \cdots & \sigma_s(\alpha_n) & \sigma_{s+1}(\alpha_n)\bar{\sigma}_{s+1}(\alpha_n) & \cdots
\end{vmatrix}.
$$

Then

$$
E^2 = \Delta[\alpha_1, \ldots, \alpha_n]
$$

by definition of the discriminant, and $E^2 \neq 0$ by Theorem 2.7. Now elementary properties of determinants (column operations) yield

$$
E = (-2i)^t D
$$

so $D \neq 0$ as required. $\square$

Corollary 8.2. $\mathbb{Q}$-*linearly independent elements of the number field* K *map under* σ *to* $\mathbb{R}$-*linearly independent elements of* $\mathbb{L}^{st}$. $\square$

Corollary 8.3. *Suppose that* G *is a finitely generated subgroup of* $(K, +)$ *with* $\mathbb{Z}$-*basis* $\{\alpha_1, \ldots, \alpha_m\}$. *Then the image of* G *in* $\mathbb{L}^{st}$ *is a lattice with generators* $\sigma(\alpha_1), \ldots, \sigma(\alpha_m)$. $\square$

The 'geometric representation' of K in $\mathbb{L}^{st}$ defined by σ, in combination with Minkowski's theorem, provides the key to several of the deeper parts of the theory in Chapters 9, 10, and Appendix B. For these applications we need a notion of 'distance' on $\mathbb{L}^{st}$. Since $\mathbb{L}^{st}$ is isomorphic to $\mathbb{R}^{s+2t}$ as a real vector space, the natural idea is to transfer the usual Euclidean metric from $\mathbb{R}^{s+2t}$ to $\mathbb{L}^{st}$. This amounts to choosing a basis in $\mathbb{L}^{st}$ and defining an inner product with respect to which this basis is orthonormal. The natural basis to pick is

$$\begin{cases} (1,0,\ldots,0;0,\ldots,0) \\ (0,1,\ldots,0;0,\ldots,0) \\ \quad\vdots \\ (0,0,\ldots,1;0,\ldots,0) \\ (0,0,\ldots,0;1,0,\ldots,0) \\ (0,0,\ldots,0;i,0,\ldots,0) \\ \quad\vdots \\ (0,0,\ldots,0;0,0,\ldots,1) \\ (0,0,\ldots,0;0,0,\ldots,i). \end{cases} \qquad (8.4)$$

With respect to this basis, the element

$$(x_1,\ldots,x_s;y_1+iz_1,\ldots,y_t+iz_t)$$

of $\mathbb{L}^{st}$ has coordinates

$$(x_1,\ldots,x_s,y_1,z_1,\ldots,y_t,z_t).$$

Changing notation slightly, if we take

$$\begin{aligned} x &= (x_1,\ldots,x_{s+2t}) \\ x' &= (x'_1,\ldots,x'_{s+2t}) \end{aligned}$$

with respect to the *new* coordinates (8.4), then the inner product is defined by

$$(x,x') = x_1 x'_1 + \ldots + x_{s+2t} x'_{s+2t}.$$

The *length* of a vector x is then

$$\|x\| = \sqrt{(x,x)}$$

and the *distance* between x and x' is $\|x-x'\|$.

Referred to our original mixture of real and complex coordinates, any element

$$x = (x_1,\ldots,x_s;y_1+iz_1,\ldots,y_t+iz_t)$$

has length

$$\|x\| = \sqrt{(x_1^2 + \ldots + x_s^2 + y_1^2 + z_1^2 + \ldots + y_t^2 + z_t^2)}.$$

8.2 Exercises

1. Find the monomorphisms $\sigma_i : K \to \mathbb{C}$ for the following fields. Determine the number s of the σ_i satisfying $\sigma_i(K) \subseteq \mathbb{R}$, and the number t of distinct conjugate pairs σ_i, σ_j such that $\bar{\sigma}_i = \sigma_j$.

   (i) $\mathbb{Q}(\sqrt{5})$

   (ii) $\mathbb{Q}(\sqrt{-5})$

   (iii) $\mathbb{Q}(\sqrt[4]{5})$

   (iv) $\mathbb{Q}(\zeta)$ where $\zeta = e^{2\pi i/7}$

   (v) $\mathbb{Q}(\zeta)$ where $\zeta = e^{2\pi i/p}$ for a rational prime p

2. For $K = \mathbb{Q}(\sqrt{d})$ where d is a squarefree integer, calculate $\sigma : K \to \mathbb{L}^{st}$, distinguishing the cases $d < 0$, $d > 0$. Compute $N(x)$ for $x \in \mathbb{L}^{st}$ and by direct calculation verify

$$N(\alpha) = N(\sigma(\alpha)) \quad (\alpha \in K)$$

3. Let $K = \mathbb{Q}(\theta)$ where the algebraic integer θ has minimum polynomial f. If f factorizes over $\mathbb{R}$ into irreducibles as

$$f(t) = g_1(t) \ldots g_q(t) h_1(t) \ldots h_r(t)$$

   where g_i is linear and h_j quadratic, prove that $q = s$ and $r = t$ in the notation of the chapter for s, t.

4. Let $K = \mathbb{Q}(\theta)$ where $\theta \in \mathbb{R}$ and $\theta^3 = 3$. What is the map σ in this case? Pick a basis for K and verify Theorem 8.1 for it.

5. Find a map from $\mathbb{R}^2$ to itself under which $\mathbb{Q}$-linearly independent sets map to $\mathbb{Q}$-linearly independent sets, but some $\mathbb{R}$-linearly independent set does not map to an $\mathbb{R}$-linearly independent set.

6. If $K = \mathbb{Q}(\theta)$ where $\theta \in \mathbb{R}$ and $\theta^3 = 3$, verify Corollary 8.3 for the additive subgroup of K generated by $1 + \theta$ and $\theta^2 - 2$.

9

Class-Group and Class-Number

We now use the geometric ideas that we have developed to build further insight into the property of unique factorization. We already know that in the ring of integers of any number field, factorization into primes is unique if and only if every ideal is principal. We now refine this statement to give a quantitative measure of how *non-unique* factorization can be. To do this we use the fractional ideals introduced in Chapter 5. The class-group of a number field is defined to be the quotient of the group of fractional ideals by the (normal) subgroup of principal fractional ideals; the class-number is the order of this group. This gives the required measure: factorization in a ring of integers is unique if and only if the corresponding class-number is 1. If the class-number is greater than 1, factorization is non-unique. Intuitively, the larger the class-number is, the more complicated the possibilities for non-uniqueness are. In a sense, prime factorization becomes 'less unique' as the class-number increases.

It turns out that the class-number is always finite, an important fact which we prove using Minkowski's theorem. Simple group-theoretic considerations then yield useful conditions for an ideal to be principal. These conditions lead to a proof that every ideal becomes principal in a suitable extension field, which is one formulation of the basic idea of Kummer's 'ideal numbers' within the ideal-theoretic framework.

The importance of the class-number can only be hinted at here. It is crucial in the proof of Kummer's special case of Fermat's Last Theorem in Chapter 11. Many deep and delicate results in the theory of numbers are related to arithmetic properties of the class-number, or to algebraic properties of the class-group.

9.1   The Class-Group

As usual let $\mathfrak{O}$ be the ring of integers of a number field K of degree n. Theorem 5.21 tells us that prime factorization in $\mathfrak{O}$ is unique if and only if every ideal of $\mathfrak{O}$ is principal. Our aim here is to find a way of measuring how far prime factorization fails to be unique in the case where $\mathfrak{O}$ contains non-principal ideals, or equivalently how far away the ideals of $\mathfrak{O}$ are from being principal.

To this end we use the group of fractional ideals defined in Chapter 5. Say that a fractional ideal of $\mathfrak{O}$ is *principal* if it is of the form $c^{-1}\mathfrak{a}$ where $\mathfrak{a}$ is a principal ideal of $\mathfrak{O}$. Let $\mathcal{F}$ be the group of fractional ideals under multiplication. It is easy to check that the set $\mathcal{P}$ of principal fractional ideals is a subgroup of $\mathcal{F}$. We define the *class-group* of $\mathfrak{O}$ to be the quotient group

$$\mathcal{H} = \mathcal{F}/\mathcal{P}$$

The *class-number* $h = h(\mathfrak{O})$ is defined to be the order of $\mathcal{H}$.

Since each of $\mathcal{F}, \mathcal{P}$ is an infinite group we have no immediate way of deciding whether h is finite. In fact it is, and we develop a proof of this deep and important fact. We begin by reformulating the definition of the class-group in terms of ideals rather than fractional ideals.

Say that two fractional ideals are *equivalent* if they belong to the same coset of $\mathcal{P}$ in $\mathcal{F}$, or in other words if they map to the same element of $\mathcal{F}/\mathcal{P}$. If a and b are fractional ideals we write

$$\mathfrak{a} \sim \mathfrak{b}$$

if $\mathfrak{a}$ and $\mathfrak{b}$ are equivalent, and use

$$[\mathfrak{a}]$$

to denote the equivalence class of $\mathfrak{a}$.

The class-group $\mathcal{H}$ is the set of these equivalence classes.

If $\mathfrak{a}$ is a fractional ideal then $\mathfrak{a} = c^{-1}\mathfrak{b}$ where $c \in \mathfrak{O}$ and $\mathfrak{b}$ is an ideal. Therefore

$$\mathfrak{b} = c\mathfrak{a} = \langle c \rangle \, \mathfrak{a}$$

and since $\langle c \rangle \in \mathcal{P}$ this means that $\mathfrak{a} \sim \mathfrak{b}$. In other words, *every equivalence class contains an ideal*.

Now let $\mathfrak{x}$ and $\mathfrak{y}$ be equivalent ideals. (These symbols are Gothic x and y, despite appearances.) Then $\mathfrak{x} = \mathfrak{c}\mathfrak{y}$ where $\mathfrak{c}$ is a principal fractional ideal, say $\mathfrak{c} = d^{-1}\mathfrak{e}$ for $d \in \mathfrak{O}$, $\mathfrak{e}$ a principal ideal. Therefore

$$\mathfrak{x} \langle d \rangle = \mathfrak{y}\mathfrak{e}.$$

Conversely if $\mathfrak{x}\mathfrak{b} = \mathfrak{y}\mathfrak{e}$ for $\mathfrak{b}$, $\mathfrak{e}$ principal ideals then $\mathfrak{x} \sim \mathfrak{y}$.

This leads to an alternative description of $\mathcal{H}$: on the set $\mathcal{F}$ of all ideals, define a relation $\sim$ by $\mathfrak{x} \sim \mathfrak{y}$ if and only if there exist *principal* ideals $\mathfrak{b}$, $\mathfrak{e}$ with $\mathfrak{x}\mathfrak{b} = \mathfrak{y}\mathfrak{e}$. This is an equivalence relation, and $\mathcal{H}$ is the set of equivalence classes $[\mathfrak{x}]$ with group operation

$$[\mathfrak{x}][\mathfrak{y}] = [\mathfrak{x}\mathfrak{y}].$$

This is why $\mathcal{H}$ is called the *class*-group.

The significance of the class-group is that it captures the extent to which factorization is not unique. In particular:

Theorem 9.1. *Factorization in $\mathfrak{O}$ is unique if and only if the class-group $\mathcal{H}$ has order 1, or equivalently the class-number $h = 1$.*

Proof: By Theorem 5.21, factorization is unique if and only if every ideal of $\mathfrak{O}$ is principal, which is true if and only if every fractional ideal is principal, which is equivalent to $\mathcal{F} = \mathcal{P}$, which is equivalent to $|\mathcal{H}| = h = 1$. ☐

The rest of this chapter proves that h is finite, and deduces a few useful consequences. In the next chapter we develop some methods whereby h, and the structure of $\mathcal{H}$, may be computed: such methods are an obvious necessity for applications of the class-group in particular cases.

9.2   An Existence Theorem

The finiteness of h rests on an application of Minkowski's theorem to the space $\mathbb{L}^{st}$. It is, in fact, possible to give a more elementary proof that h is finite, Lang [47], but Minkowski's theorem gives a better bound, and is in any case needed elsewhere. In this section we state and prove the relevant result, leaving the finiteness theorem to the next section.

Lemma 9.2. *If M is a lattice in $\mathbb{L}^{st}$ of dimension $s + 2t$ having fundamental domain of volume V, and if $c_1, \ldots, c_{s+t}$ are positive real numbers whose product*

$$c_1 \ldots c_{s+t} > \left(\frac{4}{\pi}\right)^t V$$

then there exists a non-zero $x = (x_1, \ldots, x_{s+t}) \in M$ such that

$$|x_1| < c_1, \ldots, |x_s| < c_s;$$

$$|x_{s+1}|^2 < c_{s+1}, \ldots, |x_{s+t}|^2 < c_{s+t}.$$

Proof: Let X be the set of all points $x \in \mathbb{L}^{st}$ for which the conclusion holds. Compute

$$v(X) = \int_{-c_1}^{c_1} dx_1 \ldots \int_{-c_s}^{c_s} dx_s \times \iint_{y_1^2 + z_1^2 < c_{s+1}} dy_1 dz_1 \ldots$$

$$\times \iint_{y_t^2 + z_t^2 < c_{s+t}} dy_t dz_t$$

$$= 2c_1 \cdot 2c_2 \ldots 2c_s \cdot \pi c_{s+1} \ldots \pi c_{s+t}$$

$$= 2^s \pi^t c_1 \ldots c_{s+t}.$$

Now X is a cartesian product of line segments and circular discs, so X is bounded, symmetric, and convex. Minkowski's theorem yields the required result provided

$$2^s \pi^t c_1 \ldots c_{s+t} > 2^{s+2t} V,$$

that is

$$c_1 \ldots c_{s+t} > \left(\frac{4}{\pi}\right)^t V. \qquad \square$$

Let K be a number field of degree $n = s + 2t$ as usual, with ring of integers $\mathfrak{O}$; and let $\mathfrak{a}$ be an ideal of $\mathfrak{O}$. Then $(\mathfrak{a}, +)$ is a free abelian group of rank n by Theorem 2.16, so by Corollary 8.3 its image $\sigma(\mathfrak{a})$ in $\mathbb{L}^{st}$ is a lattice of dimension n. To apply Lemma 9.2 we must know the volume of a fundamental domain for $\sigma(\mathfrak{a})$. A useful general result is:

Lemma 9.3. *Let L be an n-dimensional lattice in $\mathbb{R}^n$ with basis $\{e_1, \ldots, e_n\}$. Suppose that*

$$e_i = (a_{1i}, \ldots, a_{ni}).$$

Then the volume of the fundamental domain T of L defined by this basis is

$$v(T) = |\det a_{ij}|.$$

Proof: The volume is

$$v(T) = \int_T dx_1 \ldots dx_n.$$

Define new variables by

$$x_i = \sum_j a_{ij} y_j.$$

The Jacobian of this transformation is $\det a_{ij}$, and T is the set of points $\sum a_{ij} y_i$ with $0 \leq y_i < 1$. By the transformation formula for multiple integrals, Apostol [2] p. 271.

$$
\begin{aligned}
v(T) &= \int_T |\det a_{ij}| \, dy_1 \ldots dy_n \\
&= |\det a_{ij}| \int_0^1 dy_1 \ldots \int_0^1 dy_n \\
&= |\det a_{ij}|. \qquad\qquad \square
\end{aligned}
$$

Given a lattice L there exist many different $\mathbb{Z}$-bases for L, hence many distinct fundamental domains. However, since distinct $\mathbb{Z}$-bases are related by a unimodular matrix, Lemma 9.3 implies that the volumes of these distinct fundamental domains are all equal.

Theorem 9.4. *Let K be a number field of degree $n = s + 2t$ as usual, with ring of integers $\mathfrak{O}$, and let $0 \neq \mathfrak{a}$ be an ideal of $\mathfrak{O}$. Then the volume of a fundamental domain for $\sigma(\mathfrak{a})$ in $\mathbb{L}^{st}$ is*

$$2^{-t} N(\mathfrak{a}) \sqrt{|\Delta|}$$

where Δ is the discriminant of K.

Proof: Let $\{\alpha_1, \ldots, \alpha_n\}$ be a $\mathbb{Z}$-basis for $\mathfrak{a}$. Then, in the notation of Theorem 8.1, a $\mathbb{Z}$-basis for $\sigma(\mathfrak{a})$ in $\mathbb{L}^{st}$ is

$$(x_1^{(1)}, \ldots, x_s^{(1)}, y_1^{(1)}, z_1^{(1)}, \ldots, y_t^{(1)}, z_t^{(1)}),$$
$$\vdots$$
$$(x_1^{(n)}, \ldots, x_s^{(n)}, y_1^{(n)}, z_1^{(n)}, \ldots, y_t^{(n)}, z_t^{(n)}).$$

By Lemma 9.3, if T is a fundamental domain for $\sigma(\mathfrak{a})$ then $v(T) = |D|$, where D is as in Theorem 8.1. In the notation of that theorem,

$$D = (-2i)^{-t} E$$

so that

$$|D| = 2^{-t} |E|.$$

Now $E^2 = \Delta[\alpha_1, \ldots, \alpha_n]$ and

$$N(\mathfrak{a}) = \left| \frac{\Delta[\alpha_1, \ldots, \alpha_n]}{\Delta} \right|^{1/2}$$

by Theorem 5.9. $\square$

Lemma 9.2 and Theorem 9.4 now yield the important:

Theorem 9.5. *If* $\mathfrak{a} \neq 0$ *is an ideal of* $\mathfrak{O}$ *then* $\mathfrak{a}$ *contains an integer* α *with*

$$|N(\alpha)| \leq \left(\frac{2}{\pi}\right)^t N(\mathfrak{a})\sqrt{|\Delta|}$$

where Δ *is the discriminant of* K.

Proof: For fixed but arbitrary $\epsilon > 0$ choose positive real numbers $c_1, \ldots, c_{s+t}$ with

$$c_1 \ldots c_{s+t} = \left(\frac{2}{\pi}\right)^t N(\mathfrak{a})\sqrt{|\Delta|} + \epsilon.$$

By Lemma 9.2 and Theorem 9.4 there exists $0 \neq \alpha \in \mathfrak{a}$ such that

$$
\begin{aligned}
|\sigma_1(\alpha)| &< c_1, \ldots, |\sigma_s(\alpha)| < c_s \\
|\sigma_{s+1}(\alpha)|^2 &< c_{s+1}, \ldots, |\sigma_{s+t}(\alpha)|^2 < c_{s+t}.
\end{aligned}
$$

Multiply all these inequalities together:

$$|N(\alpha)| < c_1 \ldots c_s c_{s+1} \ldots c_{s+t} = \left(\frac{2}{\pi}\right)^t N(\mathfrak{a})\sqrt{|\Delta|} + \epsilon.$$

Since a lattice is discrete, the set A_ϵ of such α is finite. Also $A_\epsilon \neq \emptyset$, so $A = \cap_\epsilon A_\epsilon \neq \emptyset$. It we pick $\alpha \in A$ then

$$|N(\alpha)| \leq \left(\frac{2}{\pi}\right)^t N(\mathfrak{a})\sqrt{|\Delta|}.$$

$\square$

Corollary 9.6. *Every non-zero ideal* $\mathfrak{a}$ *of* $\mathfrak{O}$ *is equivalent to an ideal whose norm is* $\leq (2/\pi)^t \sqrt{|\Delta|}$.

Proof: The class of fractional ideals equivalent to $\mathfrak{a}^{-1}$ contains an ideal $\mathfrak{c}$, so $\mathfrak{a}\mathfrak{c} \sim \mathfrak{O}$. Use Theorem 9.5 to find an integer $\gamma \in \mathfrak{c}$ such that

$$|N(\gamma)| \leq \left(\frac{2}{\pi}\right)^t N(\mathfrak{c})\sqrt{|\Delta|}.$$

Since $\mathfrak{c}|\gamma$ we have

$$\langle\gamma\rangle = \mathfrak{c}\mathfrak{b}$$

for some ideal $\mathfrak{b}$. Since $\mathrm{N}(\mathfrak{b})\mathrm{N}(\mathfrak{c}) = \mathrm{N}(\mathfrak{b}\mathfrak{c}) = \mathrm{N}(\langle\gamma\rangle) = |\mathrm{N}(\gamma)|$,

$$\mathrm{N}(\mathfrak{b}) \leq \left(\frac{2}{\pi}\right)^t \sqrt{|\Delta|}.$$

We claim that $\mathfrak{b} \sim \mathfrak{a}$. This is clear since $\mathfrak{c} \sim \mathfrak{a}^{-1}$ and $\mathfrak{b} \sim \mathfrak{c}^{-1}$. $\qquad\square$

An Explicit Computation. If $K = \mathbb{Q}(\sqrt{-5})$, then $\mathfrak{O} = \mathbb{Z}[\sqrt{-5}]$ does not have unique factorization, so $h > 1$. Because the monomorphisms $\sigma_i : K \to \mathbb{C}$ are σ_1, σ_2 where $\sigma_1 \neq \sigma_2$ and $\bar{\sigma}_1 = \sigma_2$, we have $t = 1$. The discriminant Δ of K is $\Delta = -20$, so

$$\left(\frac{2}{\pi}\right)^t \sqrt{|\Delta|} = \frac{2\sqrt{20}}{\pi} < 2.85.$$

Every ideal of $\mathfrak{O}$ is then equivalent to an ideal of norm less than 2.85, which means a norm of 1 or 2. An ideal of norm 1 is the whole ring $\mathfrak{O}$, hence principal. An ideal $\mathfrak{a}$ of norm 2 satisfies $\mathfrak{a}|2$ by Theorem 5.14 (b), so $\mathfrak{a}$ is a factor of $\langle 2 \rangle$. But

$$\langle 2 \rangle = \langle 2, 1 + \sqrt{-5} \rangle^2$$

where $\langle 2, 1 + \sqrt{-5} \rangle$ is prime and has norm 2. So $\langle 2, 1 + \sqrt{-5} \rangle$ is the *only* ideal of norm 2. Hence every ideal of $\mathfrak{O}$ is equivalent to $\mathfrak{O}$ or $\langle 2, 1 + \sqrt{-5} \rangle$ which are themselves inequivalent (since $\langle 2, 1 + \sqrt{-5} \rangle$ is not principal), proving that $h = 2$.

9.3   Finiteness of the Class-Group

Theorem 9.7. *The class-group of a number field is a finite abelian group. The class-number h is finite.*

Proof: Let K be a number field of discriminant Δ and degree $n = s + 2t$ as usual. The class-group $\mathcal{H} = \mathcal{F}/\mathcal{P}$ is abelian, so it remains to prove $\mathcal{H}$ finite. This is true if and only if the number of distinct equivalence classes of fractional ideals is finite. Let $[\mathfrak{c}]$ be such an equivalence class. Then $[\mathfrak{c}]$ contains an ideal $\mathfrak{a}$, and by Corollary 9.6, $\mathfrak{a}$ is equivalent to an ideal $\mathfrak{b}$ with $\mathrm{N}(\mathfrak{b}) \leq (2/\pi)^t \sqrt{|\Delta|}$. Since only finitely many ideals have a given norm (Theorem 5.17 (c)) there are only finitely many choices for $\mathfrak{b}$. Since $[\mathfrak{c}] = [\mathfrak{a}] = [\mathfrak{b}]$ (because $\mathfrak{c} \sim \mathfrak{a} \sim \mathfrak{b}$) there are only finitely many equivalence classes $[\mathfrak{c}]$, whence $\mathcal{H}$ is a finite group and $h = |\mathcal{H}|$ is finite. $\qquad\square$

From simple group-theoretic facts we obtain the useful:

Proposition 9.8. *Let K be a number field of class-number h, and $\mathfrak{a}$ an ideal of the ring of integers $\mathfrak{O}$. Then*

(a) *$\mathfrak{a}^h$ is principal.*

(b) *If q is prime to h and $\mathfrak{a}^q$ is principal, then $\mathfrak{a}$ is principal.*

Proof: Since $h = |\mathcal{H}|$ we have $[\mathfrak{a}]^h = [\mathfrak{O}]$ for all $[\mathfrak{a}] \in \mathcal{H}$, because $[\mathfrak{O}]$ is the identity element of $\mathcal{H}$. Hence $[\mathfrak{a}^h] = [\mathfrak{a}]^h = [\mathfrak{O}]$, so $\mathfrak{a}^h \sim \mathfrak{O}$, so $\mathfrak{a}^h$ is principal. This proves (a). For (b) choose u and $v \in \mathbb{Z}$ such that $uh + vq = 1$. Then $[\mathfrak{a}]^q = [\mathfrak{O}]$, so

$$
\begin{aligned}
[\mathfrak{a}] &= [\mathfrak{a}]^{uh+vq} \\
&= \left([\mathfrak{a}]^h\right)^u \left([\mathfrak{a}]^q\right)^v \\
&= [\mathfrak{O}]^u [\mathfrak{O}]^v \\
&= [\mathfrak{O}]
\end{aligned}
$$

and again $\mathfrak{a}$ is principal.   □

9.4   How to Make an Ideal Principal

(*This section and the next are not required elsewhere and may be omitted.*)

Given an ideal $\mathfrak{a}$ in the ring $\mathfrak{O}$ of integers of a number field K, we already know that $\mathfrak{a}$ has at most two generators

$$\mathfrak{a} = \langle \alpha, \beta \rangle \quad (\alpha, \beta \in \mathfrak{O}).$$

In this section we demonstrate that there exists an extension number field $E \supseteq K$ with integers $\mathcal{D}'$, such that the extended ideal $\mathfrak{O}'\mathfrak{a}$ in $\mathfrak{O}'$ is principal. As standard notation we retain the symbols $\langle \alpha \rangle$, $\langle \alpha, \beta \rangle$ to denote the ideals in $\mathfrak{O}$ generated by α and by $\alpha, \beta \in \mathfrak{O}$. We write the ideal in $\mathfrak{O}'$ generated by $S \subseteq \mathfrak{O}'$ as $\mathfrak{O}'S$. For example $\mathfrak{O}'\kappa$ denotes the principal ideal in $\mathfrak{O}'$ generated by $\kappa \in \mathfrak{O}'$.

Lemma 9.9. *If S_1, S_2 are subsets of $\mathfrak{O}'$, then*

$$\mathfrak{O}'(S_1 S_2) = (\mathfrak{O}'S_1)(\mathfrak{O}'S_2).$$

Proof: Trivial (remembering $1 \in \mathfrak{O}'$).   □

The central result is:

Theorem 9.10. *Let K be a number field, $\mathfrak{a}$ an ideal in the ring of integers $\mathfrak{O}$ of K. Then there exists an algebraic integer κ such that if $\mathfrak{O}'$ is the ring of integers of $K(\kappa)$, then*

(a) $\mathfrak{O}'\kappa = \mathfrak{O}'\mathfrak{a}$.

(b) $(\mathfrak{O}'\kappa) \cap \mathfrak{O} = \mathfrak{a}$.

(c) *If $\mathbb{B}$ is the ring of all algebraic integers, then $(\mathbb{B}\kappa) \cap K = \mathfrak{a}$.*

(d) *If $\mathfrak{O}''\gamma = \mathfrak{O}''\mathfrak{a}$ for any $\gamma \in \mathbb{B}$, and any ring $\mathfrak{O}''$ of integers, then $\gamma = u\kappa$ where u is a unit of $\mathbb{B}$.*

Proof: By Proposition 9.8, $\mathfrak{a}^h$ is principal, say $\mathfrak{a}^h = \langle \omega \rangle$. Let $\kappa = \omega^{1/h} \in \mathbb{B}$, and consider $E = K(\kappa)$. Let $\mathfrak{O}' = \mathbb{B} \cap E$ be the ring of integers in E; clearly $\kappa \in \mathfrak{O}'$. Since $\mathfrak{a}^h = \langle \omega \rangle$ Lemma 9.9 implies that

$$(\mathfrak{O}'\mathfrak{a})^h = \mathfrak{O}'(\mathfrak{a}^h) = \mathfrak{O}'\omega = \mathfrak{O}'\kappa^h = (\mathfrak{O}'\kappa)^h.$$

Uniqueness of factorization of ideals in $\mathfrak{O}'$ easily yields

$$\mathfrak{O}'\mathfrak{a} = \mathfrak{O}'\kappa,$$

proving (a).

Since (c) implies (b), we now consider (c). The inclusion $\mathfrak{a} \subseteq \mathbb{B}\kappa \cap K$ is straightforward. Conversely, suppose $\gamma \subseteq \mathbb{B}\kappa \cap K$. Then

$$\gamma = \lambda\kappa \quad (\lambda \in \mathbb{B})$$

and we must show that $\gamma \in \mathfrak{a}$. First note that, since $\gamma \in K$, $\kappa \in E$, we have $\lambda = \gamma\kappa^{-1} \in E$, so $\lambda \in E \cap \mathbb{B} = \mathfrak{O}'$. This gives

$$\gamma^h = \lambda^h\kappa^h = \lambda^h\omega \quad (\gamma \in K, \lambda \in \mathfrak{O}', \omega \in \mathfrak{O})$$

so $\gamma^h \in \mathbb{B}$, and by Theorem 2.10, $\gamma \in \mathbb{B}$. Thus $\gamma \in \mathbb{B} \cap K = \mathfrak{O}$. Considering the equation $\gamma^h = \lambda^h\omega$ again, we find

$$\lambda^h = \gamma^h\omega^{-1} \in K$$

so $\lambda^h \in K \cap \mathbb{B} = \mathfrak{O}$. Thus we finish up with

$$\gamma^h = \lambda^h\omega \quad (\gamma, \lambda^h, \omega \in \mathfrak{O}).$$

Taking ideals in $\mathfrak{O}$,

$$\langle \gamma \rangle^h = \langle \lambda^h \rangle \langle \omega \rangle = \langle \lambda^h \rangle \mathfrak{a}^h.$$

Unique factorization in $\mathfrak{O}$ implies that $\langle \lambda^h \rangle = \mathfrak{b}^h$ for some ideal $\mathfrak{b}$, so

$$\langle \gamma \rangle^h = \mathfrak{b}^h\mathfrak{a}^h.$$

Unique factorization once more implies that

$$\langle \gamma \rangle = \mathfrak{b}\mathfrak{a},$$

so $\gamma \in \mathfrak{a}$ as required.

To prove (d), observe that by Theorem 5.20, $\mathfrak{a} = \langle \alpha, \beta \rangle$ for $\alpha, \beta \in \mathfrak{O}$. Substituting in (d) gives

$$\mathfrak{O}''\gamma = \mathfrak{O}'' \langle \alpha, \beta \rangle .$$

Thus

$$\gamma = \lambda\alpha + \mu\beta$$

where $\lambda, \mu \in \mathfrak{O}''$, so certainly $\lambda, \mu \in \mathbb{B}$. From (a), $\alpha, \beta \in \mathfrak{O}'\kappa$, so

$$\alpha = \eta\kappa, \quad \beta = \zeta\kappa \quad (\eta, \zeta \in \mathfrak{O}' \subseteq \mathbb{B}).$$

Hence $\gamma = \lambda\eta\kappa + \mu\zeta\kappa$ and $\kappa | \gamma$ in $\mathbb{B}$. Finally, interchange the roles of γ, κ to prove (d). $\qquad\square$

Theorem 9.10 can be improved, for as it stands the extension ring $\mathfrak{O}'$ in which $\mathfrak{O}'\mathfrak{a}$ is principal depends on $\mathfrak{a}$. We can actually find a single extension ring in which the extension of every ideal is principal. This depends on the following lemma and the finiteness of the class-number:

Lemma 9.11. *If $\mathfrak{a}, \mathfrak{b}$ are equivalent ideals in the ring $\mathfrak{O}$ of integers of a number field and $\mathfrak{O}'\mathfrak{a}$ is principal, then so is $\mathfrak{O}'\mathfrak{b}$.*

Proof: By the definition of equivalence, there exist principal ideals $\mathfrak{d}, \mathfrak{e}$ of $\mathfrak{O}$ such that $\mathfrak{a}\mathfrak{d} = \mathfrak{b}\mathfrak{e}$. Hence

$$(\mathfrak{O}'\mathfrak{a})(\mathfrak{O}'\mathfrak{d}) = (\mathfrak{O}'\mathfrak{b})(\mathcal{D}'\mathfrak{e})$$

where now $\mathfrak{O}'\mathfrak{a}$, $\mathfrak{O}'\mathfrak{d}$, $\mathfrak{O}'\mathfrak{e}$ are all principal. Since the set $\mathcal{P}$ of principal fractional ideals of $\mathfrak{O}'$ is a group, $\mathfrak{O}'\mathfrak{b}$ is a principal fractional ideal which is also an ideal, so $\mathfrak{O}'\mathfrak{b}$ is a principal ideal. $\qquad\square$

Theorem 9.12. *Let K be a number field with integers $\mathfrak{O}_K$. Then there exists a number field $L \supseteq K$ with ring $\mathfrak{O}_L$ of integers such that for every ideal $\mathfrak{a}$ in $\mathfrak{O}_K$:*

(a) $\mathfrak{O}_L\mathfrak{a}$ *is a principal ideal.*

(b) $(\mathfrak{O}_L\mathfrak{a}) \cap \mathfrak{O}_K = \mathfrak{a}$.

Proof: Since h is finite, select a representative set of ideals $\mathfrak{a}_1, \ldots, \mathfrak{a}_h$, one from each class. Choose algebraic integers $\kappa_1, \ldots, \kappa_h$ such that $\mathfrak{O}_i\mathfrak{a}_i$ is principal where $\mathfrak{O}_i$ is the ring of integers of $K(\kappa_i)$. Let $L = K(\kappa_1, \ldots, \kappa_h)$; its ring of integers $\mathfrak{O}_L$ contains all the $\mathfrak{O}_i$. Hence each ideal $\mathfrak{O}_L\mathfrak{a}_i$ is principal in $\mathfrak{O}_L$. Since every ideal $\mathfrak{a}$ in $\mathfrak{O}$ is equivalent to some $\mathfrak{a}_i$, the ideal $\mathfrak{O}_L\mathfrak{a}$ is principal by Lemma 9.11. That is, for some $\alpha \in \mathbb{B}$

$$\mathfrak{O}_L\mathfrak{a} = \mathfrak{O}_L\alpha.$$

This proves (a).

Clearly $\mathfrak{a} \subseteq (\mathfrak{O}_L\mathfrak{a}) \cap \mathfrak{O}_K$. For the converse inclusion, Theorem 9.10(d) implies that $\alpha = u\kappa$ where u is a unit in $\mathbb{B}$. Now

$$
\begin{aligned}
(\mathfrak{O}_L\mathfrak{a}) \cap \mathfrak{O}_K &= (\mathfrak{O}_L\alpha) \cap \mathfrak{O}_K \\
&\subseteq (\mathbb{B}\alpha) \cap K \\
&= (\mathbb{B}\kappa) \cap K \\
&= \mathfrak{a}
\end{aligned}
$$

by Theorem 9.10(c).   □

For many years it was an open question, going back to David Hilbert, whether every number field can be embedded in one with unique factorization. However, in 1964 Golod and Šafarevič [35] showed that this is not always possible, citing the explicit example

$$\mathbb{Q}(\sqrt{(-3 \cdot 5 \cdot 7 \cdot 11 \cdot 13 \cdot 17 \cdot 19)}).$$

The proof is ingenious rather than hard, but it uses ideas we have not developed.

9.5 Unique Factorization of Elements in an Extension Ring

The results of the last section can be translated from principal ideals back to elements to give the version of Kummer's theory alluded to in the introduction to Chapter 5. There we considered examples of non-unique factorization such as

$$10 = 2 \cdot 5 = (5 + \sqrt{15})(5 - \sqrt{15})$$

in the ring of integers of $\mathbb{Q}(\sqrt{15})$. Viewing this as an equation in $\mathbb{Q}(\sqrt{3}, \sqrt{5})$, we saw that the factors can be further reduced as

$$2 = (\sqrt{5} + \sqrt{3})(\sqrt{5} - \sqrt{3})$$
$$5 = \sqrt{5}\sqrt{5}$$
$$5 + \sqrt{15} = \sqrt{5}(\sqrt{5} + \sqrt{3})$$
$$5 - \sqrt{15} = \sqrt{5}(\sqrt{5} - \sqrt{3})$$

and the two factorizations of 10 found above are just regroupings of the factors

$$10 = \sqrt{5}\sqrt{5}(\sqrt{5} + \sqrt{3})(\sqrt{5} - \sqrt{3}).$$

We now show that a similar phenomenon occurs for all non-unique prime factorizations in all rings of integers.

Theorem 9.13. *Suppose K is a number field with integers $\mathfrak{O}_K$. Then there exists an extension field $L \supseteq K$ with integers $\mathfrak{O}_L$ such that every non-zero, non-unit $a \in \mathfrak{O}_K$ has a factorization*

$$a = p_1 \ldots p_r \quad (p_i \in \mathfrak{O}_L)$$

where the p_i are non-units in $\mathfrak{O}_L$, and the following property is satisfied. Given any factorization in $\mathfrak{O}_K$:

$$a = a_1 \ldots a_s$$

where the a_i are non-units in $\mathfrak{O}_K$, there exist integers

$$1 \le n_1 < \ldots < n_s = r$$

and a permutation π of $\{1, \ldots, r\}$ such that the following elements are associates in $\mathfrak{O}_L$:

$$a_1, p_{\pi(1)} \cdots p_{\pi(n_1)}$$

$$\vdots$$

$$a_s, p_{\pi(n_{s-1}+1)} \cdots p_{\pi(n_s)}.$$

Remark. What this theorem says in plain language is that the factorizations of elements into irreducibles in $\mathfrak{O}_K$ may not be unique, but all factorizations of an element in $\mathfrak{O}_K$ come from different groupings of associates of a single factorization in $\mathfrak{O}_L$. In this sense elements in $\mathfrak{O}_K$ have unique factorization into elements in $\mathfrak{O}_L$.

Proof: There is a unique factorization of $\langle a \rangle$ into prime ideals in $\mathfrak{O}_K$, say

$$\langle a \rangle = \mathfrak{p}_1 \ldots \mathfrak{p}_r.$$

Since a is a non-unit, $r \geq 1$. Let $\mathfrak{O}_L$ be a ring of integers as in Theorem 9.12 where every ideal of $\mathfrak{O}_K$ extends to a principal ideal, and suppose that

$$\mathfrak{O}_L \mathfrak{p}_i = \mathfrak{O}_L p_i \quad (p_i \in \mathfrak{O}_L).$$

Then $a = u p_1 \ldots p_r$ where u is a unit in $\mathfrak{O}_L$, and since $r \geq 1$, we may replace p_1 by $u p_1 \in \mathfrak{O}_L p_1$ to get a factorization of the form

$$a = p_1 \ldots p_r.$$

Given any factorization of elements

$$a = a_1 \ldots a_s$$

where the a_i are non-units in $\mathfrak{O}_K$, we obtain

$$\langle a \rangle = \langle a_1 \rangle \ldots \langle a_s \rangle,$$

where all the $\langle a_i \rangle$ are proper ideals. Unique factorization in $\mathfrak{O}_K$ gives us the integers n_i and the permutation π such that

$$\langle a_1 \rangle = \mathfrak{p}_{\pi(1)} \cdots \mathfrak{p}_{\pi(n_1)}$$

$$\vdots$$

$$\langle a_s \rangle = \mathfrak{p}_{\pi(n_{s-1}+1)} \cdots \mathfrak{p}_{\pi(n_s)}.$$

Now take ideals in $\mathfrak{O}_L$ generated by these ideals. $\qquad \square$

Example 9.14. From the explicit computation of Section 2, if $K = \mathbb{Q}(\sqrt{-5})$, then $h = 2$ and a representative set of ideals is $\mathfrak{O}$, and $\langle 2, 1 + \sqrt{-5} \rangle$ where $\langle 2, 1 + \sqrt{-5} \rangle^2 = \langle 2 \rangle$. Hence we may take $L = K(\sqrt{2}) = \mathbb{Q}(\sqrt{-5}, \sqrt{2})$. Theorem 9.13 tells us that *every* element of $\mathbb{Z}[\sqrt{-5}]$ factorizes uniquely in the integers of $\mathbb{Q}(\sqrt{-5}, \sqrt{2})$. The case of the factorization of the element 6 is dealt with in Exercise 7 at the end of this chapter, where

$$6 = \sqrt{2}\sqrt{2}(\tfrac{1}{2}\sqrt{2} + \tfrac{1}{2}\sqrt{-10})(\tfrac{1}{2}\sqrt{2} - \tfrac{1}{2}\sqrt{-10}).$$

That $\tfrac{1}{2}\sqrt{2} \pm \tfrac{1}{2}\sqrt{-10}$ really are integers may be dealt with by computing the explicit minimum polynomials of these elements over $\mathbb{Q}$. Granted this, it is an easy matter to check that the two alternative factorizations

$$6 = 2 \cdot 3 = (1 + \sqrt{-5})(1 - \sqrt{-5})$$

in $\mathbb{Z}[\sqrt{-5}]$ are just different groupings of the factors in the integers of $\mathbb{Q}(\sqrt{-5}, \sqrt{2})$. (Do it!)

The above example underlines a basic problem when factorizing elements in an extension ring. We have not given a general method for computing the integers of a number field. To date we have explicitly calculated only the integers in quadratic and cyclotomic fields—and even those calculations were not trivial. There is also another weakness when factorizing elements in an extension ring. The elements p_i occurring in the factorization of a in Theorem 9.13 need not be irreducible. (For instance we might work in a slightly larger ring $\mathfrak{O}_{L'}$ containing $\sqrt{p_i}$; the method of adjoining $\kappa = \omega^{1/h}$ may very well add such roots.) However, the proof of Theorem 9.13 tells us that the factorization of the element $\mathfrak{a}$ in $\mathfrak{O}_L$ which gives the unique factorization properties is given by the factorization of the ideal $\langle a \rangle$ in the ring $\mathfrak{O}_K$. For this reason we may just as well stick to ideals in the original ring rather than embellish the situation by factorizing elements outside. Our computations in future will be concerned mainly with ideals—unique factorization of ideals proves so much easier to handle!

9.6   Exercises

1. Let $K = \mathbb{Q}(\sqrt{-5})$, and let $\mathfrak{p}$, $\mathfrak{q}$, $\mathfrak{r}$ be the ideals defined in Exercise 2 of Chapter 5 (page 126). Let $\mathcal{H}$ be the class group. Show that in $\mathcal{H}$

$$[\mathfrak{p}]^2 = [\mathfrak{O}], \quad [\mathfrak{p}][\mathfrak{q}] = [\mathfrak{O}], \quad [\mathfrak{p}][\mathfrak{r}] = [\mathfrak{O}],$$

and hence show that $\mathfrak{p}$, $\mathfrak{q}$, $\mathfrak{r}$ are equivalent.

2. Verify by explicit computation that $\mathfrak{p}$, $\mathfrak{q}$, $\mathfrak{r}$ are equivalent.

3. Using Corollary 9.6, show that for $K = \mathbb{Q}(\sqrt{-6})$ every ideal is equivalent to one of norm at most 3. Verify that

$$\langle 2 \rangle = \langle 2, \sqrt{-6} \rangle^2,$$
$$\langle 3 \rangle = \langle 3, \sqrt{-6} \rangle^2,$$

and conclude that the only ideals of norm 2, 3 are $\langle 2, \sqrt{-6} \rangle$, $\langle 3, \sqrt{-6} \rangle$. Deduce $h \le 3$ and using $\langle 2, \sqrt{-6} \rangle^2 = \langle 2 \rangle$, or otherwise, show $h = 2$.

4. Find principal ideals $\mathfrak{a}$, $\mathfrak{b}$ in $\mathbb{Z}[\sqrt{-6}]$ such that

$$\mathfrak{a} \langle 2, \sqrt{-6} \rangle = \mathfrak{b} \langle 3, \sqrt{-6} \rangle .$$

5. Find all squarefree integers d in $-10 < d < 10$ such that the class-number of $\mathbb{Q}(\sqrt{d})$ is 1. (*Hint:* look up a few theorems!)

6. Using methods similar to Exercise 3, calculate the class-number of $\mathbb{Q}(\sqrt{d})$ for d squarefree and $-10 \le d \le 10$.

7. Suppose $K = \mathbb{Q}(\sqrt{-5})$, $\mathfrak{p} = \langle 2, 1+\sqrt{-5}\rangle$. Let $\mathfrak{O}'$ be the ring of integers of $\mathbb{Q}(\sqrt{-5}, \sqrt{2})$. Show $\mathfrak{O}'\mathfrak{p} = \mathfrak{O}'\sqrt{2}$. Find explicit integers $a, b \in \mathfrak{O}'$ such that

$$2 = \sqrt{2}a, \quad 1 + \sqrt{-5} = \sqrt{2}b,$$

and verify that a, b *are* integers by computing the monic polynomials which they satisfy over $\mathbb{Q}$. Using the notation of Exercise 1, find $\kappa_1, \kappa_2 \in \mathfrak{O}'$ such that

$$\mathfrak{O}'\kappa_1 = \mathfrak{O}'\mathfrak{q}, \quad \mathfrak{O}'\kappa_2 = \mathfrak{O}'\mathfrak{r}$$

and use the factorization $\langle 6\rangle = \mathfrak{p}^2\mathfrak{q}\mathfrak{r}$ to factorize the element 6 in $\mathfrak{O}'$. Explain how this factorization relates to

$$6 = 2 \cdot 3 = (1 + \sqrt{-5})(1 + \sqrt{-5})$$

in $\mathbb{Z}[\sqrt{-5}]$.

8. In $\mathbb{Z}[\sqrt{-10}]$ we have the factorizations into irreducibles

$$14 = 2 \cdot 7 = (2 + \sqrt{-10})(2 - \sqrt{-10}).$$

Find an extension ring $\mathfrak{O}_L$ of $\mathbb{Z}[\sqrt{-10}]$ and a factorization of 14 in $\mathfrak{O}_L$ such that the given factorizations are found by different groupings of the factors.

9. Factorize $6 = 2 \cdot 3 = (4 + \sqrt{10})(4 - \sqrt{10}) \in \mathbb{Z}[\sqrt{10}]$ in an extension ring to exhibit the given factors as different groupings of the new ones.

10. Relate the factorization

$$10 = \sqrt{5}\sqrt{5}(\sqrt{5} + \sqrt{3})(\sqrt{5} - \sqrt{3})$$

in the integers of $\mathbb{Q}(\sqrt{3}, \sqrt{5})$ to the factorization of $\langle 10\rangle$ into prime ideals in the integers of $\mathbb{Q}(\sqrt{15})$. Explain how this gives rise to the different factorizations

$$10 = 2 \cdot 5 = (5 + \sqrt{15})(5 - \sqrt{15})$$

into irreducibles in the integers of $\mathbb{Q}(\sqrt{15})$.

III

Number-Theoretic
Applications

10

Computational Methods

The results of this chapter, although apparently diverse, all have a strong bearing on the question of practical computation of the class-number, within the limits of the techniques now at our command. We focus on computations performed by hand. For complicated calculations, mathematicians and computer scientists have developed many software packages for algebraic computations. The website numbertheory.org provides a lengthy list of specialized packages, along with links to other useful information such as tables.

In the first section we study a special case of how a rational prime breaks up into prime ideals in a number field. The second section supplements this by showing that the distinct classes of fractional ideals may be found from the prime ideals dividing a finite set of rational primes, this set being in some sense small provided the degree of K and its discriminant are not too large. Several specific cases are studied, especially quadratic fields: in particular we complete the list of fields $\mathbb{Q}(\sqrt{d})$ with negative d and with class-number 1. We do not prove this list is complete, however, because the methods required are beyond the scope of this book.

10.1   Factorization of a Rational Prime

If p is a prime number in $\mathbb{Z}$, it is not generally true that $\langle p \rangle$ is a prime ideal in the ring of integers $\mathfrak{O}$ of a number field K. For instance, in $\mathbb{Q}(\sqrt{-1})$ we have the factorization

$$\langle 2 \rangle = \langle 1 + \sqrt{-1} \rangle^2 .$$

It is obviously useful to compute the prime factors of $\langle p \rangle$. In the case where the ring of integers is generated by a single element, which includes quadratic and cyclotomic fields, the following theorem of Dedekind is decisive.

Theorem 10.1. *Let K be a number field of degree n with ring of integers $\mathfrak{O} = \mathbb{Z}[\theta]$ generated by $\theta \in \mathfrak{O}$. Given a rational prime p, suppose the minimum polynomial f of θ over $\mathbb{Q}$ gives rise to the factorization into irreducibles over $\mathbb{Z}_p$:*

$$\bar{f} = \bar{f}_1^{e_1} \dots \bar{f}_r^{e_r}$$

where the bar denotes the natural map $\mathbb{Z}[t] \to \mathbb{Z}_p[t]$. Then if $f_i \in \mathbb{Z}[t]$ is any polynomial mapping onto $\bar{f}_i$, the ideal

$$\mathfrak{p}_i = \langle p \rangle + \langle f_i(\theta) \rangle$$

is prime and the prime factorization of $\langle p \rangle$ in $\mathfrak{O}$ is

$$\langle p \rangle = \mathfrak{p}_1^{e_1} \dots \mathfrak{p}_r^{e_r}.$$

Proof: Let θ_i be a root of $\bar{f}_i$ in $\mathbb{Z}_p[\theta_i] \cong \mathbb{Z}_p[t]/\langle \bar{f}_i \rangle$. There is a natural map $\nu_i : \mathbb{Z}[\theta] \to \mathbb{Z}_p[\theta_i]$ given by

$$\nu_i(p(\theta)) = \bar{p}(\theta_i).$$

The image of ν_i is $\mathbb{Z}_p[\theta_i]$, which is a field, so $\ker \nu_i$ is a prime ideal of $\mathbb{Z}[\theta] = \mathfrak{O}$. Clearly

$$\langle p \rangle + \langle f_i(\theta) \rangle \subseteq \ker \nu_i.$$

But if $g(\theta) \in \ker \nu_i$, then $\bar{g}(\theta_i) = 0$, so $\bar{g} = \bar{f}_i \bar{h}$ for some $\bar{h} \in \mathbb{Z}_p[t]$; this means that $g - f_i h \in \mathbb{Z}[t]$ has coefficients divisible by p. Thus

$$g(\theta) = (g(\theta) - f_i(\theta)h(\theta)) + f_i(\theta)h(\theta) \in \langle p \rangle + \langle f_i(\theta) \rangle ,$$

showing that

$$\ker \nu_i = \langle p \rangle + \langle f_i(\theta) \rangle .$$

Let

$$\mathfrak{p}_i = \langle p \rangle + \langle f_i(\theta) \rangle .$$

Then for each $\bar{f}_i$ the ideal $\mathfrak{p}_i$ is prime and satisfies $\langle p \rangle \subseteq \mathfrak{p}_i$, that is, $\mathfrak{p}_i | \langle p \rangle$.
    For any ideals $\mathfrak{a}, \mathfrak{b}_1, \mathfrak{b}_2$,

$$(\mathfrak{a} + \mathfrak{b}_1)(\mathfrak{a} + \mathfrak{b}_2) \subseteq \mathfrak{a} + \mathfrak{b}_1\mathfrak{b}_2,$$

so by induction

$$\mathfrak{p}_1^{e_1} \dots \mathfrak{p}_r^{e_r} \subseteq \langle p \rangle + \langle f_1(\theta)^{e_1} \dots f_r(\theta)^{e_r} \rangle$$
$$\subseteq \langle p \rangle + \langle f(\theta) \rangle$$
$$= \langle p \rangle.$$

Thus $\langle p \rangle \mid \mathfrak{p}_1^{e_1} \dots \mathfrak{p}_r^{e_r}$, and the only prime factors of $\langle p \rangle$ are $\mathfrak{p}_1, \dots, \mathfrak{p}_r$, showing that

$$\langle p \rangle = \mathfrak{p}_1^{k_1} \dots \mathfrak{p}_r^{k_r} \tag{10.1}$$

where

$$0 < k_i \leq e_i \quad (1 \leq i \leq r). \tag{10.2}$$

The norm of $\mathfrak{p}_i$ is, by definition, $|\mathfrak{O}/\mathfrak{p}_i|$, and the isomorphisms

$$\mathfrak{O}/\mathfrak{p}_i = \mathbb{Z}[\theta]/\mathfrak{p}_i \cong \mathbb{Z}_p[\theta_i]$$

imply that

$$N(\mathfrak{p}_i) = |\mathbb{Z}_p[\theta_i]| = p^{d_i}$$

where $d_i = \partial \bar{f}_i$, or equivalently $d_i = \partial f_i$. Also

$$N(\langle p \rangle) = |\mathbb{Z}[\theta]/\langle p \rangle)| = p^n,$$

so, taking norms in Equation (10.1),

$$p^n = N(\langle p \rangle) = N(\mathfrak{p}_1)^{k_1} \dots N(\mathfrak{p}_r)^{k_r} = p^{d_1 k_1 + \dots + d_r k_r},$$

which implies that

$$d_1 k_1 + \dots + d_r k_r = n = d_1 e_1 + \dots + d_r e_r. \tag{10.3}$$

Equation (10.2) leads to $k_i = e_i$ $(1 \leq i \leq r)$. $\qquad\qquad\square$

This result is not always applicable, since in general $\mathfrak{O}$ need not be of the form $\mathbb{Z}[\theta]$: see Section 2.6 Example 2.23. But for quadratic or cyclotomic fields we have already shown that $\mathfrak{O} = \mathbb{Z}[\theta]$, so the theorem applies in these cases—and in many others. It also has the advantage of computability. Since there is only a finite number of polynomials over $\mathbb{Z}_p$ of given degree, the factorization of $\bar{f}$ can be performed in a finite number of steps. A little native wit helps, but, if the worst comes to the worst, there is only a finite number of polynomials of lower degree than $\bar{f}$ to try as factors.

For example, in $\mathbb{Q}(\sqrt{-1})$ we have $\mathfrak{O} = \mathbb{Z}[\theta]$ where θ has minimum polynomial

$$t^2 + 1.$$

To find the factorization of $\langle 2 \rangle$ we look at this polynomial (mod 2), where

$$t^2 + 1 = (t + 1)^2.$$

Hence $\langle 2 \rangle = \mathfrak{p}^2$ where

$$\mathfrak{p} = \langle 2 \rangle + \langle \sqrt{-1} + 1 \rangle$$
$$= \langle 1 + \sqrt{-1} \rangle$$

(because $2 = (1 + \sqrt{-1})(1 - \sqrt{-1})$), and we recover the example noted at the beginning of this section.

More generally, consider the factorization in $\mathbb{Z}[\sqrt{-1}]$ of a prime $p \in \mathbb{Z}$. There are three cases to consider:

(1) $t^2 + 1$ irreducible (mod p)

(2) $t^2 + 1 \equiv (t - \lambda)(t + \lambda) \pmod{p}$, (where $\lambda^2 \equiv -1 \pmod{p}$) and $\lambda \not\equiv -\lambda$ (i.e. $p \neq 2$)

(3) $t^2 + 1 \equiv (t + 1)^2 \pmod{2}$ when $p = 2$

In case (1) $\langle p \rangle$ is prime; in case (2) $\langle p \rangle = \mathfrak{p}_1 \mathfrak{p}_2$ for distinct prime ideals $\mathfrak{p}_1, \mathfrak{p}_2$; in case (3) $\langle p \rangle = \mathfrak{p}_1^2$ for a prime ideal $\mathfrak{p}_1$.

The distinction between cases (1) and (2) is whether -1 is congruent to a square (mod p). In Appendix A on quadratic residues we show that (1) applies if p is of the form $4k - 1$ ($k \in \mathbb{Z}$), and (2) applies if p is of the form $4k + 1$ ($k \in \mathbb{Z}$).

The results in this section are, in fact, the tip of the iceberg of a large and significant portion of algebraic number theory. Given a prime ideal $\mathfrak{p}$ in the ring $\mathfrak{O}_K$ of integers in a number field K, we may consider the extension ideal $\mathfrak{O}_L \mathfrak{p}$ in the ring of integers $\mathfrak{O}_L$ of an extension algebraic number field. We find

$$\mathfrak{O}_L \mathfrak{p} = \mathfrak{q}_1^{e_1} \ldots \mathfrak{q}_r^{e_r}$$

where $\mathfrak{q}_1, \ldots, \mathfrak{q}_r$ are distinct prime ideals in $\mathfrak{O}_L$.

10.2   Minkowski Constants

The proof of Theorem 9.5 leaves room for improvement, because it is based on Lemma 9.2, which is far stronger than we really need. What we want is a point α such that

$$|\sigma_1(\alpha)| \ldots |\sigma_s(\alpha)| \, |\sigma_{s+1}(\alpha)|^2 \ldots |\sigma_{s+t}(\alpha)|^2 < c_1 \ldots c_{s+t}, \qquad (10.4)$$

but what we actually find is a point α satisfying the considerably stronger restriction

$$
|\sigma_1(\alpha)| < c_1, \ldots, |\sigma_s(\alpha)| < c_s,
$$
$$
|\sigma_{s+1}(\alpha)|^2 < c_{s+1}, \ldots, |\sigma_{s+t}(\alpha)|^2 < c_{s+t}. \tag{10.5}
$$

Certainly the inequalities (10.5) imply (10.4), but not the reverse.

The reason for using (10.5) is that we wish to employ Minkowski's theorem. For (10.5) the relevant set of points in $\mathbb{L}^{st}$ is convex and symmetric, so the theorem applies; but for (10.4) the relevant set, though symmetric, is not convex. This means we cannot use (10.5) directly. The gap between (10.4) and (10.5) is so great, however, that we might hope to find another set of inequalities, corresponding to a convex subset of $\mathbb{L}^{st}$, and implying (10.4): this would lead to improved estimates in Theorem 9.5 and Corollary 9.6.

To do this, we use the well-known inequality between arithmetic and geometric means:

$$
(a_1 \ldots a_n)^{1/n} \leq \frac{1}{n}(a_1 + \ldots + a_n). \tag{10.6}
$$

The result is:

Theorem 10.2. *If $\mathfrak{a} \neq 0$ is an ideal of $\mathfrak{O}$ then $\mathfrak{a}$ contains an element α with*

$$
|N(\alpha)| \leq \left(\frac{4}{\pi}\right)^t \cdot \frac{n!}{n^n}\sqrt{|\Delta|}N(\mathfrak{a}),
$$

where n is the degree of K and Δ is the discriminant.

Proof: Let X_c be the set of all $x \in \mathbb{L}^{st}$ such that

$$
|x_1| + \ldots + |x_s| + 2\sqrt{(y_1^2 + z_1^2)} + \ldots + 2\sqrt{(y_t^2 + z_t^2)} < c,
$$

where c is a positive real number. Then X_c is convex and centrally symmetric, and it is a routine though non-trivial exercise to compute

$$
v(X_c) = 2^s \left(\frac{\pi}{2}\right)^t \cdot \frac{1}{n!}c^n
$$

using induction and a change to polar coordinates. For details see Lang [47] page 116.

By Minkowski's theorem, X_c contains a point $\alpha \neq 0$ of $\sigma(\mathfrak{a})$ provided that

$$v(X_c) > 2^{s+2t}v(T),$$

where T is a fundamental domain for $\sigma(\mathfrak{a})$. By Theorem 9.4

$$v(T) = 2^{-t}N(\mathfrak{a})\sqrt{|\Delta|}$$

so the condition on X_c becomes

$$2^s\left(\frac{\pi}{2}\right)^t \cdot \frac{1}{n!}c^n > 2^{s+2t}2^{-t}N(\mathfrak{a})\sqrt{|\Delta|},$$

which is

$$c^n > \left(\frac{4}{\pi}\right)^t n!N(\mathfrak{a})\sqrt{|\Delta|}.$$

For such an α

$$|N(\alpha)| = |\sigma_1(\alpha)\ldots\sigma_s(\alpha)\sigma_{s+1}(\alpha)^2\ldots\sigma_{s+t}(\alpha)^2| \leq \left(\frac{c}{n}\right)^n$$

by the inequality between arithmetic and geometric means.

Using ϵ's as in Theorem 9.5, we may assume that α can be found for

$$c^n = \left(\frac{4}{\pi}\right)^t n!N(\mathfrak{a})\sqrt{|\Delta|}$$

and then

$$|N(\alpha)| \leq \left(\frac{4}{\pi}\right)^t \cdot \frac{n!}{n^n}N(\mathfrak{a})\sqrt{|\Delta|}.$$

□

The geometric considerations involved in the choice of X_c in this proof are illustrated in Figure 10.1 for the case $n = 2$, $s = 2$, $t = 0$. The three regions

A: $|xy| \leq 1$

B: $\frac{|x|+|y|}{2} \leq 1$

C: $|x| \leq 1$, $|y| \leq 1$

correspond respectively to the inequality (10.4), the region chosen in the proof of Theorem 10.2, and the inequality (10.5). Note that A is not convex, although B, C are; that $C \subseteq B \subseteq A$; and that B is much larger than C (which is why it leads to a better estimate).

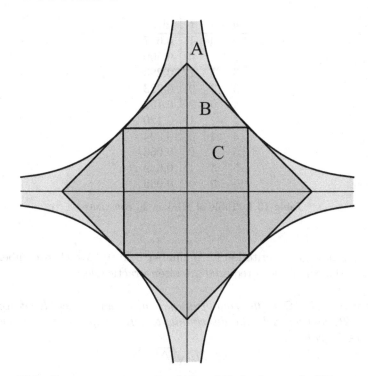

Figure 10.1. Geometry suggests the choice of X_c in the proof of Theorem 10.2, here illustrated for $n = s = 2, t = 0$. Region B is convex, lies within region A, and is larger than the more obvious region C. Therefore the use of B, in conjunction with Minkowski's theorem, yields a better bound.

Corollary 10.3. *Every class of fractional ideals contains an ideal* $\mathfrak{a}$ *with*

$$N(\mathfrak{a}) \leq \left(\frac{4}{\pi}\right)^t \cdot \frac{n!}{n^n} \sqrt{|\Delta|}.$$

Proof: As for Corollary 9.6.   □

This result suggests the introduction of *Minkowski constants*

$$M_{st} = \left(\frac{4}{\pi}\right)^t \frac{(s + 2t)!}{(s + 2t)^{s+2t}}.$$

For future use, we give a short table of their values, taken from Lang [47]. The numbers in the last column have all been rounded *upwards* in the third decimal place, to avoid underestimates.

n	s	t	M_{st}
2	0	1	0.637
2	2	0	0.500
3	1	1	0.283
3	3	0	0.223
4	0	2	0.152
4	2	1	0.120
4	4	0	0.094
5	1	2	0.063
5	3	1	0.049
5	5	0	0.039

Table 10.1. Table of Minkowski constants.

We can now give a criterion for a number field to have class-number 1, for which the calculations required are often practicable.

Theorem 10.4. *Let $\mathfrak{D}$ be the ring of integers of a number field K of degree $n = s + 2t$, and let Δ be the discriminant of K. Suppose that for every prime $p \in \mathbb{Z}$ with*

$$p \leq M_{st}\sqrt{|\Delta|},$$

every prime ideal dividing $\langle p \rangle$ is principal. Then $\mathfrak{D}$ has class-number $h = 1$.

Proof: Every class of fractional ideals contains an ideal $\mathfrak{a}$ with $N(\mathfrak{a}) \leq M_{st}\sqrt{|\Delta|}$. Now

$$N(\mathfrak{a}) = p_1 \ldots p_k$$

where $p_1, \ldots, p_k \in \mathbb{Z}$ and $p_i \leq M_{st}\sqrt{|\Delta|}$. Further, $\mathfrak{a}|N(\mathfrak{a})$, so $\mathfrak{a}$ is a product of prime ideals, each dividing some p_i. By hypothesis these prime ideals are principal, so $\mathfrak{a}$ is principal. Therefore every class of fractional ideals is equal to $[\mathfrak{D}]$, and $h = 1$. $\square$

Specific numerical applications of this theorem, and related methods, are given in the next section.

10.3   Some Class-Number Calculations

Theorem 10.4 combines with Theorem 10.1 to provide a useful computational technique for fields of small degree and with small discriminant. The following examples show what is meant by 'small' in these circumstances.

1. $\mathbb{Q}(\sqrt{-19})$: The ring of integers is $\mathbb{Z}[\theta]$ where θ is a zero of

$$f(t) = t^2 - t + 5,$$

and the discriminant is -19. Then $M_{st}\sqrt{|\Delta|} \leq 0.637\sqrt{19}$ so Theorem 10.4 applies if we know the factors of primes ≤ 2. Now we use Theorem 10.1: modulo 2, $f(t)$ is irreducible, so $\langle 2 \rangle$ is prime in $\mathfrak{O}$ (and hence every prime ideal dividing $\langle 2 \rangle$ is equal to $\langle 2 \rangle$ so is principal); modulo 3, $f(t)$ is also irreducible, so $\langle 3 \rangle$ is prime and the same argument applies.

2. $\mathbb{Q}(\sqrt{-43})$: This is similar, but now

$$f(t) = t^2 - t + 11$$

and $M_{st}\sqrt{|\Delta|} \leq 0.637\sqrt{43}$ which involves looking at primes ≤ 4. But $f(t)$ is irreducible modulo 2 or 3.

3. $\mathbb{Q}(\sqrt{-67})$: For this,

$$f(t) = t^2 - t + 17$$

and $M_{st}\sqrt{|\Delta|} \leq 0.637\sqrt{67}$ which involves looking at primes ≤ 5. But $f(t)$ is irreducible modulo 2, 3, or 5.

4. $\mathbb{Q}(\sqrt{-163})$: Now

$$f(t) = t^2 - t + 41$$

and $M_{st}\sqrt{|\Delta|} \leq 0.637\sqrt{163}$ which involves looking at primes ≤ 8. But $f(t)$ is irreducible modulo 2, 3, 5, or 7.

Combining these results with Theorem 4.17 (or using the above methods for the other values of Δ) we have:

Theorem 10.5. *The class-number of $\mathbb{Q}(\sqrt{d})$ is equal to 1 for $d = -1, -2, -3, -7, -11, -19, -43, -67, -163$.* $\square$

As remarked in Section 4.3, these are in fact the only values of $d < 0$ for which $\mathbb{Q}(\sqrt{d})$ has unique factorization, or equivalently class-number 1.

Comparing with Theorem 4.18 we obtain the interesting:

Corollary 10.6. *There exist rings with unique factorization that are not Euclidean; for example, the rings of integers of* $\mathbb{Q}(\sqrt{d})$ *for* $d = -19, -43, -67,$ $-163.$ □

We can also deal with a few cyclotomic fields by the same method. If $K = \mathbb{Q}(\zeta)$ where $\zeta^p = 1$, p prime, then the degree of K is $p - 1$, and the ring of integers is $\mathbb{Z}[\zeta]$. For $p = 3$, $K = \mathbb{Q}(\sqrt{-3})$ and we already know $h = 1$ in this case.

5. $\mathbb{Q}(\zeta)$ where $\zeta^5 = 1$: Here $n = 4$, $s = 0$, $t = 2$; and $\Delta = 125$ by Theorem 3.6. Hence $M_{st}\sqrt{|\Delta|} \leq 0.152\sqrt{125}$ so we must look at primes ≤ 1. Since there are no such primes, Theorem 10.4 applies at once to give $h = 1$.

6. $\mathbb{Q}(\zeta)$ where $\zeta^7 = 1$: Here $n = 6$, $s = 0$, $t = 3$, and $\Delta = -7^5$. We have to look at primes ≤ 3. The ring of integers is $\mathbb{Z}[\zeta]$ where ζ is a zero of

$$f(t) = t^6 + t^5 + t^4 + t^3 + t^2 + t + 1.$$

Modulo 2, this factorizes as

$$(t^3 + t^2 + 1)(t^3 + t + 1)$$

so $\langle 2 \rangle = \mathfrak{p}_1\mathfrak{p}_2$ where $\mathfrak{p}_1, \mathfrak{p}_2$ are distinct prime ideals, by Theorem 10.1. In fact

$$(\zeta^3 + \zeta^2 + 1)(\zeta^3 + \zeta + 1)\zeta^4 = 2,$$

so we have

$$\langle 2 \rangle = \langle \zeta^3 + \zeta^2 + 1 \rangle \langle \zeta^3 + \zeta + 1 \rangle$$

and $\mathfrak{p}_1, \mathfrak{p}_2$ are principal.

Modulo 3, $f(t)$ is irreducible (by trying all possible divisors, or more enlightened methods), so $\langle 3 \rangle$ is prime.

Hence by Corollary 10.4, $h = 1$.

Similar methods often allow us to compute h, even when it is not 1.

7. $\mathbb{Q}(\sqrt{10})$: The discriminant $d = 40$, $n = 2$, $s = 2$, $t = 0$. Every class of ideals contains one with norm

$$\leq M_{2,0}\sqrt{|\Delta|} \leq 0.5\sqrt{40}$$

so we must factorize the primes ≤ 3. Now $\mathfrak{O} = \mathbb{Z}[\theta]$ where θ is a zero of

$$f(t) = t^2 - 10.$$

$f(t) \equiv (t+1)(t-1) \pmod 3$, so $\langle 3 \rangle = \mathfrak{g}_1 \mathfrak{g}_2$ where $\mathfrak{g}_1 = \langle 3, 1 + \sqrt{10} \rangle$, $\mathfrak{g}_2 = \langle 3, 1 - \sqrt{10} \rangle$. Modulo 2, $f(t) = t \cdot t$, so $\langle 2 \rangle = \mathfrak{p}^2$ for a prime ideal $\mathfrak{p}$. If $\mathfrak{p}$ is principal, say $\mathfrak{p} = \langle a + b\sqrt{10} \rangle$, then the equation

$$N(\mathfrak{p})^2 = N(\langle 2 \rangle) = 4$$

implies that $N(\mathfrak{p}) = 2$. Hence

$$a^2 - 10b^2 = \pm 2.$$

The latter, considered modulo 10, is impossible; hence $\mathfrak{p}$ is not principal.

We have $\mathfrak{p}\mathfrak{g}_1 = \langle -2 + \sqrt{10} \rangle$ and $[\mathfrak{g}_1] = [\mathfrak{p}]^{-1}$. Therefore every class of fractional ideals either contains a principal ideal or $\mathfrak{p}$, hence equals $[\mathfrak{O}]$ or $[\mathfrak{p}]$. Since $\mathfrak{p}$ is not principal, these two classes are distinct, so $h = 2$. The class-group is cyclic of order 2, and as verification

$$[\mathfrak{p}]^2 = [\mathfrak{p}^2] = [\langle 2 \rangle] = [\mathfrak{O}].$$

As we said in Section 4.4, all the imaginary quadratic fields $\mathbb{Q}(\sqrt{d})$ with unique factorization are now known, verifying a conjecture of Gauss. But Gauss also stated a more general conjecture, the *Class Number Problem*. This states that for any given class number h, the set of $d < 0$ for which $\mathbb{Q}(\sqrt{d}) = h$ is finite. It was proved in 1934 by Hans Heilbronn. A stronger result was proved in 1983 by Dorian Goldfeld, Benedict Gross, and Don Zagier, and is described in a masterly survey by Goldfeld [33].

10.4 Table of Class-Numbers

To give an idea of how irregularly the class-number of $\mathbb{Q}(\sqrt{d})$ depends upon d, we give a short table (Table 10.2) showing, for squarefree d with $0 < d < 100$, the class-numbers h of $\mathbb{Q}(\sqrt{d})$ and h' of $\mathbb{Q}(\sqrt{-d})$.

Methods more suited to such computations than ours above exist, especially analytic methods which are beyond our present scope. See Borevič and Šafarevič [8] p. 342 onwards.

d	h	h'	d	h	h'	d	h	h'
1	-	1	34	2	4	69	1	8
2	1	1	35	2	2	70	2	4
3	1	1	37	1	2	71	1	7
5	1	2	38	1	6	73	1	4
6	1	2	39	2	4	74	2	10
7	1	1	41	1	8	77	1	8
10	2	2	42	2	4	78	2	4
11	1	1	43	1	1	79	3	5
13	1	2	46	1	4	82	4	4
14	1	4	47	1	5	83	1	3
15	2	2	51	2	2	85	2	4
17	1	4	53	1	6	86	1	10
19	1	1	55	2	4	87	2	6
21	1	4	57	1	4	89	1	12
22	1	2	58	2	2	91	2	2
23	1	3	59	1	3	93	1	4
26	2	6	61	1	6	94	1	8
29	1	6	62	1	8	95	2	8
30	2	4	65	2	8	97	1	4
31	1	3	66	2	8			
33	1	4	67	1	1			

Table 10.2. Class-numbers h of $\mathbb{Q}(\sqrt{d})$ and h' of $\mathbb{Q}(\sqrt{-d})$.

10.5   Exercises

1. Let $K = \mathbb{Q}(\sqrt{3})$. Use Theorem 10.1 to factorize the following principal ideals in the ring $\mathfrak{O}$ of integers of K:

$$\langle 2 \rangle, \langle 3 \rangle, \langle 5 \rangle, \langle 10 \rangle, \langle 30 \rangle.$$

2. Factorize the following principal ideals in the ring of integers of $\mathbb{Q}(\sqrt{5})$:

$$\langle 2 \rangle, \langle 3 \rangle, \langle 5 \rangle, \langle 12 \rangle, \langle 25 \rangle.$$

3. Factorize the following ideals in $\mathbb{Z}[\zeta]$ where $\zeta = e^{2\pi i/5}$:

$$\langle 2 \rangle, \langle 5 \rangle, \langle 20 \rangle, \langle 50 \rangle.$$

4. Compute the volume integral quoted in the proof of Theorem 10.2.

5. If K is a number field of degree n, prove that

$$|\Delta| \geq \left(\frac{\pi}{4}\right)^n \left(\frac{n^n}{n!}\right)^2,$$

where Δ is the discriminant.

6. Prove that there exist only finitely many number fields with any given discriminant.

7. Using the methods of this chapter, compute the class-numbers of fields $\mathbb{Q}(\sqrt{d})$ for $-20 \leq d \leq 20$.

11

Kummer's Special Case of Fermat's Last Theorem

We now have sufficient machinery at our disposal to tackle Fermat's Last Theorem in a special case, namely when the exponent n in the equation $x^n + y^n = z^n$ is a so-called 'regular prime', and when n does not divide any of x, y, or z. We begin with a short historical survey to set this version of the problem in perspective. Following this we show how elementary methods dispose of the case $n = 4$ and reduce the problem to odd prime values of n. In this chapter we do not deal with the case where one of x, y, or z is divisible by n, neither do we deal with irregular prime n. These cases are described in Chapter 14. In a final discursive section we discuss the regularity property and some related matters.

11.1 Some History

The origins of Fermat's Last Theorem have been explained in the Introduction. Useful references for background reading are Stewart [77] and Ribenboim [64]. Fermat himself is considered to have disposed of the cases $n = 3, 4$, because he issued these specific cases as mathematical challenges to others. In fact he produced only one written proof in the whole of his mathematical career. This states that the area of a right-angled triangle with rational sides cannot be a perfect square. Algebraically, this statement translates to the assertion that there are no (non-zero) integer solutions x, y, z of the equation $x^2 + y^2 = z^2$ where $xy/2$ is a square. From this it is easy to deduce Fermat's Last Theorem for $n = 4$.

Euler (1706-1783) produced his own proof for the case $n = 3$ in his *Algebra* of 1770. However, his proof contained a subtle error. He needed to find cubes of the form $p^2 + 3q^2$, and ingeniously showed that, for any integers a and b, if we define

$$p = a^3 - 9ab^2 \qquad q = 3(a^2 b - b^3)$$

then

$$p^2 + 3q^2 = (a^2 + 3b^2)^3.$$

However, he then tried to show the reverse process also works, namely, if $p^2 + 3q^2$ is a perfect cube, then there exist integers a, b satisfying the above relationships. Here he worked with algebraic numbers of the form $x + y\sqrt{-3}$, with x, y integers, believing that these numbers possess the same properties as ordinary integers—including uniqueness of factorization. (As it happens, factorization *is* unique in this case, but Euler did not realise that this needed proving.) This omission went unnoticed at the time. However, other results that Euler published gave an alternative proof for $n = 3$, without logical gaps, thus justifying giving him full credit for this case.

Sophie Germain (1776–1831) was one of the very few women doing research in mathematics at this time. As a woman, she was unable to attend the École Polytechnique when it opened in Paris in 1794. Instead she assumed the identity of a student, 'Monsieur Antoine-Auguste Le Blanc', who had left the course without giving formal notice. So elegant and insightful were her written solutions of weekly problems that her ability was noted by Lagrange. He insisted on a meeting, which revealed her subterfuge. He gave her positive encouragement, and she developed a serious interest in Fermat's Last Theorem. Early in her work, she found it convenient to divide the problem into two cases:

(1) None of x, y, z is divisible by n.

(2) Only one of x, y, z is divisible by n.

If two of x, y, z are divisible by n then all three are, and if all three are, then the common factor n^n can be divided out, so only these cases are needed for a proof. Legendre credits her with a proof of the 'first case' if there exists an auxiliary prime satisfying two technical conditions, and states that she verified these conditions for all $n \leq 97$. Thus she proved the first case of Fermat's Last Theorem for all n in that range.

Attention then turned to case (2). A partial proof of this case for $n = 5$ was presented to the Paris Academy by Dirichlet in July 1825. Legendre filled in the other details in September 1825, hence completing the full proof for $n = 5$. Dirichlet continued to work on the case $n = 7$, only to realise that the closely related case $n = 14$ was more amenable to his methods. He published the proof for $n = 14$ in 1832. The case $n = 7$ was finally proved in 1839 by Lamé (1796–1870). It required far more subtle computations than those of earlier cases and gave the impression that further progress would be unlikely unless a completely different line of attack was found. But the next major step forward was followed by an immediate retreat.

On 1 March 1847, Lamé addressed the Paris Academy and announced a complete proof of Fermat's Last Theorem. He outlined a proof which introduced the complex nth roots of unity and factorized the equation $x^n + y^n = z^n$ into linear terms

$$x^n + y^n = (x + y)(x + \zeta y) \ldots (x + \zeta^{n-1} y)$$

where $\zeta = e^{2\pi i/n}$ and n is odd. Lamé acknowledged that he was indebted to Liouville for this idea.

Then Liouville took the stage. He acknowledged his contribution, but pointed out that the argument used depended on uniqueness of factorization, and he suspected that this property might fail. Immediately the focus turned to unique factorization. A fortnight later Pierre Wantzel announced a proof of unique factorization for some cases, providing arguments for $n = 2, 3, 4$. He also stated that his method of proof failed for $n = 23$ (see Cauchy [13] p. 308). On 24 May Liouville informed the Academy that Kummer had already shown the failure of unique factorization three years before, but had developed a technical alternative that worked by introducing what he called 'ideal numbers'.

In 1850 Kummer produced his sensational proof of Fermat's Last Theorem for what he termed 'regular' primes, including all primes less than 100 except for 37, 59, 67. Kummer asserted that there is an infinite number of regular primes, but this has never been proved (although Johan Jensen proved that there is an infinite number of *irregular* primes in 1915). The same year, Kummer attended to the three cases 37, 59, 67, but made errors that went unnoticed until Harry Vandiver found them in 1920. A proof for $n = 37$ was given by Dimitri Mirimanoff in 1893, and he extended this as far as $n \leq 257$ in 1905. Vandiver laid down methods that made a computational approach possible, which led to proofs for $n \leq 25,000$ by Selfridge and Pollock [73], then $n \leq 125,000$ by Wagstaff [86]. By 1993 the record was $n \leq 4,000,000$, Buhler *et al.* [11].

The proof by Kummer therefore occupies a pivotal position in the development of Fermat's Last Theorem. It changed the focus from increasingly complicated ways of dealing with small values of n, using a variety of methods, before 1850, towards a more general proof for a wide variety of values of n in the late 19th and 20th centuries. Before moving on to entirely new techniques, it is therefore worth taking a detailed look at Kummer's proof and the methods behind it.

11.2 Elementary Considerations

We first consider what can be said about the *Fermat equation*

$$x^n + y^n = z^n \tag{11.1}$$

from an elementary point of view. If a solution to Equation (11.1) exists then there must exist one solution in which x, y, z are coprime in pairs. For if a prime q divides x and y, then $x = qx'$, $y = qy'$,

$$q^n(x'^n + y'^n) = z^n$$

so that q also divides z, say $z = qz'$, and then $x'^n + y'^n = z'^n$. Similarly if q divides x and z, or y and z. In this way we can remove all common factors from x, y, z.

Next note that if Equation (11.1) is impossible for an exponent n then it is impossible for all multiples of n. For if $x^{mn} + y^{mn} = z^{mn}$ then $(x^m)^n + (y^m)^n = (z^m)^n$. Now any integer ≥ 3 is divisible either by 4 or by an odd prime. Hence to prove (or disprove) the conjecture *it is sufficient to consider the cases $n = 4$ and n an odd prime*.

We start with Fermat's proof for $n = 4$. It is based on the (well known) general solution of the Pythagorean equation $x^2 + y^2 = z^2$, given by:

Lemma 11.1. *The solutions of $x^2 + y^2 = z^2$ with pairwise coprime integers x, y, z are given parametrically by*

$$
\begin{aligned}
\pm x &= 2rs \\
\pm y &= r^2 - s^2 \\
\pm z &= r^2 + s^2
\end{aligned}
$$

(or with x, y interchanged) where r, s are coprime and exactly one is odd.

Proof: We give the classical proof. It is sufficient to consider x, y, z positive. They cannot all be odd, for this gives the contradiction 'odd + odd = odd'. Since they are pairwise coprime, precisely one is even. It cannot be z, for then $z = 2k$, $x = 2a + 1$, $y = 2b + 1$ where k, a, b are rational integers, and

$$(2a + 1)^2 + (2b + 1)^2 = 4k^2.$$

This cannot occur since the left-hand side is clearly not divisible by 4 whilst the right-hand side is. So one of x, y is even. We can suppose that this is x. Then

$$x^2 = z^2 - y^2 = (z+y)(z-y).$$

Because x, $z+y$, $z-y$ are all even and positive, we can write $x = 2u$, $z + y = 2v$, $z - y = 2w$, whence

$$(2u)^2 = 2v \cdot 2w,$$

or

$$u^2 = vw. \tag{11.2}$$

Now v, w are coprime, for a common factor of v, w would divide their sum $v + w = z$ and their difference $v - w = y$, which have no proper common factors. Factorizing u, v, w into prime factors, we see that (11.2) implies v, w are both squares, say $v = r^2$, $w = s^2$. Moreover r, s are coprime because v, w are.

Thus

$$z = v + w = r^2 + s^2,$$
$$y = v - w = r^2 - s^2.$$

Because y, z are both odd, precisely one of r, s is odd. Finally

$$x^2 = z^2 - y^2 = (r^2 + s^2)^2 - (r^2 - s^2)^2 = 4r^2 s^2,$$

so

$$x = 2rs.$$

$\square$

Now we can prove a theorem even stronger than the impossibility of Equation (11.1) for $n = 4$, namely:

Theorem 11.2. *The equation $x^4 + y^4 = z^2$ has no integer solutions with $x, y, z \neq 0$.*

Proof: First note that this *is* stronger, since if $x^4 + y^4 = z^4$ then x, y, z^2 satisfy the above equation.

Suppose a solution of

$$x^4 + y^4 = z^2 \tag{11.3}$$

exists. We may assume x, y, z are positive. Among such solutions there exists one for which z is smallest: assume we have this one in (11.3). Then

x, y, z are coprime (or we can cancel a common factor and make z smaller) and so by Lemma 11.1

$$x^2 = r^2 - s^2, \quad y^2 = 2rs, \quad z = r^2 + s^2,$$

where x, z are odd and y is even. The first of these implies

$$x^2 + s^2 = r^2$$

with x, s coprime. Hence by Equation (11.1) again, since x is odd

$$x = a^2 - b^2, \quad s = 2ab, \quad r = a^2 + b^2.$$

But now we substitute back to get

$$y^2 = 2rs = 2 \cdot 2ab(a^2 + b^2)$$

so y is even, say $y = 2k$, and

$$k^2 = ab(a^2 + b^2).$$

Since a, b and $a^2 + b^2$ are pairwise coprime

$$a = c^2, \quad b = d^2, \quad a^2 + b^2 = e^2,$$

so that

$$c^4 + d^4 = e^2.$$

This is an equation of type (11.3), but $e \le a^2 + b^2 = r < z$, contradicting minimality of z.   □

11.3 Kummer's Lemma

This section begins the build-up to the solution of a special case of Fermat's Last Theorem, with a detailed study of the field $K = \mathbb{Q}(\zeta)$ where $\zeta = e^{2\pi i/p}$ for an odd prime p. As in Chapter 3 we write

$$\lambda = 1 - \zeta.$$

Further we define

$$\mathfrak{l} = \langle \lambda \rangle \, ,$$

the ideal generated by λ in the ring of integers $\mathbb{Z}[\zeta]$ of K. We start with some properties of $\mathfrak{l}$.

Lemma 11.3. (a) $\mathfrak{l}^{p-1} = \langle p \rangle$.
  (b) $N(\mathfrak{l}) = p$.

Proof: For $j = 1, \ldots, p-1$ the numbers $1 - \zeta$ and $1 - \zeta^j$ are associates in $\mathbb{Z}[\zeta]$. Clearly $1 - \zeta | 1 - \zeta^j$. But if we choose t such that $jt \equiv 1 \pmod{p}$ then $1 - \zeta = 1 - \zeta^{jt}$ so that $1 - \zeta^j | 1 - \zeta$. Hence they are associates.
  Now equation (3.9) of Chapter 3 leads to

$$\langle p \rangle = \prod_{j=1}^{p-1} \langle 1 - \zeta^j \rangle$$

but the above remarks show that $\langle 1 - \zeta^j \rangle = \langle 1 - \zeta \rangle = \mathfrak{l}$, so

$$\langle p \rangle = \mathfrak{l}^{p-1}$$

and (a) is proved. Part (b) is immediate on taking norms. $\square$

Part (b) has a useful consequence. It implies that $|\mathbb{Z}[\zeta]/\mathfrak{l}| = p$, from which it follows (on looking at the natural homomorphism $\mathbb{Z}[\zeta] \to \mathbb{Z}[\zeta]/\mathfrak{l}$) that every element of $\mathbb{Z}[\zeta]$ is congruent modulo $\mathfrak{l}$ to one of $0, 1, 2, \ldots, p-1$.
  The main aim of the rest of this section is to give a useful, though incomplete, description of the units of $\mathbb{Z}[\zeta]$. We start by finding which roots of unity occur, showing that there are no 'accidental' occurrences:

Lemma 11.4. *The only roots of unity in K are $\pm\zeta^s$ for integers s.*

Proof: First we show $i \notin K$ by arguing for a contradiction. If, on the contrary, $i \in K$, then $2 = i(1 - i)^2$, so

$$\langle 2 \rangle = \langle 1 - i \rangle^2 \, .$$

Hence when $\langle 2 \rangle$ is resolved into prime factors in $\mathbb{Z}[\zeta]$ it has repeated factors. Theorem 10.1 implies that the polynomial

$$f(t) = \frac{t^p - 1}{t - 1}$$

has a repeated irreducible factor modulo 2, hence that $t^p - 1$ has a repeated irreducible factor modulo 2. Then the remark following Theorem 1.5 tells us that $t^p - 1$ and $D(t^p - 1) = pt^{p-1}$ are not coprime. However, p is odd, so these polynomials modulo 2 take the form $t^p + 1$, t^{p-1} which are obviously coprime. This is a contradiction.

In exactly the same way we can show that for an odd prime $q \neq p$,

$$e^{2\pi i/q} \notin K.$$

We just use

$$\langle q \rangle = \left(1 - e^{2\pi i/q}\right)^{q-1}.$$

Next we remark that

$$e^{2\pi i/p^2} \notin K$$

because $e^{2\pi i/p^2}$ satisfies $t^{p^2} - 1 = 0$, but not $t^p - 1 = 0$, so it is a zero of

$$f(t) = (t^{p^2} - 1)/(t^p - 1) = \sum_{r=0}^{p-1} t^{rp}.$$

Applying Eisenstein's criterion to $f(t + 1)$, a little arithmetic shows that $f(t+1)$, hence also $f(t)$, is irreducible. Thus f is the minimum polynomial of $e^{2\pi i/p^2}$. Since $[K : \mathbb{Q}] = p - 1$, Theorems 1.10 and 1.11 imply that $e^{2\pi i/p^2} \notin K$.

Suppose now that $e^{2\pi i/m} \in K$ for an integer m. Then the above results show that

$$4 \nmid m, \quad q \nmid m, \quad p^2 \nmid m.$$

Hence $m | 2p$ which leads at once to the desired result. $\qquad\square$

Lemma 11.5. *For each $\alpha \in \mathbb{Z}[\zeta]$ there exists $a \in \mathbb{Z}$ such that*

$$\alpha^p \equiv a \pmod{\mathfrak{l}^p}.$$

Proof: We have already remarked on the existence of $b \in \mathbb{Z}$ such that $\alpha \equiv b \pmod{\mathfrak{l}}$. Now

$$\alpha^p - b^p = \prod_{j=0}^{p-1} (\alpha - \zeta^j b)$$

and since $\zeta \equiv 1 \pmod{\mathfrak{l}}$ each factor on the right is congruent to $\alpha - b \equiv 0 \pmod{\mathfrak{l}}$. Multiplying up, $\alpha^p - b^p \equiv 0 \pmod{\mathfrak{l}^p}$.

Next comes a curious result about polynomials and roots of unity:

Lemma 11.6. *If $p(t) \in \mathbb{Z}[t]$ is a monic polynomial, all of whose zeros in $\mathbb{C}$ have absolute value 1, then every zero is a root of unity.*

Proof: Let $\alpha_1, \ldots, \alpha_k$ be the zeros of $p(t)$. For each integer $l > 0$ the polynomial

$$p_l(t) = (t - \alpha_1^l) \ldots (t - \alpha_k^l)$$

lies in $\mathbb{Z}[t]$ by the usual argument on symmetric polynomials. Now if

$$p_l(t) = t^k + a_{k-1}t^{k-1} + \ldots + a_0$$

then

$$|a_j| \leq \binom{k}{j} \quad (j = 0, \ldots, k - 1)$$

by estimating the size of elementary symmetric polynomials in the α_j and using $|\alpha_j| = 1$. But only finitely many distinct polynomials over $\mathbb{Z}$ can satisfy this system of inequalities, so for some $m \neq l$ we must have

$$p_l(t) = p_m(t).$$

Hence there exists a permutation π of $\{1, \ldots, k\}$ such that

$$\alpha_j^l = \alpha_{\pi(j)}^m$$

for $j = 1, \ldots, k$. Inductively,

$$\alpha_j^{l^r} = \alpha_{\pi^r(j)}^{m^r}.$$

Since $\pi^{k!}(j) = j$, we have $\alpha_j^{l^{k!}} = \alpha_j^{m^{k!}}$ so

$$\alpha_j^{(l^{k!} - m^{k!})} = 1.$$

Since $l^{k!} \neq m^{k!}$ it follows that α_j is a root of unity. $\square$

Now we may prove the main result of this section, known as *Kummer's lemma*:

Lemma 11.7. Kummer's Lemma.   *Every unit of $\mathbb{Z}[\zeta]$ is of the form $r\zeta^g$ where r is real and g is an integer.*

Proof: Let ϵ be a unit in $\mathbb{Z}[\zeta]$. There exists a polynomial $e(t) \in \mathbb{Z}[t]$ such that $\epsilon = e(\zeta)$. For $s = 1, \ldots, p - 1$ the elements

$$\epsilon_s = e(\zeta^s)$$

are conjugate to ϵ. Now $1 = \pm N(\epsilon) = \pm \epsilon_1 \ldots \epsilon_{p-1}$, so each ϵ_s is also a unit. Further, if bars denote complex conjugation,

$$\epsilon_{p-s} = e(\zeta^{p-s}) = e(\zeta^{-s}) = e(\overline{\zeta^s}) = \overline{e(\zeta^s)} = \overline{\epsilon_s}.$$

Therefore

$$\epsilon_s \epsilon_{p-s} = |\epsilon_s|^2 > 0.$$

Then

$$\pm 1 = N(\epsilon) = (\epsilon_1 \epsilon_{p-1})(\epsilon_2 \epsilon_{p-2}) \ldots > 0$$

so $N(\epsilon) = 1$.

Now each $\epsilon_s/\epsilon_{p-s}$ is a unit, of absolute value 1, and by a symmetric polynomial argument

$$\prod_{s=1}^{p-1} (t - \epsilon_s/\epsilon_{p-s})$$

has coefficients in $\mathbb{Z}$. By Lemma 11.6 its zeros are roots of unity. Lemma 11.4 yields

$$\epsilon/\epsilon_{p-1} = \pm \zeta^u$$

for integer u. Since p is odd either u or $u + p$ is even, so

$$\epsilon/\epsilon_{p-1} = \pm \zeta^{2g} \qquad (11.4)$$

for $0 < g \in \mathbb{Z}$.

The crucial step now is to find out whether the sign in (11.4) is positive or negative. To do this we work out the left-hand side modulo $\mathfrak{l}$, as follows. We know that for some $v \in \mathbb{Z}$

$$\zeta^{-g}\epsilon \equiv v \quad (\mathrm{mod}\ \mathfrak{l}).$$

Taking complex conjugates,

$$\zeta^g \epsilon_{p-1} \equiv v \quad (\mathrm{mod}\ \langle \bar{\lambda} \rangle).$$

But $\bar{\lambda} = 1 - \zeta^{p-1}$ is an associate of λ, so in fact $\langle \bar{\lambda} \rangle = \mathfrak{l}$. Eliminate v to get

$$\epsilon/\epsilon_{p-1} \equiv \zeta^{2g} \quad (\mathrm{mod}\ \mathfrak{l}).$$

A negative sign in Equation (11.4) leads to

$$\mathfrak{l} | 2\zeta^{2g}.$$

Taking norms,

$$N(\mathfrak{l}) | 2^{p-1}$$

which contradicts Lemma 11.3(b). So the sign in (11.4) is positive. Hence

$$\zeta^{-g}\epsilon = \zeta^g \epsilon_{p-1}.$$

The two sides of this equation are complex conjugates, so are in fact real. Therefore $\zeta^{-g}\epsilon = r \in \mathbb{R}$. $\square$

11.4   Kummer's Theorem

In order to state Kummer's special case of Fermat's Last Theorem, we need a technical definition. A prime p is *regular* if it does not divide the class-number of $\mathbb{Q}(\zeta)$, where $\zeta = e^{2\pi i/p}$. By Section 10.3, $p = 3, 5, 7$ are regular. Further discussion of the regularity property is postponed until Section 11.5, for we are now in a position to state and prove:

Theorem 11.8. *If p is an odd regular prime then the equation*

$$x^p + y^p = z^p$$

has no solutions in integers x, y, z satisfying

$$p \nmid x, \quad p \nmid y, \quad p \nmid z.$$

Proof: Consider instead the equation

$$x^p + y^p + z^p = 0 \tag{11.5}$$

which exhibits greater symmetry. Since we can pass from this to the Fermat equation by changing z to $-z$, it suffices to work on (11.5). Assume, for a contradiction, that there exists a solution (x, y, z) of (11.5) in integers prime to p. We may as usual assume further that x, y, z are pairwise coprime. Factorize (11.5) in $\mathbb{Q}(\zeta)$ to obtain

$$\prod_{j=0}^{p-1} \left(x + \zeta^j y\right) = -z^p$$

and pass to ideals:

$$\prod_{j=0}^{p-1} \left\langle x + \zeta^j y \right\rangle = \langle z \rangle^p. \tag{11.6}$$

First we establish that all factors on the left of this equation are pairwise coprime. For suppose $\mathfrak{p}$ is a prime ideal dividing $\left\langle x + \zeta^k y \right\rangle$ and $\left\langle x + \zeta^l y \right\rangle$ with $0 \le k < l \le p - 1$. Then $\mathfrak{p}$ contains

$$(x + \zeta^k y) - (x + \zeta^l y) = y\zeta^k (1 - \zeta^{l-k}).$$

Now $1 - \zeta^{l-k}$ is an associate of $1 - \zeta = \lambda$, and ζ^k is a unit, so $\mathfrak{p}$ contains $y\lambda$. Since $\mathfrak{p}$ is prime either $\mathfrak{p} | y$ or $\mathfrak{p} | \lambda$. In the first case $\mathfrak{p}$ also divides z by (11.6). Now y and z are coprime integers, so there exist $a, b \in \mathbb{Z}$ such that

$az + by = 1$. But $y, z \in \mathfrak{p}$ so $1 \in \mathfrak{p}$, a contradiction. On the other hand, since $N(\mathfrak{l}) = p$, Theorem 5.14(a) implies that $\mathfrak{l}$ is prime, so if $\mathfrak{p}|\lambda$ then $\mathfrak{p} = \mathfrak{l}$. Then $\mathfrak{l}|z$ so

$$p = N(\mathfrak{l})|N(z) = z^{p-1}$$

and $p|z$ contrary to hypothesis.

Uniqueness of prime factorization of ideals now implies that each factor on the left of Equation (11.6) is a pth power of an ideal, since the right-hand side is a pth power and the factors are pairwise coprime. In particular there is an ideal $\mathfrak{a}$ such that

$$\langle x + \zeta y \rangle = \mathfrak{a}^p.$$

Thus $\mathfrak{a}^p$ is principal. Regularity of p means that $p \nmid h$, the class-number of $\mathbb{Q}(\zeta)$, and then Proposition 9.8(b) tells us that $\mathfrak{a}$ is principal, say $\mathfrak{a} = \langle \delta \rangle$. Therefore

$$x + \zeta y = \epsilon \delta^p$$

where ϵ is a unit.

Now we use Lemma 11.7 to conclude that

$$x + \zeta y = r\zeta^g \delta^p$$

where r is real. By Lemma 11.5 there exists $a \in \mathbb{Z}$ such that

$$\delta^p \equiv a \pmod{\mathfrak{l}^p}.$$

Hence

$$x + \zeta y \equiv ra\zeta^g \pmod{\mathfrak{l}^p}.$$

Lemma 11.3 (a) shows that $\langle p \rangle \mid \mathfrak{l}^p$, so

$$x + \zeta y \equiv ra\zeta^g \pmod{\langle p \rangle}.$$

Now ζ^{-g} is a unit, so

$$\zeta^{-g}(x + \zeta y) \equiv ra \pmod{\langle p \rangle}.$$

Take complex conjugates:

$$\zeta^g(x + \zeta^{-1}y) \equiv ra \pmod{\langle p \rangle}.$$

Eliminate ra to obtain the important congruence

$$x\zeta^{-g} + y\zeta^{1-g} - x\zeta^g - y\zeta^{g-1} \equiv 0 \pmod{\langle p \rangle}. \tag{11.7}$$

Observe that $1 + \zeta$ is a unit (put $t = -1$ in Equation (3.3)). We investigate possible values for g in Equation (11.7).

Suppose $g \equiv 0 \pmod{p}$. Then $\zeta^g = 1$, the terms with x cancel, and (11.7) becomes

$$y(\zeta - \zeta^{-1}) \equiv 0 \pmod{\langle p \rangle}$$

so

$$y(1 + \zeta)(1 - \zeta) \equiv 0 \pmod{\langle p \rangle}.$$

Since $1 + \zeta$ is a unit,

$$y\lambda \equiv 0 \pmod{\langle p \rangle}.$$

Now $\langle p \rangle = \langle \lambda \rangle^{p-1}$ and $p - 1 \geq 2$, so we have $\lambda | y$. Taking norms, $p|y$, contrary to hypothesis. Hence $g \not\equiv 0 \pmod{p}$. A similar argument shows that $g \not\equiv 1 \pmod{p}$.

Rewrite (11.7) in the form

$$\alpha p = x\zeta^{-g} + y\zeta^{1-g} - x\zeta^g - y\zeta^{g-1}$$

for some $\alpha \in \mathbb{Z}[\zeta]$. By the previous paragraph no exponent $-g, 1-g, g, g-1$ is divisible by p. Now

$$\alpha = \frac{x}{p}\zeta^{-g} + \frac{y}{p}\zeta^{1-g} - \frac{x}{p}\zeta^p - \frac{y}{p}\zeta^{g-1}. \tag{11.8}$$

Moreover, $\alpha \in \mathbb{Z}[\zeta]$ and $\{1, \zeta, \ldots, \zeta^{p-2}\}$ is a $\mathbb{Z}$-basis. Hence if all four exponents are incongruent modulo p we have $x/p \in \mathbb{Z}$, contrary to hypothesis. So some pair of exponents must be congruent modulo p. Since $g \not\equiv 0, 1 \pmod{p}$ the only possibility left is that $2g \equiv 1 \pmod{p}$.

But now (11.8) can be rewritten as

$$\begin{aligned}
\alpha p \zeta^g &= x + y\zeta - x\zeta^{2g} - y\zeta^{2g-1} \\
&= (x - y)\lambda.
\end{aligned}$$

Taking norms we get $p|(x - y)$, so

$$x \equiv y \pmod{p}.$$

By the symmetry of (11.5),

$$y \equiv z \pmod{p}$$

and hence

$$0 \equiv x^p + y^p + z^p \equiv 3x^p \pmod{p}.$$

Since $p \nmid x$ we must have $p = 3$.

It remains to deal with the possibility $p = 3$. Note that modulo 9, cubes of numbers prime to p (namely 1, 2, 4, 5, 7, 8) are congruent either to 1 or to -1. Hence modulo 9 a solution of (11.5) in integers prime to 3 takes the form

$$\pm 1 \pm 1 \pm 1 \equiv 0 \pmod{9}$$

which is impossible. Hence finally $p \neq 3$, a contradiction. □

A complete solution of Fermat's Last Theorem (for regular primes!) is thus reduced to the case where one of x, y, or z is a multiple of p. Kummer's proof of this case also depends heavily on ideal theory and, although long, would be accessible to us at this stage, except for one fact. We need to know that (still for p regular) if a unit in $\mathbb{Q}(\zeta)$ is congruent modulo p to a rational integer, then it is a pth power of another unit in $\mathbb{Q}(\zeta)$. The proof of this requires new methods. It seems best to refer the reader to Borevič and Šafarevič [8] pages 378–81 for the missing details.

11.5   Regular Primes

Theorem 11.8 is, of course, useless without a test for regularity. There is, in fact, quite a simple test, but once more the proofs are far beyond our present methods. We nonetheless sketch what is involved, and again refer the reader to Borevič and Šafarevič [8] for details.

Everything rests on a remarkable gadget known as the *analytic class-number formula*. Let K be a number field, and define the *Dedekind zeta-function*

$$\zeta_K(x) = \sum N(\mathfrak{a})^{-x}$$

where $\mathfrak{a}$ runs through all ideals of the ring of integers $\mathfrak{O}$ of K, and for the moment $1 < x < \infty$. The key formula is

$$\lim_{x \to 1}(x-1)\zeta_K(x) = \frac{2^{s+t}\pi^t R}{m\sqrt{|\Delta|}}h$$

in which s and t are the number of real, or complex conjugate pairs of, monomorphisms $K \to \mathbb{C}$; m is the number of roots of unity in K; Δ is the discriminant of K; R is a new constant called the *regulator* of K; and h is the class-number.

The point is that nearly everything on the right, except h, is quite easy to compute, though R is much harder than the rest. If we could evaluate the limit on the left we could then work out h. To evaluate this limit we first extend the definition of $\zeta_K(x)$ to allow complex values of x, and then use powerful techniques from complex function theory. These involve another gadget known as a *Dirichlet L-series*.

In the case $K = \mathbb{Q}(\zeta)$ for $\zeta = e^{2\pi i/p}$, p prime, the analysis leads to an expression for h in the form of a product

$$h = h_1 h_2.$$

In this, h_2 is the class-number of the related number field $\mathbb{Q}(\zeta + \zeta^{-1})$, and h_1 is a computable integer. This would not be very helpful, except that it can be proved that *if h_1 is prime to p, then so is h_2.* Therefore h is prime to p, or equivalently p is regular, if and only if h_1 is prime to p.

Analysis of h_1 leads to a criterion: h_1 is divisible by p if and only if one of the numbers

$$S_k = \sum_{n=1}^{p-1} n^k \quad (k = 2, 4, \ldots, p-3)$$

is divisible by p^2.

The numbers S_k have long been associated with the *Bernoulli numbers* B_k defined by the series expansion

$$\frac{t}{e^t - 1} = 1 + \sum_{m=1}^{\infty} \frac{B_m}{m!} t^m.$$

Their values behave very irregularly: for m odd $\neq 1$ they are zero, for $m = 1$ we have $B_1 = -\frac{1}{2}$, and for even m the first few are:

$$B_2 = \tfrac{1}{6} \qquad B_4 = -\tfrac{1}{30} \qquad B_6 = \tfrac{1}{42} \qquad B_8 = -\tfrac{1}{30}$$

$$B_{10} = \tfrac{5}{66} \qquad B_{12} = -\tfrac{691}{2730} \qquad B_{14} = \tfrac{7}{6} \qquad B_{16} = -\tfrac{3617}{510}$$

The connection between the S_k and the B_k may be shown to give:

Criterion 11.9. *A prime p is regular if and only if it does not divide the numerators of the Bernoulli numbers $B_2, B_4, \ldots, B_{p-3}$.* $\square$

The first 10 irregular primes, found from this criterion, are 37, 59, 67, 101, 103, 131, 149, 157, 233, 257. As a check, it is possible to compute the number h_1, with the results in Table 11.1. Observe that h_1 is divisible by p exactly in the cases $p = 37, 59, 67$ (marked in bold type) as expected.

11.6 Exercises

1. If x, y, z are integers such that $x^2 + y^2 = z^2$, prove that at least one of x, y, z is a multiple of 3, at least one is a multiple of 4, and at least one is a multiple of 5.

p	h_1	p	h_1
3	1	43	211
5	1	47	$5 \cdot 139$
7	1	53	4889
11	1	59	$3 \cdot \mathbf{59} \cdot 233$
13	1	61	$41 \cdot 1861$
17	1	67	$\mathbf{67} \cdot 12739$
19	1	71	$7^2 \cdot 79241$
23	3	73	$89 \cdot 134353$
29	2^3	79	$5 \cdot 53 \cdot 377911$
31	3^2	83	$3 \cdot 279405653$
37	$\mathbf{37}$	89	$113 \cdot 118401449$
41	11^2	97	$577 \cdot 3457 \cdot 206209$

Table 11.1. Values of the class-number h_1.

2. Show that the smallest value of z for which there exist four distinct solutions to $x^2 + y^2 = z^2$ with x, y, z pairwise coprime (not counting sign changes or interchanges of x, y as distinct) is 1105, and find the four solutions.

3. Show that there exist no solutions in non-zero integers to the equation $x^3 + y^3 = 3z^3$.

4. Show that the general solution in rational numbers of the equation

$$x^3 + y^3 = u^3 + v^3$$

is

$$
\begin{aligned}
x &= k(1 - (a - 3b)(a^2 + 3b^2)), \\
y &= k((a + 3b)(a^2 + 3b^2) - 1), \\
u &= k((a + 3b) - (a^2 + 3b^2)^2), \\
v &= k((a^2 + 3b^2)^2 - (a - 3b)),
\end{aligned}
$$

where a, b, k are rational and $k \neq 0$; or $x = y = 0$, $u = -v$; or $x = u$, $y = v$, or $x = v$, $y = u$. (*Hint*: write $x = X - Y$, $y = X + Y$, $u = U - V$, $v = U + V$, and factorize the resulting equation in $\mathbb{Q}(\sqrt{-3})$.)

5. For p an odd prime, show that if $\zeta = e^{2\pi i/p}$, then

$$\sqrt{\left(\frac{1 - \zeta^s}{1 - \zeta} \cdot \frac{1 - \zeta^{-s}}{1 - \zeta^{-1}} \right)}$$

is a real unit in $\mathbb{Q}(\zeta)$ for $s = 1, 2, \ldots, p - 1$.

6. Let p be an odd prime, $\zeta = e^{2\pi i/p}$. Kummer's lemma says that the units of $\mathbb{Z}[\zeta]$, thought of in the complex plane $\mathbb{C}$, lie on equally spaced radial lines through the origin, passing through the vertices of a regular p-gon (namely the powers ζ^s). Now $1 + \zeta$ is a unit, so why does Figure 11.1 not contradict Kummer's lemma?

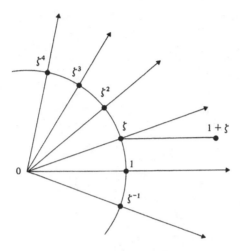

Figure 11.1. Why doesn't this contradict Kummer's lemma?

12

The Path to the
Final Breakthrough

In the late 19th and early 20th centuries, the study of Fermat's Last Theorem built mainly on Kummer's methods, with the notion of ideal numbers being supplanted by Dedekind's theory of ideals in a commutative ring. The techniques required a high degree of mathematical and computational facility, and were applied to more and more special cases. For instance, in 1905 Mirimanoff extended Kummer's results as far as $n \leq 257$. In 1908 Dickson generalized the theories of Germain and Legendre by investigating $x^n + y^n = z^n$ in the case where (n is prime and) none of x, y, z is divisible by n. Fermat's Last Theorem was proving to have a nasty sting in its tail. Despite the apparently simple statement of the problem, the proofs of special cases were becoming ever more complex, requiring the highly specialized activity of mathematical experts.

12.1 The Wolfskehl Prize

In 1908 the situation changed dramatically, and the problem was opened up to other mathematicians and to a wider world of amateurs. The agent of change was Paul Friedrich Wolfskehl, the son of a wealthy Jewish banker; he was born in Darmstadt in 1856. He first studied medicine, obtaining his doctorate in 1880. However, debilitating multiple sclerosis made it impossible for him to practice surgery, and in 1880 he turned instead to mathematics. He began this activity in Bonn, but moved to Berlin the following year, where he attended the lectures of the 72-year old Kummer. So fascinated did Wolfskehl become with the still unproved Last Theorem of Fermat that he left 100,000 marks in his will, to be awarded to the first person either to prove the theorem or to give a counterexample, see Barner [4]. In today's currency, this would have been worth around $1.7 million. The prize was announced by the Royal Society of Science in Göttingen, on 13 September 1908, exactly two years after Wolfskehl's death: it was to be claimed on or before 13 September 2007. Although hyperinflation in Germany in the 1920s greatly diminished the value of the bequest, it was still valued at 75,000 Deutsche marks in modern currency at the end of the 20th century, thanks to judicious investment. (The mark was replaced by the euro and euro banknotes were first available in 2002.)

Wolfskehl's act of altruism proved a mixed blessing to the mathematical community. In the first year alone, 621 solutions were submitted, and although the frequency slowly decreased, attempted solutions continued to flow in for the next ninety years. The total number sent to the Göttingen Academy has been estimated at over 5,000, and each attempt had to be read and considered by one of the judges. The endless succession of 'proofs' of Fermat's Last Theorem kept the staff and assistants involved continually busy. Not only did they have to deal with problems regularly: they could also become involved in protracted correspondence in addition to the initial reply. One correspondence on record extended to over sixty communications.

Other universities did not escape the burden. At the Royal Society of Science in Berlin the numerous attempted proofs were dealt with by a single individual, Albert Fleck, who courteously replied to each aspirant, highlighting the error in the manuscript and succinctly explaining the mistake.

Sometimes the solutions were put forward by eminent mathematicians. Ferdinand Lindemann (1852–1939), who is famous for his proof of the transcendence of π, published a fallacious proof of Fermat's Last Theorem in 1901. He soon withdrew it, but he continued his efforts with a 64-page paper in 1908. Fleck showed him his error on pages 23 and 24, rendering the remainder of the enterprise worthless. Fleck was a true 'amateur' who loved his work: his 'Fermat Clinic' at the Berlin Academy consisted solely of himself, at his desk in his room in the Mathematics Department. For these efforts, Fleck was awarded the Leibniz silver medal of the Berlin Mathematical Society in 1915, and he continued in this task until his death in 1943. As a Jew in Nazi Germany, his final years were blighted by persecution and humiliation. As the 20th century continued, the volume of solutions diminished, but they continued to arrive at intervals from all corners of the world. When the Berlin wall was removed, the number of solutions from Eastern Europe suddenly increased, because academics from the former Soviet Union were once more able to communicate freely with the West.

Despite the large number of attempted proofs, the actual advances in the early 20th century were prosaic and highly technical. In 1909, Wieferich focused on the case where p has no factors in common with x, y, z, and proved that if there is a solution then the condition

$$2^{p-1} \equiv 1 \pmod{p^2}$$

must be satisfied. Relationships like this became much more useful with the arrival of computers in the middle of the 20th century, because it was then practical to check them for large p. The American mathematician Harry Schultz Vandiver (1882–1973) introduced methods that made a computational approach to the full theorem possible for any specific p (not *too* large). He had little formal education, and left school early to work in his father's firm. In 1904 he collaborated with the 20-year old George Birkhoff in a paper on the factorization of integers of the form $a^n - b^n$, becoming yet another in the long line of amateurs who were fascinated with number theory. He took a university appointment in 1919, and worked extensively on Fermat's Last Theorem: he was awarded the Cole Prize of the American Mathematical Society in 1931 for this work. His findings built on the work of Kummer, and were particularly amenable to computation. In 1952, at the age of seventy, he used a computer to prove Fermat's Last Theorem for $n \leq 2,000$. The value of n continued to be raised at intervals over the years. In 1976, Wagstaff proved the theorem for $n \leq 125,000$, and by 1993 subsequent computations by others had raised this to $n \leq 4,000,000$, see Buhler *et al.* [11].

The methods continued to involve heavy calculations, making painful step-by-step progress without any simple fundamental insight that addressed the whole problem in a truly conceptual way. The proof was remarkably elusive. It seemed that the Wolfskehl Prize would be unclaimed in the few years left before time ran out in 2007.

12.2 Other Directions

Meanwhile, mathematics was continung to grow in other directions, which seemed at the time to have nothing whatsoever to do with Fermat's Last

Theorem. However, history is littered with cases where mathematicians attempting to solve one problem ended up formulating and proving something quite different. Indeed, Kummer's original breakthrough in his proof of many cases of Fermat's Last Theorem occurred when he was working on a totally different problem in the generalized theory of quadratic reciprocity. In the same manner, the ingredients that were to lead Wiles to the final proof of Fermat's Last Theorem arose in areas which, at first, seemed to have no possible link with it.

In the last decade of the 19th century, Henri Poincaré developed the new theory of algebraic topology in his book *Analysis Situs* (1895). He invented ways to translate topological problems into algebraic form. He classified surfaces in terms of their 'fundamental group' which, among other things, gives information about the number of 'holes' in the surface and relates this number to an integer called the 'genus'. A sphere with no holes has genus 0, a torus has genus 1, and other surfaces with 'more holes' have genus $g \geq 2$.

Initially, this idea seemed to have no relationship with Fermat's Last Theorem. However, there is a connection. An integer solution of Fermat's equation, say $a^n + b^n = c^n$, corresponds to the rational solution $x = a/c, y = b/c$ of the polynomial equation

$$x^n + y^n - 1 = 0. \tag{12.1}$$

Therefore Fermat's Last Theorem is equivalent to showing that this polynomial equation has no rational solutions. The Cambridge mathematician Louis Mordell had the bright idea of looking not only at the rational solutions of a polynomial equation $Q(x, y) = 0$ with rational coefficients, but also at its complex solutions. Topologically, the complex solutions of (12.1) are related to a surface whose genus happens to be $(n - 1)(n - 2)/2$. For $n \geq 4$, the genus is therefore 2 or more. In 1922 Mordell formulated what is now called the Mordell Conjecture: a polynomial equation $Q(x, y) = 0$ with rational coefficents and genus $g \geq 2$ has only finitely many rational solutions. If this could be proved, then it would immediately follow that the Fermat equation $a^n + b^n = c^n$ $(n \geq 4)$ has at most a finite number of integer solutions.

At first this seemed not to carry the Fermat quest very far forward. To start with, it was an unproved conjecture. Even if it were proved, it would only show that the equation has a finite number of solutions, when what we actually wish to show is that there are none. Nevertheless, the Mordell Conjecture turned out to be an important step towards the final proof of Fermat's Last Theorem. Early work of Weil [87] led to significant progress in special cases, and the full Mordell Conjecture was finally proved by Gerd Faltings in 1983, see Bloch [7]. The proof was immediately followed by new results. In 1985, two different papers were published to confirm that Fermat's Last Theorem is true for 'almost all' n. Andrew Granville and 'Roger' Heath-Brown showed that the proportion of those n for which Fermat's Last Theorem is true tends to 1 as n becomes large. It is a remarkable result, but still not the full proof that was so eagerly being sought.

12.3 Modular Functions and Elliptic Curves

Other ideas of Poincaré also proved to be seminal in the proof of Fermat's Last Theorem, although again the link was not obvious when they were first introduced. As a visual thinker, Poincaré loved to study systems that have symmetry. An area of particular interest was that of symmetries in complex function theory. The context for this work requires:

Definition 12.1. Let $a, b, c, d \in \mathbb{C}$ with $ad - bc \neq 0$. The function

$$g(z) = \frac{az + b}{cz + d}$$

is called a *Möbius map* or *bilinear map*.

Classically these maps are also called Möbius transformations or bilinear transformations.

Poincaré studied complex functions $f(z)$ that remain invariant when their domains are operated on by a Möbius map (12.1) for integers a, b, c, d. That is, functions such that

$$f\left(\frac{az + b}{cz + d}\right) = f(z) \quad \forall z.$$

It can be checked that these maps form a group under composition. When $z = -d/c$, the image under the transformation is infinite, so that to obtain a more satisfactory theory, it is best to adjoin the point at infinity to the complex plane to give a surface that is topologically like the surface of a sphere. Functions that are invariant under a countably infinite group of Möbius maps are called *automorphic*.

Poincaré went further, and considered those functions transforming the upper half plane ($z = x + iy$ where $y > 0$) to itself that remain invariant under the same kinds of map. Adding one or two technical conditions, he developed a theory of *modular functions*. These will be discussed in

Chapter 13. For the moment, it is sufficient to know that modular functions have properties that eventually made them a pivotal idea in the proof of Fermat's Last Theorem.

The introduction of complex numbers into the study of Fermat's Last Theorem—particularly the study of polynomial equations with rational coefficients, as in the Mordell Conjecture—played another important role. This relates to *elliptic curves*—curves defined by the formula

$$y^2 = Ax^3 + Bx^2 + Cx + D$$

where A, B, C, D are all rational. The trick that opens up a route to a proof of Fermat's Last Theorem can be formulated as follows: imagine this equation for complex x and y, and to attempt to parametrize it with functions $x = f(z), y = g(z)$ satisfying the equation

$$g(z)^2 = Af(z)^3 + Bf(z)^2 + Cf(z) + D.$$

However, this point of view on the ideas involved is a fairly recent one: the early formulations were stated in more technical ways. See Rubin and Silverberg [70].

12.4   The Taniyama–Shimura–Weil Conjecture

In 1955, a highly significant step was taken by two Japanese mathematicians who were planning a conference in Tokyo on algebraic number theory. Yutaka Taniyama was interested in elliptic curves. He had a powerful intuitive grasp of mathematics but was prone to making errors. But his friend Goro Shimura, a much more formal mathematician, realised that Taniyama had an instinctive ability to imagine new relationships that were not available to more careful thinkers. At their conference they presented a number of problems for consideration by the participants. Four of these, proposed by Taniyama, dealt with possible relationships between elliptic curves and modular functions. From these developed what became known as the Taniyama–Shimura–Weil Conjecture. This conjecture hypothesized that every elliptic curve can be parametrized by modular functions. (Its technical statement was different, and only much later was it reinterpreted in this way as a result of other discoveries in the area.)

At the time this was a surprising idea to most workers in the field, who saw elliptic curves and modular functions as inhabiting quite different parts of mathematics, so at first the conjecture was not taken seriously. Shimura left Tokyo for Princeton in 1957, resolving to return in two years to continue work with his colleague. His plans were not realised: in November 1958

Taniyama committed suicide. A letter left beside his body explained that he did not really know why he had decided on this action: simply that he was in a frame of mind where he had lost confidence in his future. He was due to be married within a month. A few weeks later his fiancée also took her own life. They had promised each other they would never be parted and she chose to follow him in death.

Shimura reacted to this double tragedy by devoting his energies to understanding the relationship between elliptic curves and modular functions. Over the years he gathered so much supporting evidence that the Taniyama–Shimura–Weil Conjecture became more widely appreciated. It occupies a pivotal position between two different areas of mathematics. Both of these areas had been studied intensely, but had remained separate. If the conjecture were true, then unsolved problems in one area could be translated into the language and concepts of the other, and perhaps solved by the novel methods available there.

In the 1960s and 1970s, hundreds of mathematical papers appeared which showed that if the Taniyama–Shimura–Weil Conjecture were true, then other—very important—results would follow. A whole mathematical industry was being built on a principle that still eluded proof.

12.5   Frey's Elliptic Equation

In the depths of the Black Forest in Germany, near the town of Oberwolfach, is a retreat for mathematical researchers, where they can gather in a relaxed environment to share their ideas. In the summer of 1984 a group of number theorists assembled to discuss their latest ideas on elliptic equations. In a lecture at the meeting Gerhard Frey, from Saarbrucken, formulated an idea that forever changed the landscape in the search for a proof of Fermat's Last Theorem.

In common with almost all of the great breakthroughs in number theory, Frey's idea depended on an ingenious calculation. He made the assumption that a genuine solution to Fermat's equation exists, so that $a^n + b^n = c^n$ where a, b, c are integers and $n > 2$. Such a solution would, of course, be a counterexample to Fermat's Last Theorem. He then wrote the following elliptic equation on the board:

$$
\begin{aligned}
y^2 &= x(x + a^n)(x - b^n) \\
&= x^3 + (a^n - b^n)x^2 - a^n b^n x
\end{aligned}
$$

which later became known as the *Frey curve*. He explained that this equation has very special properties. For instance, the 'discriminant' of such an equation is defined to be

$$(x_1 - x_2)^2 (x_2 - x_3)^2 (x_3 - x_1)^2$$

where x_1, x_2, x_3 are the roots of the right-hand side. In this case $x_1 = 0, x_2 = -a^n, x_3 = b^n$, so the discriminant is

$$(-a^n - b^n)^2 (b^n - 0)^2 (0 - (-a)^n)^2 = c^{2n} b^{2n} a^{2n} = (abc)^{2n}$$

(using $a^n + b^n = c^n$). Frey remarked that it is highly unusual for a discriminant to be a perfect power in this way, and went on to suggest that the equation has other equally strange properties which mean that it contradicts the Taniyama–Shimura–Weil Conjecture. He was unable to prove this in full, but he offered convincing evidence for such a connection. So, if the Taniyama–Shimura–Weil Conjecture is true, then there cannot be any solution of the Fermat equation ... so Fermat's Last Theorem must be true.

12.6   The Amateur who Became a Model Professional

Andrew Wiles now enters the story. His love of mathematics dated from his childhood in Cambridge. As he recalled in the BBC Television Programme *Horizon* on 27 September 1997:

> I was a 10-year-old, and one day I happened to be looking in my local public library and I found a book on math and it told a bit about the history of this problem—that someone had resolved this problem 300 years ago, but no one had ever seen the proof, no one knew if there was a proof, and people ever since have looked for the proof. And here was a problem that I, a 10-year-old, could understand, but none of the great mathematicians in the past had been able to resolve. And from that moment of course I just tried to solve it myself. It was such a challenge, such a beautiful problem.

This problem was Fermat's Last Theorem. It became an obsession. As a teenager, Wiles reasoned that Fermat would have had only limited resources, which did not include the more subtle theories that came after him. Wiles therefore felt that it was worthwhile to attack Fermat's Last Theorem using only the knowledge that he already had from school. As his interest developed, though, he began to read the literature on the subject, and to delve more and more deeply into it.

In 1971 he went to Merton College, Oxford, to study mathematics. After graduating in 1974 he moved to Clare College, Cambridge, to study for a doctorate. At the time he wanted to pursue his quest for a proof of Fermat's Last Theorem, but his PhD supervisor John Coates advised against this, because it was possible to spend many years working on the problem but getting nowhere. Instead, Wiles worked in his supervisor's area of expertise which happened to be the Iwasawa theory of elliptic curves—a fortuitous choice, given how the story would later turn out.

Wiles was a Junior Research Fellow at Clare College from 1977 to 1980, spending a period during this time at Harvard. In 1980 he was awarded his doctorate, and then he spent a time at Bonn before taking a post at the Princeton Institute for Advanced Study in 1981. He became a professor at Princeton University in 1982. He was awarded a Guggenheim Fellowship to visit the Institut des Hautes Études Scientifiques and the École Normale Supérieure in Paris during 1985–86, and it was here that events occurred which were to change his life.

In 1986, Ken Ribet completed a chain of arguments that began with the Frey curve and used ideas of Jean-Pierre Serre on modular Galois groups, to prove Frey's contention that the Taniyama–Shimura–Weil Conjecture implies Fermat's Last Theorem. Wiles took this as an opportunity to begin work in earnest. If he could prove the Taniyama–Shimura–Weil Conjecture, then he would finally crack the problem that had defeated the entire mathematical community for nearly three hundred and fifty years.

He soon learned that so many people continued to have an interest in Fermat's Last Theorem that talking about it would lead to wide-ranging discussions that would use up valuable time. So for the next seven years he worked on the problem in secret. As he worked, only his wife, and later his young children and his Head of Department, were aware of what he was doing. He spent his life on his mathematics and with his family. When he got stuck, he took a walk down the road to the lake near the Princeton Institute. Sometimes the combination of relaxation and deep incubation of ideas would suddenly come together in a new revelation. He found it necessary always to have a pencil and paper with him, to write down anything that occurred before it slipped his mind.

In 1988 he was stunned to see an announcement in the *Washington Post* and the *New York Times* that Fermat's Last Theorem had been proved by Yoichi Miyaoka of Tokyo University. Miyaoka also translated the number-theoretic problem into one in another area of mathematics, but he related the problem to differential geometry, not elliptic curves. He presented the first outline of his proof at a seminar in Bonn. Two weeks later he released a five-page algebraic proof, and close scrutiny by other mathematicians began. But soon his 'theorem' was seen to contradict a result in geometry that had been proved conclusively several years before. A fortnight later, Gerd Faltings pinpointed the fatal flaw in Miyaoka's proof. Within two months the consensus was that Miyaoka had failed. Wiles could breathe again and continue his work.

In the next three years he made considerable progress with various parts of the proof. As he explained later,

> Perhaps I can best describe my experience of doing mathematics in terms of a journey through a dark unexplored mansion. You enter the first room of the mansion, and it's completely dark. You stumble around bumping into the furniture, but gradually you learn where each piece of furniture is. Finally, after six months or so, you find the light switch, you turn it on, and suddenly it's all illuminated. You can see exactly where you were. Then you move into the next room and spend another six months in the dark. So each of these breakthroughs, while sometimes they're momentary, sometimes over a period of a day or so, they are the culmination of—and couldn't exist without—the many months of stumbling around in the dark that preceded them.

He tried to use the Iwasawa theory that he had studied for his PhD. He knew that the theory as it stood would be of little help, so he tried to generalize it and fix it up to attack his difficulties. It didn't work. In 1991, after a period of getting nowhere, he met Coates at a conference, who told him about something that appeared to bridge the gap. A brilliant young student, Mattheus Flach, had just written a beautiful paper analysing elliptic equations. Wiles took a look at the work and concluded that it was exactly what he needed. Progress thereafter was more rapid.

In 1993 Wiles gave a series of three lectures at the Isaac Newton Institute in Cambridge, England, on Monday, Tuesday, and Wednesday 21–23 June. The title of the series was 'Modular forms, elliptic curves and Galois representations'. With typical modesty, he made no advance announcements of his recent activity. Even so, many of the giants of number theory realised that something special was about to happen, and they attended the lectures—some with cameras ready to record the event for posterity. In the course of his lectures, Wiles proved a partial version of the Taniyama–Shimura–Weil Conjecture. It was sufficiently powerful to have a very special corollary. At 10:30 am, at the end of his third lecture, he wrote this corollary on the blackboard. It was the statement of Fermat's Last Theorem. At this point he turned to the audience, and as he sat down he said 'I will stop here.'

12.7 Technical Hitch

Wiles was not allowed to stop there. His proof now had to be subject to the usual reviewing process, and soon doubts began to arise. In response to a query from a colleague, Nick Katz, he realised that there was a hole in his use of the Flach technique, employed in the final stages of his proof. On 4 December 1993 he issued a statement that in the reviewing process a number of issues had arisen, most of which had been resolved. However, in view of the speculation buzzing around at the time, he acknowledged that a certain problem had occurred, and he wished to withdraw his claim that he had a proof. Despite this, he said that he remained confident that he could repair the difficulty using the methods he had announced in his Cambridge lectures.

His life was suddenly in turmoil. Instead of being able to work in secret, his difficulties were now public knowledge. In view of the many false proofs of Fermat's Last Theorem that had preceded Wiles's announcement, many mathematical colleagues began to voice doubts about the validity of his proof. In March 1994, in *Scientific American*, Faltings wrote:

> If it were easy, he would have solved it by now. Strictly speaking, it was not a proof when it was announced.

In the same magazine, André Weil was even more damning:

> I believe he has had some good ideas in trying to construct the proof, but the proof is not there. To some extent, proving Fermat's Theorem is like climbing Everest. If a man wants to climb Everest and falls short of it by 100 yards, he has not climbed Everest.

From the beginning of 1994, Wiles began to collaborate with his former student Richard Taylor in an attempt to fill the gaps in the proof. They concentrated on the step based on Flach's method, which was now seen to be inadequate, but they were unable to find an alternative argument. In August, Wiles addressed the International Congress of Mathematicians and had to announce that he was no nearer to a solution. Taylor suggested that they revisit Flach's method to see if another approach were possible, but Wiles was sure it would never work. Nevertheless, he agreed to give it another try to convince Taylor that it was hopeless.

12.8 Flash of Inspiration

They worked on alternative approaches for a couple of weeks, with no result. Then Wiles suddenly had a blinding inspiration as to why the Flach technique failed:

> In a flash I saw that the thing that stopped it working was something
> that would make another method I had tried previously work.

His inspiration cut through the final difficulties. On 6 October he sent
the new proof to three mathematicians primed for the job, and all three
reviewers found the new ideas satisfactory. The new method was even
simpler than his earlier failed attempt, and one of the three, Faltings,
even suggested a further simplification of part of the argument. By the
following year there was general agreement that the proof was correct and
complete. When Taylor lectured at the British Mathematical Colloquium
in Edinburgh in April 1995, there were no longer any real doubts about its
validity.

The proof was finally published in May 1995 in two papers in *Annals
of Mathematics*. The first, from page 443 to 551, was Wiles's paper on
'Modular elliptic curves and Fermat's Last Theorem', the second, from page
553 to 572, was the final step by Taylor and Wiles, entitled 'Ring theoretic
properties of Hecke algebras'. See Wiles [90], Taylor and Wiles [82], and
the survey by Darmon *et al.* [21].

In the years that followed, Wiles was fêted around the world. In 1995 he
received the Schock Prize in Mathematics from the Royal Swedish Academy
of Sciences and the Prix Fermat from the Université Paul Sabatier. The
American Mathematical Society awarded him the Cole Prize in Number
Theory, worth $4,000. He was presented with a $50,000 share in the 1995/6
Wolf Prize by the Israeli President Ezer Weizman for his 'spectacular con-
tributions to number theory and related fields, major advances on fundamental conjectures, and for settling Fermat's Last Theorem'. The other
recipient, Robert Langlands, was honoured for his own work in number
theory, automorphic forms, and group representations. In 1996 Wiles received the National Academy of Sciences Award ($5,000) followed in 1997
by a five year MacArthur Fellowship ($275,000). On 27 June 1997, after
his proof had been published for the statutory two years laid down in the
rules, he received the Wolfskehl Prize. A decade later, the hundred-year
period laid down in the original bequest would have run out.

There is no Nobel Prize in mathematics: the equivalent honour is the
Fields Medal, awarded to up to four mathematicians every four years at the
International Congress of Mathematicians. But by tradition the medal is
limited to individuals under the age of forty, and Wiles was just over this age
when he proved Fermat's Last Theorem. So in August 1998 the Congress
celebrated this event at the Fields Medal Ceremony by awarding Wiles a
special Silver Plaque—a unique honour in the history of the organization.
In 1999 he won the King Faisal International Prize for Science ($200,000),
being nominated for this honour by the London Mathematical Society.
In addition he has been awarded honorary degrees at many universities

around the world and, in the New Years Honours List in 2000, he became Sir Andrew Wiles, Knight Commander of the British Empire. The ten-year-old boy had grown to achieve his lifetime ambition, and had been lionized around the world for his success. He had conquered a problem that had foiled the world of mathematicians for 358 years.

12.9 Exercises

These exercises really are intended to be taken seriously. Don't just be amused (or not, according to taste): do them.

1. Think about how creative mathematics needs hard work (first) and relaxation. Note how the great insights mentioned in this chapter occurred. Does this suggest any way to help yourself understand difficult mathematics?

2. Find a lake or other idyllic setting; relax and think great thoughts.

3. Prepare yourself for the rigours to come. Subtler details will be outlined in the next two chapters.

13

Elliptic Curves

In this chapter we introduce the important notion of an 'elliptic curve'. Elliptic curves are a natural class of plane curves that generalise the straight lines and conic sections studied in nearly all university mathematics courses (and many high school courses). However, the study of elliptic curves involves two new ingredients. First, it is useful to consider *complex* curves, not just real ones. Second, for some purposes it is more satisfactory to work in complex projective space rather than the complex plane $\mathbb{C}^2$. (Algebraic geometers call $\mathbb{C}$ the complex *line* because it is 1-dimensional over $\mathbb{C}$. So $\mathbb{C}^2$ becomes the complex *plane*.) We introduce these refinements in simple stages.

The main topics dicussed in this chapter are:

- Lines and conic sections in the plane.

- The 'secant process' on a conic section and its relation to Diophantine equations.

- The definition and elementary properties of elliptic curves.

- The 'tangent/secant' process on an elliptic curve and the associated group structure.

Our point of view emphasises analogies between conic sections, where the key ideas take on an especially familiar form, and elliptic curves. This should help to explain the origin of the ideas involved in the theory of elliptic curves, and make them appear more natural.

13.1   Review of Conics

The simplest real plane curves are straight lines, which can be defined as
the set of solutions $(x, y) \in \mathbb{R}^2$ to a *linear* (or degree 1) polynomial equation

$$Ax + By + C = 0 \tag{13.1}$$

where $A, B, C \in \mathbb{R}$ are constants and $(A, B) \neq 0$.

Next in order of complexity come the *conic sections* or *conics*, defined
by a general quadratic (or degree 2) polynomial equation

$$Ax^2 + Bxy + Cy^2 + Dx + Ey + F = 0 \tag{13.2}$$

where $A, B, C, D, E, F \in \mathbb{R}$ are constants and $A, B, C \neq 0$.

It is well-known that conic sections can be classified into seven different
types: ellipse, hyperbola, parabola, two distinct lines, one 'double' line,
a point, or empty. A good way to see this is to transform (13.2) into a
simpler form, usually known as a *normal form*, by a change of coordinates.
In fact, a general invertible linear change of coordinates

$$\begin{aligned} X &= ax + by \\ Y &= cx + dy \end{aligned}$$

(with $ad - bc \neq 0$ for invertibility) transforms (13.2) into one or other of
the forms

$$\begin{aligned} \epsilon_1 X^2 + \epsilon_2 Y^2 + P &= 0 \\ X^2 + Y + Q &= 0 \end{aligned}$$

where $P, Q \in \mathbb{R}$ and $\epsilon_1, \epsilon_2 = 0, 1$, or -1.

The usual proof of this (see for example Loney [48] page 323, Anton [1]
page 359, or Roe [68] page 251) begins by rotating coordinates orthogonally
to diagonalise the quadratic form $Ax^2 + Bxy + Cy^2$, which changes (13.2)
to the slightly simpler form

$$\lambda_1 x'^2 + \lambda_2 y'^2 + \alpha x' + \beta y' + \gamma = 0.$$

If $\lambda_1 \neq 0$ then the term $\alpha x'$ can be eliminated by 'completing the square',
and similarly if $\lambda_2 \neq 0$ then the term $\beta y'$ can be eliminated. The coefficients
of x'^2 and y'^2 can be scaled to 0, 1, or -1 by multiplying them by a nonzero
constant; furthermore, x' and y' can be interchanged if necessary. Finally,
the entire equation can be multiplied throughout by -1. The result is the
following catalogue of normal forms:

Theorem 13.1. *By an invertible linear coordinate change, every conic can
be put in one of the following normal forms:*

(1) $X^2 + Y^2 + P = 0$

(2) $X^2 - Y^2 + P = 0$

(3) $X^2 + Y + Q = 0$

(4) $X^2 + Q = 0$ □

In case (1) we get an ellipse (indeed a circle) if $P < 0$, a point if $P = 0$, and the empty set if $P > 0$. In case (2) we get a (rectangular) hyperbola if $P \neq 0$ and two distinct intersecting lines if $P = 0$. Case (3) is a parabola. Case (4) is a pair of parallel lines if $Q < 0$, a 'double' line if $Q = 0$, and empty if $Q > 0$.

Transforming back into the original (x, y) coordinates, circles transform into ellipses, rectangular hyperbolas transform into general hyperbolas, parabolas transform into parabolas, lines transform into lines, and points transform into points.

Even the conics, then, exhibit a rich set of possibilities when viewed as curves in the real plane $\mathbb{R}^2$. The situation simplifies somewhat if we consider the same equations, but in *complex* variables; it simplifies even more if we work in projective space. In complex coordinates, the map $Y \mapsto iY$ sends Y^2 to $-Y^2$, a scaling that cannot be performed over the reals. This coordinate transformation sends normal form (1) to normal form (2) and thereby abolishes the distinction between hyperbolas and ellipses.

13.2 Projective Space

We now show that in projective space, all the different types of conic section other than the double line and the point can be transformed into each other. This is the case even in *real* projective space. First, we recall the basic notions of projective geometry. For further details, see Coxeter [19], Loney [48], or Roe [68].

Definition 13.2. The *real projective plane* $\mathbb{RP}^2$ is the set of lines L through the origin in $\mathbb{R}^3$. Each such line is referred to as a *projective point*. Each plane through the origin in $\mathbb{R}^3$ is called a *projective line*. A projective point is *contained in* a projective line if and only if the corresponding line through the origin is contained in the corresponding plane through the origin.

A *projective transformation* or *projection* is a map from $\mathbb{RP}^2$ to itself of the form $L \mapsto \phi(L)$, where ϕ is an invertible linear transformation of $\mathbb{R}^3$.

Two configurations of projective lines and projective points are *projectively equivalent* if one can be mapped to the other by a projection.

This definition may seem strange when first encountered, but it represents the distillation of a considerable effort on the part of geometers to 'complete' the ordinary (or *affine*) plane $\mathbb{R}^2$ by adding 'points at infinity' at which parallel lines can be deemed to meet. We explain this idea in a moment, but first we record:

Proposition 13.3. *In the projective plane, any two projective lines meet in a unique projective point, and any two projective points can be joined by a unique projective line.*

Proof: These properties follow from the analogous properties of lines and planes through the origin in $\mathbb{R}^3$.   □

Now we describe the interpretation of the projective plane in terms of points at infinity. One way to see how this comes about is to consider the plane $\mathcal{P} = \{(x, y, z) : z = 1\} \subseteq \mathbb{R}^3$. Each point $(x, y, 1) \in \mathcal{P}$ can be identified with a point (x, y) in the affine plane $\mathbb{R}^2$. We write $(x, y) \equiv (x, y, 1)$. Alternatively, the point $(x, y, 1)$ can be identified with the line through the origin in $\mathbb{R}^3$ that passes through it. Nearly every line through the origin in $\mathbb{R}^3$ is of this form: the exceptions are precisely the lines that lie in the plane $\mathcal{Q} = \{(x, y, z) : z = 0\}$, that is, the lines parallel to $\mathcal{P}$.

In the same way, any straight line M in $\mathcal{P}$ can be identified with either a straight line in $\mathbb{R}^2$, or with a plane through the origin in $\mathbb{R}^3$ — namely, the unique plane that contains both the origin of $\mathbb{R}^3$ and M. Precisely *one* plane through the origin of $\mathbb{R}^3$ is not of this form, namely, the plane $\mathcal{Q}$ that is parallel to $\mathcal{P}$. See Figure 13.1.

These identifications therefore *embed* the affine plane $\mathbb{R}^2$ in the projective plane $\mathbb{RP}^2$, in such a way that points embed as projective points and lines embed as projective lines. However, $\mathbb{RP}^2$ contains exactly one extra projective line, called the *line at infinity*: namely, the projective line that corresponds to the plane $\mathcal{Q}$ through the origin of $\mathbb{R}^3$. Moreover, $\mathbb{RP}^2$ contains extra projective points that do not correspond to points in $\mathbb{R}^2$; indeed, these are precisely the projective points that lie on the line at infinity, since they correspond to lines through the origin in $\mathbb{R}^3$ that lie in the plane $\mathcal{Q}$. These are called 'points at infinity'.

Each point at infinity corresponds to a unique direction in the plane $\mathbb{R}^2$, that is, a set of parallel lines, because any such set is parallel to precisely one line in the plane $\mathcal{Q}$. In these terms, a direction and its exact opposite, a 180° rotation, are identical. See Figure 13.2.

The key feature of this set-up is:

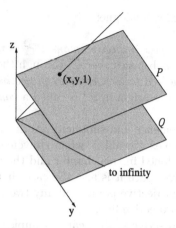

Figure 13.1. Construction of the projective plane.

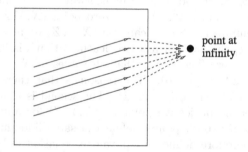

Figure 13.2. Points at infinity correspond to directions in the affine plane.

Lemma 13.4. *Any two parallel lines in* $\mathbb{R}^2$ *meet in* $\mathbb{RP}^2$ *at exactly one point at infinity.*

Proof: Suppose J, K are parallel lines in $\mathbb{R}^2$. They correspond to projective lines, namely, the planes J', K' through the origin that contain J, K respectively. But these meet in a unique projective point, namely the line in Q that is parallel to J and K. And this is a point at infinity in $\mathbb{RP}^2$.

$\square$

There is also a complex analogue:

Definition 13.5. The *complex projective plane* $\mathbb{CP}^2$ is the set of lines (that is, 1-dimensional vector subspaces over $\mathbb{C}$) through the origin in $\mathbb{C}^3$.

Each such line is referred to as a *complex projective point*.

Each plane through the origin in $\mathbb{R}^3$ is called a *complex projective line*.

For notational convenience and simplicity we break with tradition and use x, y to denote complex variables, when convenient. The convention that $z = x + iy$ is abandoned in this chapter and the next.

In complex projective space it is also the case that any two projective lines meet in a unique projective point, and any two projective points can be joined by a unique projective line.

The geometry of projective space, real or complex, is richer than mere lines. Any curve in $\mathbb{R}^2$ or $\mathbb{C}^2$ can be embedded in the corresponding $\mathbb{RP}^2$ or $\mathbb{CP}^2$. If the equation of the curve is polynomial, then this can be done in a systematic manner, so that 'points at infinity' on the curve can also be defined. The easiest way to achieve this is to introduce 'homogeneous coordinates'. Again, the idea is straightforward. A point in $\mathbb{RP}^2$ is a line through the origin in $\mathbb{R}^3$. Any nonzero point (X, Y, Z) on that line defines the line uniquely. So we can use (X, Y, Z) as a system of coordinates. However, this system has two features that distinguish it from Cartesian coordinates. The first is that all values of X, Y, Z are permitted *except* $(X, Y, Z) = (0, 0, 0)$. The reason is that there is no unique line joining $(0, 0, 0)$ to the origin in $\mathbb{R}^3$. The second is that (aX, aY, aZ) represents that same projective point as (X, Y, Z) for any nonzero constant a, since clearly both points define the same line through the origin of $\mathbb{R}^3$. We therefore define an equivalence relation $\sim$ by $(X, Y, Z) \sim (aX, aY, aZ)$ for any nonzero constant a. In other words, it is not the *values* of (X, Y, Z) that determine the corresponding projective point, but their ratios.

The embedding of $\mathbb{R}^2$ into $\mathbb{RP}^2$ defined above, which identifies $(x, y) \in \mathbb{R}^2$ with $(x, y, 1) \in \mathcal{P} \subseteq \mathbb{R}^3$, also identifies the usual coordinates (x, y) on $\mathbb{R}^2$ with the corresponding coordinates $(x, y, 1)$ on $\mathbb{RP}^2$. This represents the same projective point as (ax, ay, a) for any $a \neq 0$. In other words, when $Z \neq 0$ the projective point (X, Y, Z) is the same as $(X/Z, Y/Z, 1) \in \mathcal{P}$. On the other hand, when $Z = 0$ the projective point (X, Y, Z) lies in the plane $\mathcal{Q}$ and hence represents a point at infinity.

This system of coordinates (X, Y, Z) on $\mathbb{RP}^2$ is known as *homogeneous coordinates*. Homogeneous coordinates are really $\sim$-equivalence classes of triples (X, Y, Z), but it is more convenient to work with representatives and remember to take the equivalence relation $\sim$ into account. The choice of the line at infinity is conventional: in principle any line in $\mathbb{RP}^2$ can be deemed to be the line at infinity, and there is then a corresponding embedding of $\mathbb{R}^2$ in $\mathbb{RP}^2$. Indeed, any projective line in $\mathbb{RP}^2$ can be mapped to any other projective line by a projection, since any plane through the origin in $\mathbb{R}^3$ can be transformed into any other plane by an invertible linear map. For the purposes of this book, however, we employ the convention that $Z = 0$ defines the line at infinity.

The way to transform a polynomial equation in affine coordinates (x, y) into homogeneous coordinates is to replace x by X/Z and y by Y/Z, and then to multiply through by the smallest power of Z that makes the result a polynomial. For example the Cartesian equation $y - x^2 = 0$ becomes:

$$\begin{aligned} (Y/Z) - (X/Z)^2 &= 0, \\ YZ^{-1} - X^2 Z^{-2} &= 0, \\ YZ - X^2 &= 0. \end{aligned}$$

Notice that as well as the usual points $(x, x^2) \equiv (x, x^2, 1)$, this projective curve also contains the point at infinity given by $Z = 0$, which forces $X = 0$ but any nonzero Y. Since $(0, Y, 0) \sim (0, 1, 0)$ the parabola contains exactly one new point at infinity, in addition to the usual points in $\mathbb{R}^2$. It is easy to check that this point lies in the direction towards which the arms of the affine parabola 'diverge', namely the y-axis. See Figure 13.3.

Moreover, adding this point at infinity to the parabola causes it to close up (since the point at infinity lies on *both* arms). It is now plausible that the parabola is just an ellipse in disguise—that is, that they are projectively equivalent. We can verify this by means of the projection $\phi(X, Y, Z) = (X, Y + Z, Y - Z)$, which transforms $YZ - X^2 = 0$ into $Y^2 - Z^2 - X^2 = 0$ or $X^2 + Z^2 = Y^2$. Compose with $\psi(X, Y, Z) = (X, Z, Y)$ to turn this into $X^2 + Y^2 = Z^2$. Finally restrict back to the plane $\mathbb{R}^2$ by setting

point at
infinity

Figure 13.3. Adding a point at infinity to a parabola.

$(X, Y, Z) = (x, y, 1)$ and we get $x^2 + y^2 = 1$, a circle. Which, of course, is just a special type of ellipse.

If we had not interchanged Y and Z, the result would have been $x^2 - y^2 = 1$, a hyperbola. So in fact the ellipse, hyperbola, and parabola are all projectively equivalent over $\mathbb{R}$. So in real projective space the list of conics collapses to a smaller one. Namely: ellipse ($=$ parabola $=$ hyperbola), intersecting lines ($=$ pair of parallel lines), double-line, point.

What about complex projective space? Think about it. Hint: the real surprise is 'point'.

13.3   Rational Conics and the Pythagorean Equation

There is an interesting link between the geometry of conics and solutions of quadratic Diophantine equations.

Definition 13.6. A *rational line* in $\mathbb{R}^2$ or $\mathbb{C}^2$ is a line

$$ax + by + c = 0 \tag{13.3}$$

whose coefficients a, b, c are rational numbers.

A *rational conic* in $\mathbb{R}^2$ or $\mathbb{C}^2$ is a conic

$$f(x, y) = ax^2 + bxy + cy^2 + dx + ey + f = 0 \tag{13.4}$$

whose coefficients a, b, c, d, e, f are rational numbers.

A *rational point* in $\mathbb{R}^2$ or $\mathbb{C}^2$ is a point whose coordinates are rational numbers.

There are similar definitions for real and complex projective planes.

An intersection point of two rational lines is obviously a rational point. However, an intersection point of a rational line and a rational conic need not be rational—for example, consider the intersection of $x - y = 0$ with $x^2 + y^2 - 2 = 0$, which consists of the two points $(\pm\sqrt{2}, \pm\sqrt{2})$.

Not all rational conics possess rational points. For an example see Exercise 1 at the end of this chapter. A necessary and sufficient condition for a rational conic to possess at least one rational point was proved by Legendre, and can be found in Goldman [34] page 318. However, many rational conics do possess rational points, and from now on we work with such a conic.

Proposition 13.7. *Let p be a rational point on a rational conic C. Then any rational line through p intersects C in rational points.*

Proof: We discuss real conics: the complex case is similar. Let $f(x, y)$ be a rational conic as in (13.4) and let $Ax + By + C = 0$ be a rational line. Suppose that $B \neq 0$; if not, then $A \neq 0$ and a similar argument applies. Their intersection is the set of all (x, y) for which $y = (Ax - C)/B$ and x satisfies the quadratic equation $f(x, (Ax - C)/B) = 0$. Suppose that $f(x, (Ax - C)/B) = Kx^2 + Lx + M$: then K, L, M are rational. This equation has at least one real root, given by p, so it has two real roots (which are identical if and only if the line is tangent to the conic at p). The sum of those roots is $-L/K \in \mathbb{Q}$, and the root given by p is rational; therefore the second root is also rational.   □

This result immediately leads to a method for parametrizing all rational points on a rational conic (provided it possesses at least one rational point). Let C be a rational conic with a rational point p and let L be any rational line. For any point $q \in L$ the line joining q to p meets C at p, and at some other point (which is distinct from p unless the line concerned is tangent to C). Define a map $\pi : L \to C$ by letting $\pi(q)$ be this second point of intersection of the line joining q to p. (See Figure 13.4.)

Theorem 13.8. *With the above notation, $\pi(q)$ is rational if and only if q is rational.*

Proof: Clearly $\pi(q)$ is rational if and only if the line joining p to q is rational. But this is the case if and only if q is a rational point.   □

Example 13.9. Suppose that C is the unit circle $x^2 + y^2 - 1 = 0$, which is a rational conic. It contains the rational point $p = (-1, 0)$. Let L be the rational line $x = 0$. The rational points on L are the points $(0, t)$ where $t \in \mathbb{Q}$.

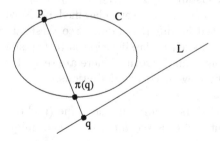

Figure 13.4. Parametrization of the rational points of a rational conic in terms of the rational points on a rational line.

The line joining p to $(0, t)$ has equation

$$y = t(x + 1)$$

and this meets the circle at

$$(x, y) = \left(\frac{1 - t^2}{1 + t^2}, \frac{2t}{1 + t^2} \right).$$

Thus we have the identity

$$\left(\frac{1 - t^2}{1 + t^2} \right)^2 + \left(\frac{2t}{1 + t^2} \right)^2 = 1$$

or equivalently

$$(1 - t)^2 + (2t)^2 = (1 + t)^2$$

providing solutions of the Diophantine equation $x^2 + y^2 = z^2$ in rational numbers. Indeed by Theorem 13.8 every rational solution is of this form.

This is very close to the parametrization of Pythagorean triples obtained in Lemma 11.1. Indeed, if we set $t = r/s$ we can easily obtain the result of that lemma.

13.4   Elliptic Curves

Elliptic curves arise from the study of plane cubic curves

$$\sum_{i+j \leq 3} A_{ij} x^i y^j = 0 \tag{13.5}$$

where the A_{ij} are constants and $A_{30} A_{21} A_{12} A_{03} \neq 0$.

Over the reals, such curves were classified by Newton in (probably) 1668: he distinguished 58 different kinds. See Westfall [89] page 200. As for conics, the key to such a classification is to transform coordinates so that (13.5) takes some simpler form. There are several ways to do this. In order to state the first, we need two definitions:

Definition 13.10. Let C be a curve in the plane (real or complex, affine or projective). A point $x \in C$ is *regular* if there is a unique tangent to C at x. Otherwise, x is *singular*.

The curve C is *non-singular* if it has no singular points; that is, every point $x \in C$ is regular.

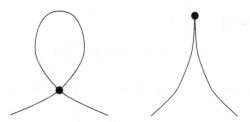

Figure 13.5. Typical singular points: (left) self-intersection, (right) cusp.

Typical singular points are *self-intersections* and *cusps* as in Figure 13.5. Although it looks as if the tangent at a cusp point is unique, actually there are three 'coincident' tangents—that is, a tangent of multiplicity 3.

Definition 13.11. Two curves $C, D \in \mathbb{CP}^2$ (or $\mathbb{RP}^2$) are *projectively equivalent* if there is a projection ϕ such that $\phi(C) = D$.

Theorem 13.12. (Weierstrass Normal Form.) *Every nonsingular cubic curve in $\mathbb{CP}^2$ is projectively equivalent to a curve which in affine coordinates takes the* Weierstrass normal form

$$y^2 = 4x^3 - g_2 x - g_3 \tag{13.6}$$

where $g_2, g_3 \in \mathbb{C}$ are constants.

Proof: We sketch the proof. The first step is to establish that every nonsingular cubic curve C has at least one *inflexion point*. This is a point at which the tangent line has *triple contact* with the curve, in the following sense. Suppose that the equation of the curve is $f(x, y) = 0$, and let $(\xi, \eta) \in C$. A general line through (ξ, η) has equation $a(x - \xi) + b(y - \eta) = 0$ for $a, b, \in \mathbb{C}$. This line meets C at (ξ, η), and in general (since the equation is cubic) it meets it at two other points. However, the cubic equation that determines these intersection points may have multiple zeros. The line is a tangent at (ξ, η) if that point corresponds to a double zero. The point (ξ, η) is an inflexion if it corresponds to a triple zero.

By writing down the equations for an inflexion point, it can be shown that any nonsingular cubic curve in $\mathbb{CP}^2$ has exactly nine inflexion points, if multiplicities are taken into account. In particular, it has at least one. See Brieskorn and Knörrer [9] page 291 for details.

By a projection, we may assume that $(0, 0, 1)$ is an inflexion point and that the tangent there has equation $X = 0$. This implies that the cubic curve has a homogeneous equation of the form

$$Y^2 Z + AXYZ + BYZ^2 + CX^3 + DX^2 Z + EXZ^2 + FZ^3 = 0$$

which in affine coordinates becomes

$$y^2 + (Ax + B)y + g(x) = 0 \tag{13.7}$$

where $g(x)$ is a cubic polynomial. Define new affine coordinates (x', y') by

$$y' = y + \frac{A}{2}x + \frac{B}{2}$$
$$x' = x.$$

Then (13.7) transforms into

$$y'^2 + h(x') = 0$$

where h is a cubic polynomial. There exists a linear change of coordinates $x'' = px' + q$ that puts $h(x')$ into the form $4x''^3 - g_2 x'' - g_3$ while leaving y' unchanged. $\qquad\square$

The coefficient 4 on the x^3 term in Weierstrass normal form is traditional: it could be made equal to 1, but we will shortly see that the 4 is more convenient in some circumstances. The notation g_2, g_3 for the linear and constant coefficients is also traditional.

We may now define an elliptic curve.

Definition 13.13. An *elliptic curve* is the set of points $(x, y) \in \mathbb{C}^2$ that satisfy the equation

$$y^2 = Ax^3 + Bx^2 + Cx + D \tag{13.8}$$

where $A, B, C, D \in \mathbb{C}$ are constants.

Strictly speaking, this defines a complex affine elliptic curve. There is projective analogue $Y^2 Z = AX^3 + BX^2 Z + CXZ^2 + DZ^3$, where (X, Y, Z) are homogeneous coordinates on $\mathbb{CP}^2$. When $A, B, C, D \in \mathbb{R}$ we can restrict attention to real variables, getting a *real elliptic curve*. Moreover, in the real case we can draw the graph of (13.8) in the plane to illustrate certain features of the geometry: we do this frequently below. Figure 13.6 shows the real elliptic curve $y^2 = 4x^3 - 3x + 2$ for which $g_2 = 3, g_3 = -2$.

The most important elliptic curves are those for which A, B, C, D are rational. We call these *rational elliptic curves*, and omit 'rational' whenever the context permits.

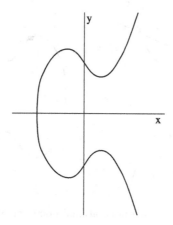

Figure 13.6. The real elliptic curve $y^2 = 4x^3 - 3x + 2$.

13.5   The Tangent/Secant Process

In Section 13.1 we showed that the rational points on a rational conic can be parametrized by the rational points on a rational line, once we know *one* rational point on the conic. A similar approach to rational points on a rational elliptic curve does not lead to such a definitive result, but in some respects the partial result that is thereby obtained is more interesting.

Proposition 13.14. *Over $\mathbb{CP}^2$ a rational line cuts a rational elliptic curve in three points (counting multiplicities). If two of these points are rational, then so is the third.*

Proof: Suppose that the affine equation of the elliptic curve is $f(x, y) = 0$ and let the line have affine equation $ax + by + c = 0$. Without loss of generality assume that $b \neq 0$, and solve for y, to get $y = -(ax + c)/b$. Substitute in f to get $f(x, -(ax + c)/b) = 0$. This is a cubic polynomial $px^3 + qx^2 + rx + s$ with rational coefficients, and its roots determine the x-coordinates of the intersection points. The corresponding y-coordinates are equal to $-(ax + c)/b$, hence are rational if and only if x is rational. Since the sum of the roots of the cubic is equal to the rational number $-p/q$, if two roots are rational then so is the third.   □

Incidentally, we stated Proposition 13.14 in projective form because in the affine case there are occasions when the third root of the cubic is at infinity. That is, the cubic actually reduces to a quadratic. We slid over this point in the proof. The proposition implies that once we have found

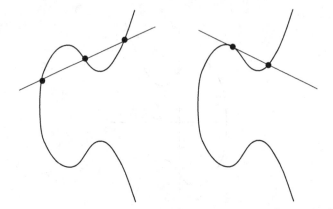

Figure 13.7. Constructing new rational points on an elliptic curve.

two rational points on a rational elliptic curve, we can find another by drawing the line through those points and seeing where else it cuts the curve, as in Figure 13.7 (left). In fact, we can do slightly better: find *one* rational point and see where else the tangent at that point cuts the curve, Figure 13.7 (right).

13.6   Group Structure on an Elliptic Curve

We now show that the rational points on a rational elliptic curve form an abelian group, under an operation of 'addition' closely related to Proposition 13.14. This remarkable fact forms the basis of the arithmetical theory of elliptic curves.

Assume that an elliptic curve C in $\mathbb{CP}^2$ contains at least one rational point, which we denote by $\mathbb{O}$ for reasons soon to become apparent.

Definition 13.15. Let P and Q be rational points on C. Define $P * Q$ to be the third point in which the line through P and Q meets C.

Let $\mathcal{G}$ be the set of all rational points in C. For some fixed but arbitrary choice $\mathbb{O}$ of a rational point on C, define the operation $+$ on $\mathcal{G}$ by

$$P + Q = (P * Q) * \mathbb{O} \tag{13.9}$$

See Figure 13.8. We now prove that the operation $+$ gives $\mathcal{G}$ the structure of an abelian group.

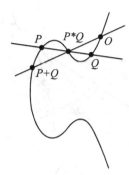

Figure 13.8. The group operation on the rational points of an elliptic curve.

In order to achieve this, we require the following fundamental theorem in algebraic geometry.

Theorem 13.16. (Bézout's Theorem.) *Let $P(X, Y, Z)$ be a homogeneous polynomial of degree p over $\mathbb{C}$, let $Q(X, Y, Z)$ be a homogeneous polynomial of degree q over $\mathbb{C}$, and suppose that P, Q have no common factor of degree > 1. Then the number of intersection points of the curves in $\mathbb{CP}^2$ defined by $P = 0, Q = 0$ is precisely pq (provided multiplicities are taken into account).*

Proof: A detailed proof, along with a careful discussion of how to count multiplicities, can be found in Brieskorn and Knörrer [9] page 227. □

Next we state without proof a lemma from algebraic geometry:

Lemma 13.17. *Let two curves of degree n meet in exactly n^2 distinct points, and let $0 \leq m \leq n$. If exactly mn of these points lie on an irreducible curve of degree mn, then the remaining $n(n - m)$ lie on a curve of degree $n - m$.*

Proof: See Brieskorn and Knörrer [9] page 245. □

We may now prove:

Theorem 13.18. *The set $\mathcal{G}$ of rational points on a rational elliptic curve forms an abelian group under the operation $+$. The identity element is $\mathbb{O}$.*

Proof: First, observe that $P * Q = Q * P$, since the process of constructing the third point on the line through P and Q does not depend on the order

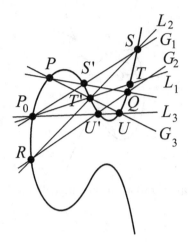

Figure 13.9. Geometry for the proof of the associative law on $\mathcal{G}$.

in which we consider P and Q. So

$$P + Q = (P * Q) * \mathbb{O} = (Q * P) * \mathbb{O} = Q + P$$

and the operation $+$ is commutative.

We claim that $P + \mathbb{O} = P$, so $\mathbb{O}$ is the identity element. This follows since

$$P + \mathbb{O} = (P * \mathbb{O}) * \mathbb{O}.$$

If $Q = P * \mathbb{O}$, then $P, \mathbb{O}, Q$ are collinear. Assume for a moment that these are distinct. Then $(P * \mathbb{O}) * \mathbb{O} = Q * \mathbb{O}$, and this must be P. If they are not distinct, then either $P = \mathbb{O}$ or $P = Q$, and in either case it is easy to complete the calculation.

The inverse of P is easily seen to be $P * (\mathbb{O} * \mathbb{O})$.

The most complicated part of the proof is the associative law

$$(P + Q) + R = P + (Q + R).$$

Figure 13.9 indicates the associated geometry.

First, we define

$$
\begin{aligned}
S &= P + Q, \\
T &= S + R, \\
U &= Q + R.
\end{aligned}
$$

We have to prove that $P + U = T$. Denote the auxiliary points used to construct R, S, T by R', S', T', as in the figure. It suffices to show that

P, U, T' are collinear. To do so, let L_1 be the line through P, Q, S', let L_2 be the line through S, R, T', let L_3 be the line through $\mathbb{O}, U', U$, let G_1 be the line through $\mathbb{O}, S', S$, and let G_2 be the line through Q, R, U'.

Recall from Lemma 13.17 that if two curves C, D of order n meet in exactly n^2 points, nm of which lie on an irreducible curve E of order m, then the remainder lie on a curve F of order $n - m$.

Apply this to the cubic curves C, D, where $D = L_1 \cup L_2 \cup L_3$, and take $E = G_1$. First, suppose $C \cap D$ contains exactly 9 distinct points. Then it follows that Q, R, U', P, T', U lie on a conic. But Q, R, U' lie on the line G_2, so G_2 is a component of this conic. Let the other component, also a line, be G_3. Then if P, T', U do not lie on G_2, they must lie on G_3, and the proof is complete.

If $C \cap D$ contains less than 9 distinct points, then some intersection points are multiple. A suitable small perturbation of the curve C splits these apart, and a limiting argument completes the proof. For a more algebro-geometric approach, replacing the limiting procedure by Zariski continuity, see Brieskorn and Knörrer [9] page 310. □

The operation $+$ can be defined in exactly the same way for any elliptic curve, rational or not. The same proof shows that $(C, +)$ is an abelian group. When C is rational, $\mathcal{G}$ is a subgroup.

One of the most important theorems in this area is:

Theorem 13.19. (Mordell's Theorem.) *Suppose that C is a nonsingular rational cubic curve in $\mathbb{CP}^2$ having a rational point. Then the group $\mathcal{G}$ of rational points is finitely generated.*

Proof: The original proof is due to Mordell [58]. A sketch, based on a version due to Weil [87], is described in Goldman [34]. The main idea is to define a function $H(x)$, for $x \in \mathcal{G}$, that measures the 'complexity' of x, and use H as the basis of an inductive argument. This function has the following properties:

- For any $K > 0$ the set $\{P \in \mathcal{G} : H(P) < K\}$ is finite.

- For each $Q \in \mathcal{G}$ there exists a constant c depending only on Q such that $H(P + Q) \leq c(H(P))^2$.

- There exists a constant d such that $H(P) \leq d(H(2P))^{1/4}$.

- The quotient group $\mathcal{G}/2\mathcal{G}$ is finite.

In fact, if $P = (x, y)$ and $x = m/n$ in lowest terms, we take $H(P) = \max(|m|, |n|)$. □

Recall from Proposition 1.18 that a finitely generated abelian group is of the form

$$F \oplus \mathbb{Z}^k$$

where F is a finite abelian group, hence a direct sum of finite cyclic groups. The group F, which is unique, consists of the elements of finite order, and is called the *torsion subgroup*. The groups $\mathcal{G}$ determined by elliptic curves are very special, as is shown by the following theorem of Mazur:

Theorem 13.20. *Let $\mathcal{G}$ be the group of rational points on an elliptic curve. Then the torsion subgroup of $\mathcal{G}$ is isomorphic either to $\mathbb{Z}_l$ where $1 \leq l \leq 10$, or $\mathbb{Z}_2 \oplus \mathbb{Z}_{2l}$ where $1 \leq l \leq 4$.*

Proof: The proof is very technical: see Mazur [52, 53].   □

13.7   Applications to Diophantine Equations

We now describe an application of the above ideas to an equation very similar to Fermat's. This application is due to Elkies [26].

We know that it is impossible for two cubes to sum to a cube, but might it be possible for three cubes to sum to a cube? It is; in fact $3^3 + 4^3 + 5^3 = 6^3$. Euler conjectured that in general n nth powers can sum to an nth power, but not $n - 1$. It has been proved that Euler's conjecture is false. In 1966 Lander and Parkin [46] found the first counterexample to Euler's conjecture: four fifth powers whose sum is a fifth power. In fact

$$27^5 + 84^5 + 110^5 + 133^5 = 144^5.$$

As a check:

$$
\begin{aligned}
27^5 &= 14348907 \\
84^5 &= 4182119424 \\
110^5 &= 16105100000 \\
133^5 &= 41615795893 \\
\hline
144^5 &= 61917364224.
\end{aligned}
$$
(13.10)

They found this example by exhaustive computer search.

In 1988 Noam Elkies found another counterexample by applying the theory of elliptic curves: three fourth powers whose sum is a fourth power.

$$
\begin{aligned}
2682440^4 &= 51774995082902409832960000 \\
15365639^4 &= 55744561387133523724209779041 \\
18796760^4 &= 124833740909952854954805760000 \\
\hline
20615673^4 &= 180630077292169281088848499041
\end{aligned}
$$
(13.11)

Instead of looking for integer solutions to the equation $x^4 + y^4 + z^4 = w^4$, Elkies divided out by w^4 and looked at the surface $r^4 + s^4 + t^4 = 1$ in coordinates (r, s, t). An integer solution to $x^4 + y^4 + z^4 = w^4$ leads to a rational solution $r = x/w, s = y/w, z = t/w$ of $r^4 + s^4 + t^4 = 1$. Conversely, given a rational solution of $r^4 + s^4 + t^4 = 1$, we can assume that r, s, t all have the same denominator w by putting them over a common denominator, and that leads directly to a solution to $x^4 + y^4 + z^4 = w^4$. Demjanenko [22] had found a rather complicated condition for a rational point (r, s, t) to lie on the closely related surface $r^4 + s^4 + t^2 = 1$. Namely, such a rational point exists if and only if there exist x, y, u such that

$$
\begin{aligned}
r &= x + y \\
s &= x - y \\
(u^2 + 2)y^2 &= -(3u^2 - 8u + 6)x^2 - 2(u^2 - 2)x - 2u \\
(u^2 + 2)t &= 4(u^2 - 2)x^2 + 8ux + (2 - u^2)
\end{aligned}
$$

To solve Elkies's problem it is enough to show that t can be made a square. A series of simplifications shows that this can be done provided the equation

$$
Y^2 = -31790X^4 + 36941X^3 - 56158X^2 + 28849X + 22030
$$

has a rational solution. This equation defines an elliptic curve. (Despite the presence of a fourth power on the right hand side, it can be transformed into a cubic. A similar transformation can be found in Section 14.2. See also McKean and Moll [56] page 254.) Conditions are known under which no solution can exist, but these conditions did not hold in this case, which showed that such a solution might possibly exist. At this stage Elkies tried a computer search, and found the solution

$$
\begin{aligned}
X &= -\frac{31}{467} \\
Y &= \frac{30731278}{467^2}.
\end{aligned}
$$

From this he deduced the rational solution

$$
\begin{aligned}
r &= -\frac{18796760}{20615673}, \\
s &= \frac{2682440}{20615673}, \\
t &= \frac{2682440}{20615673}.
\end{aligned}
$$

This led directly to the counterexample to Euler's conjecture for fourth powers, namely

$$
2682440^4 + 15365639^4 + 187960^4 = 20615673^4.
$$

In fact, there are infinitely many solutions. The theory of elliptic curves provides a general procedure for constructing new rational points from old ones—the tangent/secant construction. Using a version of this, Elkies proved that infinitely many rational solutions exist. In fact he proved that rational points are dense on the surface $r^4 + s^4 + t^4 = 1$, that is, any patch of the surface, however small, must contain a rational point. The second solution generated by the tangent/secant construction is

x = 1439965710648954492268506771833175267850201426615300442218292336336633,

y = 4417264698994538496943597489754952845854672497179047898864124209346920,

z = 9033964577482532388059482429398457291004947925005743028147465732645880,

w = 9161781830035436847832452398267266038227002962257243662070370888722169.

After Elkies had discovered there was a solution, Roger Frye of the Thinking Machines Corporation did an exhaustive computer search. He found a smaller solution, indeed the smallest possible solution:

$$
\begin{aligned}
95800^4 &= 84229075969600000000 \\
217519^4 &= 2238663363846304960321 \\
414560^4 &= 29535857400192040960000 \\
\hline
422481^4 &= 31858749840007945920321.
\end{aligned}
\tag{13.12}
$$

13.8 Exercises

1. Prove that the rational conic $x^2 + y^2 - 3 = 0$ contains no rational points.

   (*Hint:* Rational solutions correspond to integer solutions of the equation $X^2 + Y^2 = 3Z^2$. Without loss of generality X, Y, and Z have no common factor > 1. Now consider the equation mod 3.)

2. Consider a cubic curve C in $\mathbb{CP}^2$ whose equation is

$$
X^3 + Y^3 + Z^3 - AXYZ = 0 \tag{13.13}
$$

   where $A^3 = 27$. Show that C is singular.

   (*Hint:* When $A = 3$ there is a factorization $X^3 + Y^3 + Z^3 - 3XYZ = (X + Y + Z)(X^2 + Y^2 + Z^2 - XY - YZ - ZX)$. Draw the real affine versions of the two curves $X + Y + Z = 0$ and $X^2 + Y^2 + Z^2 - XY - YZ - ZX = 0$.)

14

Elliptic Functions

So far our discussion has been algebraic. We now introduce methods from complex analysis and the concept of a modular function. In the next chapter, we show how these classical ideas lead to the Taniyama–Shimura–Weil Conjecture, which forms the centrepiece of Wiles's approach to—and proof of—Fermat's Last Theorem.

The main topics dicussed in this chapter are:

- Trigonometric functions as a link between conic sections, Diophantine equations, and complex analysis.

- Weierstrassian elliptic functions and their connection with elliptic curves.

- Elliptic modular functions and their connection with elliptic curves.

An excellent reference for the material covered in this chapter, and many related topics, is McKean and Moll [56]. See also King [45] for the basic material, plus some fascinating connections with the solution of polynomial equations.

14.1 Trigonometry Meets Diophantus

In this section we explore a rich area of interconnections between trigonometric functions, complex analysis, algebraic geometry, and the Pythagorean equation

$$X^2 + Y^2 = Z^2. \tag{14.1}$$

We take the point of view that we do not yet have the machinery of trigonometric functions available, and show how to derive these functions from the aforementioned interconnections. This is a useful 'dry run' for subsequent generalizations to elliptic functions. Of course it helps to bear the standard trigonometric functions in mind throughout, since then the manipulations we perform make more sense.

An Approach to Trigonometric Functions

We can consider (14.1) as the projective form of the affine equation

$$x^2 + y^2 = 1 \tag{14.2}$$

by setting $x = X/Z, y = Y/Z$. Clearly integer or rational solutions of (14.1) correspond to rational solutions of (14.2). Solving (14.2) for y we get $y = \pm\sqrt{1 - x^2}$, which suggests looking at the integral

$$S(x) = \int \frac{dx}{y} = \int \frac{dx}{\sqrt{1 - x^2}}. \tag{14.3}$$

In order to evaluate this integral we assume that there exists a function $s(u)$, with derivative $c(u) = ds/du$, such that

$$c^2(u) + s^2(u) = 1. \tag{14.4}$$

To define these functions uniquely we impose 'initial conditions' $s(0) = 0, c(0) = 1$.

Given such a function, we can evaluate (14.3) by substituting $x = s(u)$. Then $dx = c(u)du$, so

$$\int \frac{dx}{\sqrt{1 - x^2}} = \int \frac{c(u)}{\sqrt{1 - s^2(u)}} = \int du = u.$$

Therefore

$$\int \frac{dx}{\sqrt{1 - x^2}} = s^{-1}(x). \tag{14.5}$$

Turning all this round, we can use (14.5) as the definition of the function s, by means of the equation

$$\int_0^{s(u)} \frac{dx}{\sqrt{1 - x^2}} = u \tag{14.6}$$

and we deduce that s and its derivative c satisfy (14.4), together with the initial conditions. Similar arguments (Exercise 1 at the end of this chapter) show that

$$s(-u) = -s(u), \quad c(-u) = c(u)$$

for all u for which $s(u), c(u)$ are defined.

The abstract theory of the Riemann integral guarantees that $s(u)$ is defined for u in some neighbourhood of 0. We can extend the definitions of s, c to the whole of $\mathbb{R}$ by making use of some of their special properties. So we now deduce the standard properties of trigonometric functions from our definition. Differentiate (14.4) to get

$$2c(u)c'(u) + 2s(u)c(u) = 0$$

proving that

$$\frac{d}{du}c(u) = -s(u) \tag{14.7}$$

We are now in a position to derive the standard power series for sine and cosine by invoking Taylor's Theorem. By induction, successive derivatives of s, c at the origin are given by

$$s^{(n)}(0) = \begin{cases} 0 \text{ if } n \equiv 0 \ (\mathrm{mod}\ 4) \\ 1 \text{ if } n \equiv 1 \ (\mathrm{mod}\ 4) \\ 0 \text{ if } n \equiv 2 \ (\mathrm{mod}\ 4) \\ -1 \text{ if } n \equiv 3 \ (\mathrm{mod}\ 4) \end{cases}$$

and

$$c^{(n)}(0) = \begin{cases} 1 \text{ if } n \equiv 0 \ (\mathrm{mod}\ 4) \\ 0 \text{ if } n \equiv 1 \ (\mathrm{mod}\ 4) \\ -1 \text{ if } n \equiv 2 \ (\mathrm{mod}\ 4) \\ 0 \text{ if } n \equiv 3 \ (\mathrm{mod}\ 4) \end{cases}$$

so that

$$\begin{aligned} s(x) &= \sum_{n=0}^{\infty}(-1)^n \frac{x^{2n+1}}{(2n+1)!}, \\ c(x) &= \sum_{n=0}^{\infty}(-1)^n \frac{x^{2n}}{(2n)!}. \end{aligned} \tag{14.8}$$

These series are absolutely convergent for all $x \in \mathbb{R}$ and therefore define s, c on the whole of $\mathbb{R}$. Indeed, we can replace $x \in \mathbb{R}$ by $z \in \mathbb{C}$ and extend the definitions to the complex plane: now s, c are complex analytic. The identity (14.4) now holds when x is replaced by any complex z, because two power series that are equal on an open set of real values x are equal throughout $\mathbb{C}$. At this stage we are entitled to replace s by $\sin$ and c by $\cos$, but to emphasise the logical line of development we continue to use the notation s, c.

Addition Formulas and Parametrization of the Circle

We next seek formulas for $s(u + v)$ and $c(u + v)$. Working backwards from (14.7) we see that Equation (14.4) is equivalent to $X(U) = s(u)$ and $X(u) = c(u)$ being independent solutions of the second order linear differential equation

$$\frac{d^2 X}{du^2} + X = 0. \qquad (14.9)$$

The general solution of (14.9), with arbitrary initial conditions, is

$$X(u) = As(u) + Bc(u)$$

for constants A, B. Let v be any constant. Clearly $X(u) = s(u + v)$ is a solution of (14.9), so

$$s(u + v) = As(u) + Bc(u) \qquad (14.10)$$

and by differentiation with respect to u we also have

$$c(u + v) = Ac(u) - Bs(u) \qquad (14.11)$$

where the constants A, B may depend on v. Letting $u = 0$ we see that $A = -c(v), B = s(v)$. Thus we have the addition formulas

$$s(u + v) = s(u)c(v) + c(u)s(v),$$
$$c(u + v) = c(u)c(v) - s(u)s(v).$$

We are now ready to return to the Pythagorean equation (14.2), which defines a curve in $\mathbb{R}^2$, the unit circle $\mathbb{S}^1$. By Equation (14.4) the point $(c(u), s(u))$ lies in $\mathbb{S}^1$ for any $u \in \mathbb{R}$. That is, there is a map

$$\Omega : \mathbb{R} \to \mathbb{S}^1$$
$$u \mapsto (c(u), s(u)).$$

This map is continuous, indeed infinitely differentiable.

The real line $\mathbb{R}$ has a natural abelian group operation $+$, and we can use the map Ω to transport this to $\mathbb{S}^1$ if we define $\oplus$ as follows:

$$(x_1, y_1) \oplus (x_2, y_2) = (x_1 x_2 - y_1 y_2, x_1 y_2 + y_1 x_2).$$

Under this operation $\mathbb{S}^1$ is an abelian group with identity element $(1, 0)$. The inverse of (c, s) is $(c, -s)$. The addition formulas for s, c now say that Ω is a group homomorphism, that is,

$$\Omega(u + v) = \Omega(u) \oplus \Omega(v). \tag{14.12}$$

The map Ω cannot be a group isomorphism since $\mathbb{S}^1$ is compact but $\mathbb{R}$ is not. Therefore Ω has a non-trivial kernel $\mathcal{K}$. We claim that there is a unique real number $\varpi > 0$ such that

$$\mathcal{K} = \varpi \mathbb{Z}$$

To see this, observe that the derivative of Ω is nonsingular near $u = 0$, so there exists $\epsilon > 0$ such that $\Omega(u) \neq (1, 0)$ whenever $0 \neq u$ and $|u| < \epsilon$. It follows that there exists a smallest real number $\varpi > 0$ for which $\Omega(\varpi) = (1, 0)$. Then $\Omega(n\varpi) = (1, 0)$ for all $n \in \mathbb{Z}$ since $\mathcal{K}$ is a subgroup. We claim that $\mathcal{K} = \varpi \mathbb{Z}$. If not, there exists $k \in \mathcal{K}$ such that

$$n\varpi < k < (n + 1)\varpi$$

for some $n \in \mathbb{Z}$. But then $k - n\varpi \in \mathcal{K}$ and $0 < k - n\varpi < \varpi$, a contradiction.

Numerical computations show that $\varpi \sim 6.28$, and of course $\varpi = 2\pi$.

Since Ω is a homomorphism and $\varpi \in \mathcal{K}$, we deduce that $\Omega(u + \varpi) = \Omega(u)$ for all $u \in \mathbb{R}$ (hence also for all $u \in \mathbb{C}$). That is, both s and c are ϖ-periodic.

All of the usual properties of the trigonometric functions now follow by standard methods. In particular we can show (Exercise 2) that Ω is onto. Geometrically, Ω wraps $\mathbb{R}$ round $\mathbb{S}^1$ infinitely many times, in such a manner that the usual distance in $\mathbb{R}$ becomes $1/2\pi$ times arc-length in $\mathbb{S}^1$.

The Pythagorean Equation

The group structure on $\mathbb{S}^1$ defined by $\oplus$ has an interesting implication for the Pythagorean Equation. Namely, given two solutions

$$x^2 + y^2 = 1,$$
$$u^2 + v^2 = 1$$

it implies that there exists a further solution

$$(xu - yv)^2 + (xv + yu)^2 = 1. \tag{14.13}$$

For example, from the standard $(3,4,5)$ right triangle we know that $x = 3/5, y = 4/5$ is a solution, and so is $u = 3/5, v = 4/5$. By (14.13) we compute a new solution $(7/25, 24/25)$, so that $7^2 + 24^2 = 25^2$. In this manner we can obtain an infinite number of rational solutions of (14.2), hence of integer solutions of (14.1).

We now seek to characterise those $u \in \mathbb{R}$ for which $\Omega(u) \in \mathbb{Q}^2$, that is, both $s(u)$ and $c(u)$ are rational. To this end we introduce

$$t = t(u) = \tan \frac{u}{2}.$$

From the identities

$$\cos u = \cos^2 \frac{u}{2} - \sin^2 \frac{u}{2}$$
$$1 = \cos^2 \frac{u}{2} + \sin^2 \frac{u}{2}$$

we find that

$$\cos u = \frac{1 - t^2}{1 + t^2} \qquad \sin u = \frac{2t}{1 + t^2}.$$

Proposition 14.1. $\Omega(u) \in \mathbb{Q}^2$ if and only if $t \in \mathbb{Q}$.

Proof: Clearly $t \in \mathbb{Q}$ implies $\Omega(u) \in \mathbb{Q}^2$.
    If $\Omega(u) \in \mathbb{Q}^2$ then

$$\frac{1 + t^2}{2t} = p \in \mathbb{Q},$$
$$\frac{1 - t^2}{2t} = q \in \mathbb{Q}$$

so that

$$1 + t^2 = 2pt,$$
$$1 - t^2 = 2qt.$$

Adding,

$$2 = 2(p + q)t$$

so that $p + q \neq 0$ and $t = \frac{1}{p+q} \in \mathbb{Q}$. $\qquad\qquad\square$

In other words, $\Omega(u) \in \mathbb{Q}^2$ if and only if $u \in 2 \arctan \mathbb{Q}$.

The identity (14.4) is now equivalent to the rational identity

$$\left(\frac{1-t^2}{1+t^2}\right)^2 + \left(\frac{2t}{1+t^2}\right)^2 = 1$$

which we have already encountered in Example 13.9. Putting $t = u/v$ where $u, v \in \mathbb{Z}$ this yields

$$(u^2 - v^2)^2 + (2uv)^2 = (u^2 + v^2)^2.$$

In Lemma 11.1 we showed that all primitive integer solutions of the Pythagorean equation (that is, solutions without common factors) are of this form. Thus we have found a link between the trigonometric functions (especially $t(u)$), the Pythagorean equation, and the unit circle in the plane.

A Curious Series

Later in this chapter we develop a profound generalization of the above to elliptic curves. We could continue to explore the circle and its links for some time, but we content ourselves with one curious formula that makes the 2π-periodicity of the trigonometric functions 'obvious'. Its analogue for doubly periodic complex functions forms the basis of Weierstrass's approach to elliptic functions. The relevant identity is

$$\csc z = \frac{1}{z} + \sum_{n \in \mathbb{Z} \setminus \{0\}} (-1)^n \left(\frac{1}{z - n\pi} + \frac{1}{n\pi}\right). \qquad (14.14)$$

This series is absolutely convergent and so may be differentiated term by term, yielding the simpler identity

$$\csc z \cot z = \sum_{n \in \mathbb{Z}} \frac{(-1)^n}{(z - n\pi)^2}. \qquad (14.15)$$

The series in (14.14) would itself be simpler if we could use

$$\sum_{n \in \mathbb{Z}} \frac{(-1)^n}{(z - n\pi)} \qquad (14.16)$$

but unfortunately this series fails to converge, which is why the more complicated (14.14) replaces it.

If we replace z in (14.15) by $z + 2\pi$ then the entire series shifts one place along, term by term. This makes it obvious that the function $\csc z \cot z$

is 2π-periodic in z. It is straightforward to parlay this result into 2π-periodicity of $\sin z$ and $\cos z$. So it is possible to *define* the trigonometric functions in terms of a series whose 2π-periodicity is immediately apparent from its form.

We outline the derivation of (14.14). We rely on standard ideas from complex analysis about residues. Suppose $f(z)$ is a complex analytic function whose only singularities in $\mathbb{C}$ are poles $z = a_j, j = 1, 2, 3, \ldots$ with residues b_j, where $0 < |a_1| \le |a_2| \le |a_3| \le \cdots$. Suppose that there exists a sequence of circles C_j, with centre the origin and of radius R_j which tends to infinity with j, not passing through any poles, with $f(z)$ uniformly bounded on the circles C_j; that is, $|f(z)| < M$ for all $z \in \cup_j C_j$.

An example is $f(z) = \csc z$, with $R_j = (j + \tfrac{1}{2})\pi$, to which we return shortly.

By the residue theorem (Stewart and Tall [80] Chapter 12) if x is not a pole of f then

$$\frac{1}{2\pi i} \int_{C_j} \frac{f(z)}{z - x} dz = f(x) + \sum_r \frac{b_r}{a_r - x}$$

where the sum is over all poles interior to C_j. Now

$$\frac{1}{2\pi i} \int_{C_j} \frac{f(z)}{z - x} dz = \frac{1}{2\pi i} \int_{C_j} \frac{f(z)}{z} dz + \frac{x}{2\pi i} \int_{C_j} \frac{f(z)}{z(z - x)} dz$$

$$= f(0) + \sum_r \frac{b_r}{a_r} + \frac{x}{2\pi i} \int_{C_j} \frac{f(z)}{z(z - x)} dz.$$

Let $j \to \infty$. Then the integral

$$\int_{C_j} \frac{f(z)}{z(z - x)} dz$$

tends to zero, and we get

$$f(x) = f(0) + \sum_{n=1}^{\infty} b_n \left(\frac{1}{x - a_n} + \frac{1}{a_n} \right). \tag{14.17}$$

To prove (14.14) we now set $f(z) = \csc z - \frac{1}{z}$. The singularities are at $z = a_j = j\pi, j \ne 0$, with residues $b_j = (-1)^j$. The conditions of the above calculation are easily checked, so (14.14) follows.

Weierstrass uses a very similar series to define a class of doubly periodic functions (and has a similar problem with convergence of the simplest form of the series, which he solves in the same manner). We now begin the development of Weierstrass's theory.

14.2 Elliptic Functions

We have seen that the trigonometric functions have a number of striking properties, including:

- Periodicity: $\sin(\theta + 2\pi) = \sin(\theta), \cos(\theta + 2\pi) = \cos(\theta)$

- Algebraic differential equation: $u'(\theta)^2 = 1 - u(\theta)^2$ where $u(\theta) = \sin(\theta)$ or $\cos(\theta)$

- Parametrization of circle: $(\cos(\theta), \sin(\theta))$ lies on the unit circle $x^2 + y^2 = 1$ and every point on the circle is of this form

- Addition theorems:

$$\cos(\theta + \phi) = \cos(\theta)\cos(\phi) - \sin(\theta)\sin(\phi)$$
$$\sin(\theta + \phi) = \sin(\theta)\cos(\phi) + \cos(\theta)\sin(\phi)$$

- Integration of algebraic functions: for example,

$$\int \frac{dx}{\sqrt{1 - x^2}} = \sin^{-1}(x).$$

These properties are all interconnected; moreover, they illuminate the theory of quadratic Diophantine equations, and in particular the Pythagorean Equation $X^2 + Y^2 - Z^2 = 0$, which in affine form is $x^2 + y^2 - 1 = 0$.

In 1811 Legendre published the first of a series of three volumes initiating a profound generalization of trigonometric functions, known—for the rather peripheral reason that they lead to a formula for the arc length of an ellipse—as 'elliptic functions'. In fact Legendre worked only with 'elliptic integrals', of which the most important is

$$F(k, v) = \int_0^v \frac{dz}{\sqrt{(1 - z^2)(1 - k^2 z^2)}} \qquad (14.18)$$

for a complex variable z and a complex constant k. This particular integral is the *elliptic integral of the first kind*: in Legendre's theory there are also elliptic integrals of the second and third kind. A detailed treatment can be found in Hancock [36] page 187.

Gauss, Abel, and Jacobi all noticed something that had eluded Legendre. (The history is complicated. Gauss never published the idea; Jacobi made it the basis of his monumental and influential work published in 1829; and a manuscript by Abel was submitted to the French Academy of Sciences in 1826, mislaid by Cauchy, and not published until 1841, by which time Abel was long dead. Abel also published work on elliptic functions in 1827, however.) Their common idea was to consider the integral (14.18) not as defining a function, but as defining its *inverse function*. Denote this inverse function by sn u: it satifies the equation

$$F(k, \operatorname{sn} u) = u.$$

Strictly speaking, sn is a family of functions parametrized by k. Associated with it are two other functions

$$\operatorname{cn} u = \sqrt{1 - \operatorname{sn}^2 u},$$
$$\operatorname{dn} u = \sqrt{1 - k^2 \operatorname{sn}^2 u}.$$

These functions have remarkable properties reminiscent of the trigonometric functions: a sample is

$$\operatorname{sn}(x + y) = \frac{\operatorname{sn} x \operatorname{cn} y \operatorname{dn} y + \operatorname{sn} y \operatorname{cn} x \operatorname{dn} x}{1 - k^2 \operatorname{sn}^2 x \operatorname{sn}^2 y}.$$

A vast range of similar identities can be found in Cayley [14] and Hancock [36]. The most remarkable property of all, though, is that the functions sn, cn, dn are all *doubly periodic*. That is, there exist two complex constants ω_1, ω_2 (depending on k) that are linearly independent over $\mathbb{R}$, such that

$$\operatorname{sn}(z + \omega_1) = \operatorname{sn}(z + \omega_2) = \operatorname{sn} z,$$
$$\operatorname{cn}(z + \omega_1) = \operatorname{cn}(z + \omega_2) = \operatorname{cn} z,$$
$$\operatorname{dn}(z + \omega_1) = \operatorname{dn}(z + \omega_2) = \operatorname{dn} z.$$

Legendre had recognised the equivalent property of his elliptic integrals, but its expression is far more cumbersome. Moreover, sn, cn, dn are all *meromorphic* functions of a complex variable, meaning that they are analytic except for isolated singularities, which are all poles. (See Stewart and Tall [80] or any other text on complex analysis for terminology.)

In 1882 Weierstrass developed a somewhat different approach to the whole topic of doubly periodic functions, based on a function denoted $\wp(z)$. (The symbol $\wp$ is pronounced 'pay' and is a stylized old German 'p'.) The Weierstrass $\wp$-function is closely connected with elliptic curves, and the remainder of this section is devoted to this connection.

The starting-point is to consider an arbitrary doubly periodic meromorphic function $f(z)$, where $z \in \mathbb{C}$. Then there are two complex constants ω_1, ω_2 that are linearly independent over $\mathbb{R}$, such that

$$f(z + \omega_1) = f(z + \omega_2) = f(z). \tag{14.19}$$

This implies that for all $(m, n) \in \mathbb{Z}^2$

$$f(z + m\omega_1 + n\omega_2) = f(z) \tag{14.20}$$

and opens up some interesting geometry of $\mathbb{C}$.

Definition 14.2. Let $\omega_1, \omega_2 \in \mathbb{C}$ be linearly independent over $\mathbb{R}$. Then the set

$$\mathcal{L} = \mathcal{L}_{\omega_1, \omega_2} = \{z \in \mathbb{C} : z = m\omega_1 + n\omega_2 \text{ where } m, n \in \mathbb{Z}^2\}$$

is the *lattice* generated by ω_1, ω_2.

We studied lattices in $\mathbb{R}^n$ in Chapter 6. The above definition is a special case, and arises when we identify $\mathbb{C}$ with $\mathbb{R}^2$. When (14.20) holds, we say that f is $\mathcal{L}$-*periodic*.

Suppose that T is the fundamental domain of $\mathcal{L}$, see Chapter 6 just before Lemma 6.2. Then by Lemma 6.2, every $z \in \mathbb{C}$ lies in exactly one of the sets $T + l$ for $l \in \mathcal{L}$. Therefore $f(z)$ is uniquely determined once we know $f(t)$ for all $t \in T$. Classically, the topological closure $\overline{T}$ of T is called the *period parallelogram*. We can consider f to be a function on the quotient torus $\mathbb{O}^2 = \mathbb{C}/\mathcal{L}$.

The main problem is to define 'interesting' doubly periodic functions for a given lattice $\mathcal{L}$. Weierstrass's idea is a generalization of (14.14, 14.15), namely that these can be obtained by *summing over translates by the lattice*. That is, take some (initially arbitrary) function $g(z)$ and consider

$$\hat{g}(z) = \sum_{l \in \mathcal{L}} g(z - l).$$

Then $\hat{g}$ is obviously doubly periodic, because for any $l' \in \mathcal{L}$

$$\hat{g}(z + l') = \sum_{l \in \mathcal{L}} g(z - l + l') = \sum_{l'' \in \mathcal{L}} g(z + l'') = \hat{g}(z)$$

where $l'' = l' - l$.

This is all very well, but it is necessary to choose $g(z)$ with care. Three things can go wrong:

- The series defining $\hat{g}$ fails to converge.

- The series converges but $\hat{g}$ is not meromorphic.

- The series converges but $\hat{g}$ turns out to be constant.

Avoiding these pitfalls requires a certain amount of foresight. Initially it led Weierstrass to the choice $g(z) = \frac{1}{z^3}$, for which $\hat{g}(z) = -\wp'(z)$, and from this to a suitable choice of (some constant multiple of) the integral of this particular $\hat{g}$. It turns out that taking $g(z) = \frac{1}{z^2}$ is not a good idea: the series for $\hat{g}$ fails to converge. However, by adding a suitable 'arbitrary constant' (which, taken on its own, *also* fails to converge, but in the 'opposite' way) he could obtain a meromorphic function. His final choice was:

Definition 14.3. Let $\mathcal{L} \subseteq \mathbb{C}$ be a lattice. Then the associated *Weierstrass $\wp$-function* is

$$\wp(z) = \frac{1}{z^2} + \sum_{l \in \mathcal{L} \backslash 0} \left(\frac{1}{(z-l)^2} - \frac{1}{l^2} \right). \tag{14.21}$$

This series is absolutely convergent provided $z \notin \mathcal{L}$ (Exercise 3). Evidently $\wp$ is an *even* function, that is, $\wp(-z) = \wp(z)$. We may therefore differentiate term by term to get

$$\wp'(z) = -2 \sum_{l \in \mathcal{L}} \frac{1}{(z-l)^3}.$$

This series is also absolutely convergent provided $z \notin \mathcal{L}$ (Exercise 3).

Lemma 14.4. *The function $\wp$ is meromorphic, and doubly periodic on the lattice $\mathcal{L}$.*

Proof: By the above discussion, $\wp'$ is meromorphic, and doubly periodic on the lattice $\mathcal{L}$. Therefore

$$\wp'(z + \omega_j) = \wp'(z).$$

Integrate, and remember to include an arbitrary constant:

$$\wp(z + \omega_j) = \wp(z) + c_j \quad (j = 1, 2) \tag{14.22}$$

for constants $c_j \in \mathbb{C}$. Since $\wp$ is an even function, its derivative is odd, so $\wp'(-z) = -\wp(z)$. Setting $u = -\omega_1$ in (14.22) we have

$$\wp(\omega_1) = \wp(-\omega_1) + c_1 = \wp(\omega_1) + c_1$$

so $c_1 = 0$. Similarly $c_2 = 0$, so $\wp$ is $\mathcal{L}$-periodic.

By absolute convergence, $\wp(z)$ is analytic for all $z \notin \mathcal{L}$. When $z \in \mathcal{L}$ we may without loss of generality take $z = 0$. Then (14.21) shows that $\wp(z)$

has a simple pole at 0, of order 2. Therefore all singularities are poles, and $\wp$ is meromorphic. $\qquad\qquad\square$

We now prove a useful lemma:

Lemma 14.5. *Suppose that f is a doubly periodic function that is analytic throughout $\mathbb{C}$ (that is, no poles or other singularities). Then f is constant.*

Proof: Let $\overline{T}$ be the closure of the fundamental domain of $\mathcal{L}$ (the period parallelogram). Since $\overline{T}$ is compact and f is analytic on $\overline{T}$, there exists a real constant M such that $|h(z)| < M$ for all $z \in \overline{T}$. By double periodicity, $|h(z)| < M$ for all $z \in \mathbb{C}$. By Liouville's Theorem (Stewart and Tall [80] page 184) $h(z)$ is constant. $\qquad\qquad\square$

Define

$$\begin{aligned} g_2 &= 60 \sum_{l \in \mathcal{L} \setminus 0} \tfrac{1}{l^4} \\ g_3 &= 140 \sum_{l \in \mathcal{L} \setminus 0} \tfrac{1}{l^6}. \end{aligned} \qquad (14.23)$$

(These and similar expressions are called *Eisenstein series*.) Then it may be shown (Hancock [36] page 324) that the Laurent series (Stewart and Tall [80] page 195) of $\wp, \wp'$ take the following form:

$$\wp(z) = \frac{1}{z^2} + \frac{g_2}{20}z^2 + \frac{g_3}{140}z^4 + \ldots \qquad (14.24)$$

$$\wp'(z) = \frac{-2}{z^3} + \frac{g_2}{10}z + \frac{g_3}{7}z^3 + \ldots. \qquad (14.25)$$

Theorem 14.6. *The Weierstrass $\wp$-function satisfies the differential equation*

$$\wp'(z)^2 = 4\wp(z)^3 - g_2\wp(z) - g_3 \qquad (14.26)$$

Proof: By direct computation,

$$\wp'(z)^2 = \frac{4}{z^6} - \frac{2g_2}{5z^2} + \frac{4g_3}{7} + O(z^2) \qquad (14.27)$$

$$\wp(z)^3 = \frac{1}{z^6} + \frac{3g_2}{20z^2} + \frac{3g_3}{28}z^3 + O(z^2) \qquad (14.28)$$

where $O(z^2)$ denotes a function whose Laurent series begins with terms in z^2 or higher—which, in particular, is an analytic function for all $z \in \mathbb{C}$. Therefore

$$\wp'(z)^2 - [4\wp(z)^3 - g_2\wp(z) - g_3] = O(z^2).$$

The left-hand side, which we denote by $h(z)$, is doubly periodic; the right-hand side is analytic throughout $\mathbb{C}$. Lemma 14.5 implies that $h(z)$ is constant. Since $h(z) = O(z^2)$, it follows that $h(z) = 0$.   □

Corollary 14.7. *Let C be an elliptic curve in Weierstrass normal form $y^2 = 4x^3 - g_2x - g_3$ and let $\wp$ be the corresponding Weierstrass $\wp$-function. Then $(x, y) = (\wp'(z), \wp(z))$ lies on C for all $z \in \mathbb{C}$.*

In fact, every point on C is of the above form, so the function $\wp$ *parametrizes* C. There is a close analogy with the parametrization of a circle by $(\cos(\theta), \sin(\theta))$, especially since $\sin'(\theta) = \cos(\theta)$. Moreover, the parametrization of C behaves naturally with respect to the group operation (13.9) on C, which for clarity we now rename $\oplus$ instead of $+$. (This is not the same $\oplus$ that arose earlier in this chapter in connection with trigonometric functions.) More precisely, let C be an elliptic curve with equation $y^z = 4x^3 - g_2x - g_3$. Then C is nonsingular if and only if the cubic $4x^3 - g_2x - g_3$ has distinct zeros, which happens if and only if the *discriminant*

$$9g_3^3 + 32g_2^2 \neq 0$$

There is one point at infinity on C, namely $(0, 1, 0)$, and we choose this as the identity $\mathbb{O}$ of the group $\mathcal{G}$. Straightforward calculations in coordinate geometry then show that if

$$(x_3, y_3) = (x_1, y_1) \oplus (x_2, y_2)$$

then

$$x_3 = \frac{1}{4}\left(\frac{y_1 - y_2}{x_1 - x_2}\right)^2 - (x_1 + x_2) \tag{14.29}$$

$$y_3 = \frac{y_1 - y_2}{x_1 - x_2}x_3 + \frac{x_1y_2 - x_2y_1}{x_1 - x_2}. \tag{14.30}$$

We now compare this with the addition theorem for the functions $\wp, \wp'$:

Theorem 14.8. *If $u \neq v \in \mathbb{C}$ then*

$$\wp(u + v) = \frac{1}{4}\left(\frac{\wp'(u) - \wp'(v)}{\wp(u) - \wp(v)}\right)^2 - (\wp(u) + \wp(v)) \tag{14.31}$$

$$\wp'(u + v) = \frac{\wp'(u) - \wp'(v)}{\wp(u) - \wp(v)}\wp(u + v) + \frac{\wp(u)\wp'(v) - \wp(v)\wp'(u)}{\wp(u) - \wp(v)}. \tag{14.32}$$

Proof: We sketch a proof: see Hancock [36] page 351 for details. Consider the function

$$h(u) = \wp(u+v) - \frac{1}{4}\left(\frac{\wp'(u) - \wp'(v)}{\wp(u) - \wp(v)}\right)^2$$

This is doubly periodic and meromorphic: its only poles are at points where $u + v \in \mathcal{L}$. However, by construction it is finite at $u = -v$, hence at $u = -v + l$ for all $l \in \mathcal{L}$. At $u = 0$ the function $h(u)$ goes to infinity like $-\frac{1}{u^2}$. It follows that $h(u) + \wp(u)$ is doubly periodic and analytic for all $u \in \mathbb{C}$. Lemma 14.5 implies that $h(u) + \wp(u)$ is constant. Setting $u = 0$ it follows that the constant is $-\wp(v)$.

The addition formula for $\wp'$ can be obtained by differentiation and further manipulations. $\square$

These formulas may appear complicated, but they lead immediately to a very elegant theorem:

Theorem 14.9. *The group structure on C has the property*

$$(\wp(u), \wp'(u)) \oplus (\wp(v), \wp'(v)) = (\wp(u+v), \wp'(u+v)). \qquad (14.33)$$

Proof: Compare formulas (14.29) and (14.31), and (14.30) and (14.32).

$\square$

Another way to say this is that the map $u \mapsto (\wp(u), \wp'(u))$ is a homomorphism between $(\mathbb{C}, +)$ and $(C, \oplus)$.

14.3 Legendre and Weierstrass

Legendre's theory of elliptic integrals concerns the square root of a quartic polynomial $(1 - z^2)(1 - k^2 z^2)$. Weierstrass's theory revolves around the square root of a cubic polynomial $4z^3 - g_2 z - g_3$. We briefly describe the connection between the two approaches, which is essentially that suitable birational maps transform the Legendre normal form into the Weierstrass normal form, and conversely. The discussion is summarized from Hancock [36] page 190.

Consider the integral

$$\int \frac{dz}{\sqrt{4z^3 - g_2 z - g_3}}. \qquad (14.34)$$

By Theorem 14.26 the substitution $z = \wp(t)$ transforms this into

$$\int dt = t \tag{14.35}$$

so that (14.34) is equal to $\wp^{-1}(z)$. The similarity with (14.18), which we restate here for convenience,

$$F(k,v) = \int_0^v \frac{dz}{\sqrt{(1-z^2)(1-k^2z^2)}} \tag{14.36}$$

is striking. In fact, elementary computations show that if we make the substitution

$$w = \frac{a_3 + a_2}{2} + \frac{a_3 - a_2}{2}\frac{z - k}{1 - kz}$$

in (14.36), where

$$\left(\frac{1-k}{1+k}\right)^2 = \frac{a_1 - a_2}{a_1 - a_3}$$

then (14.36) becomes

$$\frac{1}{2}\sqrt{\frac{a_2 - a_3}{k}} \int \frac{dw}{\sqrt{(w - a_1)(w - a_2)(w - a_3)}}.$$

Moreover, we can change variables from w to $u = w + c$, for a suitable constant c, and eliminate the quadratic term in $(w - a_1)(w - a_2)(w - a_3)$, reducing the cubic to Weierstrass normal form:

$$\int \frac{dw}{\sqrt{(w - a_1)(w - a_2)(w - a_3)}} = 2 \int \frac{du}{\sqrt{4u^3 - g_2 u - g_3}}$$

where

$$g_2 = -4(a_1 a_2 + a_2 a_3 + a_3 a_1)$$
$$g_3 = 4a_1 a_2 a_3.$$

The transformation is invertible, so we can also change an elliptic integral in Weierstrass normal form to one in Legendre normal form. In short, the two theories are equivalent.

This brief description conceals some beautiful mathematics that explains the relation between quartics and cubics that is exploited in the above transformation. This involves the invariants of cubic and quartic curves, the 'resolvent cubic' of a quartic equation, and the cross-ratio invariant from projective geometry. See Hancock [36] Chapter VIII.

14.4 Modular Functions

The link between elliptic curves and Fermat's Last Theorem stems from a profound generalization of doubly periodic complex functions. Liouville proved that every single-valued doubly periodic meromorphic function on a given lattice can be expressed as a rational function of Weierstrass's $\wp$-function and its derivative: see Hancock [36] page 437. This theorem classifies all such doubly periodic functions, and at first sight leaves little room for generalizations. However, translations are not the only interesting transformations of the complex plane.

Suppose that Γ is some group of invertible maps $\mathbb{C} \to \mathbb{C}$. Then we can seek complex functions f that are *invariant* under Γ, meaning that

$$f(\gamma(z)) = f(z)$$

for all $z \in C, \gamma \in \Gamma$. Doubly periodic functions arise in this way if we take Γ to be the group of all translations

$$z \mapsto z + m\omega_1 + n\omega_2$$

by elements $m\omega_1 + n\omega_2$ of the lattice $\mathcal{L}$. That is, $m, n \in \mathbb{Z}$.

The question is: what groups Γ lead to interesting results? Translations are a special case of an important class of transformations of $\mathbb{C}$, namely, the Möbius maps

$$g(z) = \frac{az + b}{cz + d}$$

of Definition 12.1.

As remarked earlier, there is a technical problem: Möbius maps take the value ∞ when $z = -d/c$. The usual way to get round this is to extend $\mathbb{C}$ to the *Riemann sphere* $\mathbb{C} \cup \{\infty\}$, see Stewart and Tall [80] page 207. If this is done, the set of Möbius maps forms a group under composition (Exercise 4).

Straight lines and circles in $\mathbb{C}$ correspond to circles on $\mathbb{C} \cup \{\infty\}$. Moreover, every Möbius map sends circles on $\mathbb{C} \cup \{\infty\}$ to circles. Being complex-analytic, every Möbius map is *conformal*: it preserves the angles at which curves meet. So Möbius maps have several remarkable properties. Translations are Möbius maps: take $a = 1, c = 1, d = 0$ to get $z \mapsto z + b$. So we may hope to find generalizations of doubly periodic functions among the Möbius maps.

The trick now is to choose fruitful subgroups of the group of Möbius maps. Taking the whole group leads to nothing of interest, because it is easy to see that a function that is invariant under all translations $z \mapsto z + b$ must be constant. As a clue, the translations by a lattice are defined in terms of a pair of integers (m, n). This leads to the following choice of group:

Definition 14.10. The *modular group* is the group of all Möbius maps

$$g(z) = \frac{az + b}{cz + d}$$

where $a, b, c, d \in \mathbb{Z}$ and $ad - bc = 1$.

It is easily seen to be a group (Exercise 5). Abstractly, it can be described in terms of the group $\mathbb{SL}_2(\mathbb{Z})$ of all matrices

$$\begin{bmatrix} a & b \\ c & d \end{bmatrix}$$

with $a, b, c, d \in \mathbb{Z}$ and $ad - bc = 1$. In fact, if Z is the subgroup comprising $\pm I$ where I is the identity matrix, the modular group is isomorphic to

$$\mathbb{PSL}_2(\mathbb{Z}) = \mathbb{SL}_2(\mathbb{Z})/Z$$

which is known as the *projective special linear group*. See Exercise 6.

For any lattice $\mathcal{L}$, the group of lattice-translations has a fundamental domain, as describe earlier. The defining property of a fundamental domain is that every point of $\mathbb{C}$ lies in exactly one of its translates by the lattice. The modular group also has a fundamental domain, but this has more subtle geometry than a parallelogram. Its construction is assisted by defining two particular elements of the modular group, namely

$$S = \begin{bmatrix} 0 & -1 \\ 1 & 0 \end{bmatrix}$$

$$T = \begin{bmatrix} 1 & 1 \\ 0 & 1 \end{bmatrix}.$$

(14.37)

These correspond to the functions

$$S(z) = -\frac{1}{z}$$
$$T = z + 1.$$

It is easy to see that $S^2 = -I$ which in $\mathbb{SL}_2(\mathbb{Z})$ represents the same element as I, that T has order ∞, and $(ST)^3 = I$. See Exercise 7.

Let $\mathbb{H} = \{z : \text{Im}(z) > 0\}$ be the upper half-plane in $\mathbb{C}$. (Despite our choice of terminology earlier, the pictures make it difficult to call this the

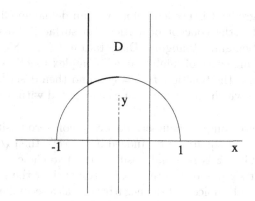

Figure 14.1. Fundamental domain for the modular group.

upper half-line.) It is easy to check that the modular group maps $\mathbb{H}$ to itself. We define the *modular domain* $\mathbb{D}$ by

$$\mathbb{D} = \{z : -\tfrac{1}{2} \leq \operatorname{Re}(z) \leq 0, |z| = 1 \text{ or } -\tfrac{1}{2} \leq \operatorname{Re}(z) < \tfrac{1}{2}, |z| > 1\}.$$

See Figure 14.1.

Theorem 14.11. $\mathbb{D}$ *is a fundamental domain for the modular group acting on* $\mathbb{H}$.

Proof: See Goldman [34] page 184. $\qquad\qquad\qquad\qquad\qquad\square$

The proof shows more: the effects of S and T on $\mathbb{D}$ are as in Figure 14.2.

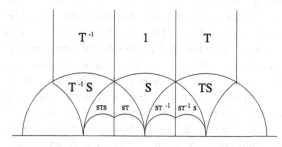

Figure 14.2. Tesselation of the upper half-plane by images of the fundamental domain.

One final ingredient is needed before we can define an elliptic modular function, namely: the concept of a Riemann surface. This concept was introduced in Riemann's Inaugural Dissertation [67] of 1851, as a general method for making sense of 'multi-valued' complex functions. We explain the idea briefly for the function $f(z) = \sqrt{z}$, and then describe an abstract generalization in which the surface is not associated with a previously defined function.

Every non-zero complex number, like every non-zero positive real number, has two distinct square roots. Indeed if $z = re^{i\theta}$, then $\sqrt{z} = +\sqrt{r}e^{i\theta/2}$ or $-\sqrt{r}e^{i\theta/2}$. When z is real and positive, the two choices can be distinguished by their sign, and it is reasonable to consider the positive square root as the 'natural' choice and the negative one as a secondary alternative. In the complex case, no straightforward distinction of this kind is possible, for the following reason. The complex setting reveals exactly why there are two choices of sign: the key point is that $-\sqrt{r}e^{i\theta/2} = +\sqrt{r}e^{i\theta/2+\pi} = +\sqrt{r}e^{i(\theta+2\pi)/2}$. That is, there are two alternatives *not* because the modulus r has two real square roots, but because the argument θ is defined only modulo 2π, and different choices of argument lead to two different values for the square root. The role of the argument is clearer if we consider the cube root $\sqrt[3]{z}$, which takes any of three values:

$$\sqrt[3]{r}e^{i\theta/3} \qquad \sqrt[3]{r}e^{i\theta/3+2\pi/3} \qquad \sqrt[3]{r}e^{i\theta/3+4\pi/3}$$

because the choices of argument $\theta, \theta + 2\pi, \theta + 4\pi$ lead to different results. (However, $\theta + 6\pi$ leads to the same cube root as θ.)

The impossibility of defining one choice to be the 'natural' square or cube root becomes obvious if we consider what happens as z moves along a continuous path in $\mathbb{C}$. For example, suppose that $z = e^{it}$ and t runs from 0 to 2π. When $t = 0$ the two square roots are $+1, -1$. The same is true when $t = 2\pi$. However, if we require $\sqrt{z}$ to vary continuously with z, then 1 lies on the path of square roots given by $\sqrt{e^{it}} = e^{it/2}$, and -1 lies on the path of square roots given by $\sqrt{e^{it}} = -e^{it/2}$. As t increases from 0 to 2π, the choice 1 changes continuously into -1, while the choice -1 changes continuously into 1. That is, the choices $1, -1$ swap places. Therefore neither can be considered more natural than the other.

Prior to Riemann, such phenomena were handled by declaring the function to be 'multi-valued', and prescribing rules for how choices of values should be made. Riemann's idea for coping with such behaviour is radically different: define the function to be single-valued (as is now conventional whenever the word 'function' is used), but specify a domain that is different from the usual complex plane $\mathbb{C}$. For the function $\sqrt{z}$ Riemann's construction can be described in terms of two superposed copies $\mathbb{C}_1$ and $\mathbb{C}_2$ of $\mathbb{C} \setminus \{0\}$, slit from 0 to $-\infty$ along the real axis. The top left-hand quadrant of $\mathbb{C}_1$ is glued to the bottom left-hand quadrant of $\mathbb{C}_2$, and the top left-hand quadrant of $\mathbb{C}_2$ is glued to the bottom left-hand quadrant of $\mathbb{C}_1$. If we try to draw the resulting surface $\mathbb{S}$ in $\mathbb{R}^3$ then it is forced to intersect itself, but abstractly no such self-intersection is implied by the gluing recipe.

There is a canonical projection $\rho : \mathbb{S} \to \mathbb{C} \setminus \{0\}$ which identifies each copy $\mathbb{C}_j$ of $\mathbb{C} \setminus \{0\}$ with $\mathbb{C} \setminus \{0\}$. Any continuous path $\gamma(t)$ in $\mathbb{C} \setminus \{0\}$ lifts to a continuous path $\hat{\gamma}(t)$ in $\mathbb{S}$, by which we mean that $\rho(\hat{\gamma}(t)) = \gamma(t)$. Suppose, for example, that $\gamma(t)$ describes the unit circle in $\mathbb{C} \setminus \{0\}$, anticlockwise, so that $\gamma(t) = e^{it}$ for $0 \leq t \leq 2\pi$. As t increases from 0 to π, this curve lifts to

$$\hat{\gamma}(t) = e^{it} \in \mathbb{C}_1.$$

However, because the two sheets $\mathbb{C}_1, \mathbb{C}_2$ are cross-connected along their negative real axes, the curve lifts to

$$\hat{\gamma}(t) = e^{it} \in \mathbb{C}_2$$

as t increases from π to 2π. So the lifted path, unlike the original, is not a closed loop: instead, it returns to a different sheet, $\mathbb{C}_2$ rather than $\mathbb{C}_1$.

If the argument of z describes the path $\gamma(t)$ in $C \setminus \{0\}$, and we choose a continuously varying argument for $\sqrt{z}$, then the choices $\pm\sqrt{z}$ change in the same manner as the sheets of the surface $\mathbb{S}$. That is, we can define a single-valued square root on $\mathbb{S}$. This is Riemann's idea. The surface $\mathbb{S}$ is one of the simplest examples of a Riemann surface in the classical sense: for further examples and proofs see Stewart and Tall [80] page 268 onwards.

The modern treatment of a Riemann surface is based on the geometry of the surface and the 'complex structure' given by its relation to $\mathbb{C}$:

Definition 14.12. A *surface* S is a topological space, covered by a countable collection of open subsets U called *patches*. Each patch U is equipped with a *local coordinate map* $\alpha_U : U \to \mathcal{D}$ where $\mathcal{D} \subseteq \mathbb{R}^2$ is the open unit disc. Finally, if $x \in S$ lies in the overlap $U \cap V$ of two patches, then the *overlap map* $\alpha_V^{-1}\alpha_U$ must be continuous where it is defined.

A *Riemann surface* is defined in the same way, but now we consider $\mathcal{D}$ to be the open unit disc in $\mathbb{C}$, and the overlap maps are required to be conformal, that is, to preserve the angles at which curves cross.

For further details see McKean and Moll [56] pages 3–5.

We can now define an elliptic modular function:

Definition 14.13. Let $N > 0$ be an integer, and define a subgroup

$$\Gamma_0(N) \subseteq \mathbb{SL}_2(\mathbb{Z})$$

by

$$\Gamma_0(N) = \left\{ \begin{bmatrix} a & b \\ c & d \end{bmatrix}, a, b, c, d \in \mathbb{Z}, ad - bc = 1, N|c \right\}.$$

This group acts on $\mathbb{H}$ and there is a compact Riemann surface $X_0(N)$ such that

$$\mathbb{H}/\Gamma_0(N) = X_0(N) \setminus \mathcal{K}$$

where $\mathcal{K}$ is some finite set of points. The points in $\mathcal{K}$ are the *cusps* of the *modular curve* $X_0(N)$ of *level N*.

Definition 14.14. An *(elliptic) modular function of level N on $\mathbb{H}$* is a function $f(z)$ that is invariant under $\Gamma_0(N)$ and descends to a function that is meromorphic on $X_0(N)$, even at the cusps.

Here 'descends' just means that f has the same value on each orbit of $\Gamma_0(N)$ on $\mathbb{H}$, and hence defines a function on the quotient space $\mathbb{H}/\Gamma_0(N)$.

14.5   Exercises

1. Prove that

$$s(-u) = -s(u), \quad c(-u) = c(u)$$

for all u for which $s(u), c(u)$ are defined.

(*Hint*: $s(-u)$ and $c(-u)$ satisfy the differential Equation (14.9), hence are of the form $As(u) + Bc(u)$ for constants A, B. Which?)

2. Prove that the map $\Omega : \mathbb{R} \to \mathbb{S}^1$ defined by $\Omega(u) = (c(u), s(u))$ is onto.

(*Hint*: Show that the image of c is the interval $[-1, 1]$ and that both signs for $\sqrt{1 - c^2(u)}$ can be realised by $s(u), s(-u)$.)

3. Prove that the series

$$\wp'(z) = -2 \sum_{l \in \mathcal{L}} \frac{1}{(z - l)^3}$$

and

$$\wp(z) = \frac{1}{z^2} + \sum_{l \in \mathcal{L} \setminus 0} \left[\frac{1}{(z - l)^2} - \frac{1}{l^2} \right]$$

are absolutely convergent provided $z \notin \mathcal{L}$.

(*Hint*: Break the sum up in terms of lattice points that lie on the parallelograms P_j with vertices $\pm 2j\omega_1 \pm 2j\omega_2$. Estimate the sums over the points in these parallelograms.) See Hancock [36] page 311.

4. Let

$$f(z) = \frac{az + b}{cz + d}$$

$$g(z) = \frac{Az + B}{Cz + D}$$

be Möbius maps, so that $ad - bc \neq 0, AD - BC \neq 0$. Show that

$$g(f(z)) = \frac{(Aa + Bc)z + (Ab + Bd)}{(Ca + Dc)z + (Cb + Dd)}.$$

Deduce that the set of Möbius maps forms a group under composition.

5. Continuing Exercise 3, if $ad - bc = 1$ and $AD - BC = 1$, prove that $(Aa + Bc)(Cb + Dd) - (Ab + Bd)(Ca + Dc) = 1$. Deduce that the modular group really is a group.

6. Prove that the modular group is isomorphic to

$$\mathbb{PSL}_2(\mathbb{Z}) = \mathbb{SL}_2(\mathbb{Z})/Z.$$

7. Let S, T be as defined in (14.37). Prove that $S^2 = -I$, T has order ∞, and ST has order 3.

15

Wiles's Strategy and Recent Developments

We continue sketching Wiles's proof of Fermat's Last Theorem. The previous chapter discusses the classical theory of modular functions. This theory provides the context required to state the Taniyama–Shimura–Weil Conjecture, which forms the centrepiece of Wiles's approach to—and proof of—Fermat's Last Theorem. The present chapter is about more recent discoveries, and develops the circle of new ideas that leads to the proof of a special case of the Taniyama–Shimura–Weil Conjecture. Wiles, building on the work of Frey and others, realised that this special case, the semistable Taniyama–Shimura–Weil Conjecture, immediately implies the truth of Fermat's Last Theorem. A vital first step is the definition of the Frey elliptic curve, which links Fermat's Last Theorem to elliptic curves.

We conclude with a summary of several recent results and conjectures that either develop the theory further, or provide insight into related questions. The main topics discussed are the full Taniyama–Shimura–Weil Conjecture, the Catalan Conjecture, the Fermat-Catalan Conjecture, the ABC Conjecture, and the Beal Conjecture.

15.1 The Frey Elliptic Curve

A major reason why problems like Fermat's Last Theorem remain unsolved for centuries is that it is difficult to find a reasonable line of attack—a place to start from. As we have seen, the 'big idea' of the 1840s and 1850s was to reformulate the problem in a cyclotomic field. This idea led to significant progress, and although in the end it failed to prove Fermat's Last Theorem, it left a legacy that was far more important than the theorem itself: the whole machinery of ideals in algebraic number theory. This kind of development is quite common in mathematics: the significance of a notorious unsolved problem often lies not in its answer (nothing of great importance would follow easily or directly from knowing whether Fermat's Last Theorem is true or false) but in the methods that the search for an answer can open up. Such problems serve as glorious reminders of areas of massive ignorance, and quell any belief that mathematics is 'pretty much worked out'. As it turned out, Fermat's Last Theorem has stimulated the creation of *several* major mathematical theories, whose far-reaching consequences are still being discovered. Wiles's ideas, leading to a complete proof of the Taniyama–Shimura–Weil Conjecture—which *is* important in its own right because it opens up new lines of attack on all kinds of questions—is the latest addition to the list.

When the approach to Fermat's Last Theorem by way of cyclotomic fields ground to a halt in the 1980s, no plausible line of attack seemed to be visible. That situation changed dramatically with the work of Helle-gouarch [41] and Frey [29, 30], who indulged in some major lateral thinking that revealed a startling link between Fermat's Last Theorem and the the-ory of elliptic curves. Bearing in mind that elliptic curves form one of the deepest areas of number theory, one equipped with a vast array of powerful machinery, the significance of this breakthrough was immediately evident to number-theorists.

It was Frey, above all, who made this link solid and complete. He started in the obvious way: assume there is a counterexample and derive a contradiction. That is, assume there exist three pairwise coprime non-zero integers a, b, c that satisfy the Fermat Equation

$$a^p + b^p = c^p$$

and then...

A lot of people have tried this, but the big problem is to derive a contradiction. This is where everyone either made a mistake or got stuck. Frey's key idea is to consider what we now call the *Frey elliptic curve* $\mathcal{F}$, defined by

$$y^2 = x(x - a^p)(x + b^p).$$

Since the above solution gives rise to two further solutions $b^p + a^p = c^p$ and $a^p + (-c)^p = (-b)^p$ we can arrange for b to be even and $a \equiv -1$ mod 4. These conditions are needed to make $\mathcal{F}$ 'semistable', a notion that we discuss below. For technical reasons, it is also useful to assume $p > 3$, which involves no loss of generality since both Fermat and Euler proved Fermat's Last Theorem for $p = 3$.

Frey's chief contribution was to recognise that the curve $\mathcal{F}$ has such strange properties that, intuitively, it cannot possibly exist. If this could be *proved*, then by contradiction there are no solutions to the Fermat equation, and Fermat's Last Theorem is proved.

Moreover, Frey [30] provided strong but incomplete evidence *why* $\mathcal{F}$ cannot exist. Namely, if it does, then it contradicts the Taniyama–Shimura–Weil Conjecture. The main gap in his argument was filled in by Serre, but only by invoking a conjecture of his own, the Special Level Reduction Conjecture. In 1986 Ribet [65] proved Serre's Special Level Reduction Conjecture (the proof was not published until 1990), at which stage the hoped-for proof of Fermat's Last Theorem rested only on the Taniyama–Shimura–Weil Conjecture. It was a sufficiently powerful special case of this conjecture that Wiles attacked, over a period of seven years, and (not without hiccups) demolished.

15.2   The Taniyama–Shimura–Weil Conjecture

The Taniyama–Shimura–Weil Conjecture (often also attributed to some subset of those three mathematicians) can be stated in numerous forms, which look very different but are equivalent given the state of knowledge in the field in the 1980s. In essence, it is this:

Conjecture 15.1. (Taniyama–Shimura–Weil Conjecture.) *Every elliptic curve over $\mathbb{Q}$ is modular.*

However, in order for this to make sense we have to explain what it means for an elliptic curve to be modular, and this is where the different alternatives arise.

The best known approach involves 'reduction modulo p', where $p \in \mathbb{Z}$ is a prime. Suppose that E is an elliptic curve over $\mathbb{Q}$, and let $\mathbb{F}_p$ be the field with p elements. We can write E in projective form as a homogeneous cubic with integer coefficients, and we can then reinterpret those coefficients as integers modulo p. We then get a cubic equation over $\mathbb{F}_p$, also in projective form; and we define b_p to be the number of distinct solutions over $\mathbb{F}_p$, including any that lie at infinity. For instance, suppose that E has affine equation

$$y^2 = x^3 + 22$$

so that in projective form it becomes

$$y^2 z = x^3 + 22 z^3$$

and let $p = 5$. In $\mathbb{F}_5$ we have $22 \equiv 2$, so we are trying to count the projectively distinct solutions of

$$y^2 z = x^3 + 2z^3$$

with $x, y, z \in \mathbb{F}_5$. By trial and error we find that there are exactly 6 of them, namely

$$
\begin{aligned}
(x, y, z) = \quad &(0, 1, 0), \\
&(1, 0, 3), \\
&(1, 1, 2), \\
&(1, 1, 4), \\
&(1, 4, 2), \\
&(1, 4, 4).
\end{aligned}
$$

The first two involve 0 so are at infinity. For example, when $(x, y, z) = (1, 4, 4)$ then $y^2 z - (x^3 + 2z^3) = -65$, which is congruent to 0 modulo 5.

Remember that if any solution is multiplied throughout by a non-zero constant, then projectively this is the same solution—so that, for instance, $(1, 1, 2)$ is the same as $(3, 3, 1)$.

We conclude that for this E,

$$b_5 = 6.$$

The numbers b_p, for various p, encode useful information about E. When E is modular, there is a formula relating *all* the numbers b_p, for all primes p, to a single function. This function is called an *eigenform*. It can be written

$$f(z) = \sum_{n=1}^{\infty} a_n e^{2\pi i n z} \tag{15.1}$$

and has some very specific properties (technically, it is a normalized cusp form of weight 2 for the $\Gamma_0(N)$ of Definition 14.14, and it is an eigenfunction for all Hecke operators). We then have:

Definition 15.2. An elliptic curve E over $\mathbb{Q}$ is *modular* if there exists an eigenform (15.1) such that for all but finitely many primes p,

$$b_p = p + 1 - a_p.$$

It is now known that this definition leads to an alternative, equivalent formulation of the Taniyama–Shimura–Weil Conjecture, which states that E can be parametrized by modular functions of a certain kind—much as the circle can be parametrized by trigonometric functions and every elliptic curve can be parametrized by the Weierstrass $\wp$-function and its derivative:

Conjecture 15.3. (Taniyama–Shimura–Weil Conjecture, alternative formulation.) *Let $y^2 = Ax^3 + Bx^2 + Cx + D$ be an elliptic curve, where $A, B, C, D \in \mathbb{Q}$. Then there exist modular functions $f(z), g(z)$, both of the same level N, such that*

$$g(z)^2 = Af(z)^3 + Bf(z)^2 + Cf(z) + D.$$

See Cox [18] and Mazur [54] for further information on this alternative formulation.

Wiles did not prove the full Taniyama–Shimura–Weil Conjecture, although this has now been done as a consequence of Wiles's ideas. He realised that a more accessible special case would be sufficient: this is known as the Semistable Taniyama–Shimura–Weil Conjecture. In order to explain what 'semistable' means, we must discuss some numerical invariants of elliptic curves. Suppose, then, that we have a rational elliptic curve $y^2 = Ax^3 + Bx^2 + Cx + D$. Over the complex numbers, the cubic equation $p(x) = Ax^3 + Bx^2 + Cx + D = 0$ has three roots x_1, x_2, x_3. Classically, the discriminant of $p(x)$ is defined to be $(x_1 - x_2)^2 (x_1 - x_3)^2 (x_2 - x_3)^2$. It is so named because it vanishes if and only if $p(x)$ has a multiple root, so it 'discriminates' between the roots. It is the forerunner of the discriminant of an algebraic number field, defined in Chapter 2 Section 2. We may now define four invariants of the Frey curve.

- The *discriminant*, which equals $a^{2p}b^{2p}c^{2p}$. Already we see that the Frey curve is special, since this is a perfect $2p$th power, a highly unusual circumstance.

- The *minimal discriminant*, equal to $2^{-8}a^{2p}b^{2p}c^{2p}$. Since b is even and $p \geq 5$, this is an integer.

- The *conductor*, which is the product of all primes that divide a, b, or c. If an elliptic curve is modular, then it can be parametrised by modular functions whose level is equal to the conductor of the curve.

- The *j-invariant*, which equals $\frac{2^8(a^{2p}+b^{2p}+a^pb^p)^3}{a^{2p}b^{2p}c^{2p}}$ and is a complete invariant in the sense that any two elliptic curves with the same j-invariant are isomorphic over $\mathbb{C}$.

Now we can define semistability.

Definition 15.4. An elliptic curve is *semistable* if whenever a prime $l > 3$ divides the discriminant, only two of the three roots of $p(x)$ are congruent modulo l; *and* similar but more technical conditions on the primes 2 and 3 hold.

Lemma 15.5. *The Frey curve $\mathcal{F}$ is semistable.*

Proof: The discriminant is $a^{2p}b^{2p}c^{2p}$ and the roots are $0, a^p, -b^p$ where a^p, b^p are coprime. When $l = 2, 3$ the conditions b even, $a \equiv -1 \pmod{4}$ are also required in the proof.   □

   Wiles's main result is:

Theorem 15.6. (Semistable Taniyama–Shimura–Weil Conjecture.) *Every semistable elliptic curve over $\mathbb{Q}$ is modular.*   □

   Thus every semistable elliptic curve over $\mathbb{Q}$ can be parametrized by modular functions of some level N (equal to the conductor). This leads to:

Lemma 15.7. *If l is an odd prime dividing N, then the j-invariant of $\mathcal{F}$ can be written in the form $r^{-mp}q$ where $m > 0$ and r is a rational number whose numerator and denominator in lowest terms are not divisible by l.*

Proof: The j-invariant of $\mathcal{F}$ is

$$\frac{2^8(a^{2p} + b^{2p} + a^p b^p)^3}{a^{2p}b^{2p}c^{2p}} = \frac{2^8(c^{2p} - b^p c^p)^3}{(abc)^{2p}}.$$

The power of l dividing the denominator is a multiple of p. Since a, b, c are pairwise coprime, the above fraction is in lowest terms. The result follows since N is the product of the primes dividing abc.   □

   This lemma fails for $l = 2$ because of the factor 2^8.

15.3   Sketch Proof of Fermat's Last Theorem

We can now sketch Wiles's proof of Fermat's Last Theorem. We need one further ingredient from complex analysis: that of a modular form of weight 2. We begin with an elliptic integral of the first kind

$$\int \frac{dx}{\sqrt{Ax^3 + Bx^2 + Cx + D}}.$$

Setting $y^2 = Ax^3 + Bx^2 + Cx + D$, defining an elliptic curve, this becomes $\int \frac{dx}{y}$. If the elliptic curve is modular then there exist modular functions $f(z), g(z)$ such that $x = f(z), y = g(z)$ parametrizes the curve. In this case

$$\frac{dx}{y} = \frac{df}{g} = \frac{f'(z)dz}{g(z)} = F(z)dz$$

where

$$F(z) = \frac{f'(z)}{g(z)}.$$

It is not hard to see that although F is not a modular function, it comes close:

$$F\left(\frac{az + b}{cz + d}\right) = (cz + d)^2 F(z).$$

See Exercise 2.

In this case we say that F is a *modular form* of *weight* 2 and *level* N. Modular forms of this type have some distinctive features: in particular, if the choice of parametrization $(f(z), g(z))$ is chosen carefully, the function $F(z)$ is analytic and vanishes at the cusps. Moreover, it is possible to work out $F(z)$ from arithmetic information about the elliptic curve, namely, the number of solutions to the congruences $y^2 \equiv Ax^3 + Bx^2 + Cx + D \pmod{p}$ for all primes p. It is this connection, rather than anything to do with Fermat's Last Theorem, that makes the Taniyama–Shimura–Weil Conjecture so important.

And now for the climax of this book:

Theorem 15.8. (Fermat's Last Theorem.) *If p is an odd prime then the Fermat equation*

$$x^p + y^p = z^p$$

has no solutions in nonzero integers x, y, z.

Proof: The full proof can be found in Wiles [90] and Wiles and Taylor [82] and is several hundred pages long. We can, however, sketch how the proof follows from the concepts introduced above. For a longer, more technical sketch, see Ribet [66].

We aim for a proof by contradiction, and suppose that there exists a solution $a^p + b^p = c^p$ for nonzero integers a, b, c. Let $\mathcal{F}$ be the corresponding Frey elliptic curve. By Theorem 15.6 the curve $\mathcal{F}$ is modular, hence has a cusp form F of weight 2 and level N, where N is the conductor of $\mathcal{F}$.

Lemma 15.7 now allows us to invoke Serre's Level Reduction Conjecture, proved by Ribet, and this implies that for any odd prime l dividing N there exists a cusp form F' of weight 2 and level N/l, which inherits various useful properties of F. (It would be too technical to say which). Inductively, we can consider an odd prime l' dividing N/l and repeat the argument to get a cusp form of level N/ll', and so on. The conductor is divisible by 2 since b is even, and by definition the conductor is a product of distinct primes. We may therefore remove all odd prime factors of N and deduce that there exists a cusp form of level 2.

The dimension of the space of such cusp forms is equal to the genus of the compact Riemann surface $X_0(N)$. But by direct calculation, the genus of $X_0(2)$ can be shown to equal 0. (Indeed, the genus of $X_0(N)$ is zero for $N \leq 10$.) That is, there are no cusp forms of weight 2 and level 2. This is a contradiction, and Fermat's Last Theorem is therefore true. $\square$

15.4   Recent Developments

Wiles's proof of Fermat's Last Theorem is not important because it solved that problem and thereby closed down that line of research. On the contrary, it is important because it introduced new methods and made new connections, opening up many new areas for future work. In this section we indicate some of these developments, including appropriate background material.

Full Taniyama–Shimura–Weil Conjecture

Perhaps the most significant development of all occurred six years after Wiles's breakthrough, when Christophe Breuil, Brian Conrad, Fred Diamond, and Richard Taylor announced a proof of the full Taniyama–Shimura–Weil Conjecture (Darmon [20]). Recall that Wiles required (and proved) only the semistable case of this conjecture. Their methods are firmly in the spirit of Wiles's pioneering work, and we content ourselves with two observations. The first is that they prove a more general theorem than the Taniyama–Shimura–Weil Conjecture by rephrasing it in algebraic form. This more general conjecture has technical advantages, which make the proof possible. The second is that their methods make heavy use of Galois Theory (Stewart [78]), which was introduced by Évariste Galois around 1830 as a way to decide whether a polynomial equation can be solved in terms of radicals—expressions involving nth roots of algebraic formulas.

One consequence of Galois's work is a simple and conceptual proof that the quintic equation (the general polynomial equation of the fifth degree) cannot be solved by radicals—a theorem proved slightly earlier by Abel, using different methods. However, contrary to common myth, no explicit proof of the insolubility of the general quintic occurs in any surviving document by Galois: see Neumann [60]. The reason is presumably that as far as Galois was concerned, this theorem had already been proved by Abel. Galois wanted to go further. He knew that some quintics were soluble by radicals, but others were not. The obvious question is: how do you tell which kind you are looking at? Galois gave a necessary and sufficient condition for solubility, and it is easy to see that the general quintic does not satisfy it. But to Galois the key result was this condition, which he (justifiably) believed improved on Abel's result.

The Catalan Conjecture

Fermat's Last Theorem is just one of many famous questions in number theory about integer powers. In 1844 the Belgian mathematician Eugène Catalan published a short letter in the *Journal für die Reine und Angewandte Mathematik* (Journal for Pure and Applied Mathematics), universally known as Crelle's journal after its founder and first editor August Crelle. It read:

> I beg you, sir, to please announce in your journal the following theorem that I believe true although I have not yet succeeded in completely proving it; perhaps others will be more successful. Two consecutive whole numbers, other than 8 and 9, cannot be consecutive powers; otherwise said, the equation $x^m - y^n = 1$ in which the unknowns are positive integers only admits a single solution.

This statement became known as the Catalan Conjecture.

The earliest significant explicit application of algebraic numbers to the Catalan Conjecture occurred in 1850, when Victor Lebesgue proved the result when the smaller of the two powers is a square; that is, the equation $x^a - y^2 = 1$ has no solutions. It took another 111 years before anyone could prove that there are no integer solutions to the deceptively similar equation $x^a - y^2 = -1$. Very little was known about this equation: for example in 1961 it was proved that x must have at least 3 billion digits. In the same year Chao Ko proved that no solutions exist except the one that started the whole game: $2^3 - 3^2 = -1$. However, it took another three years before his proof became known to the mathematical community.

The upshot of these discoveries is a useful simplification of the Catalan Conjecture. Its proof or disproof reduces to a special case, in which both x and y occur to odd prime powers. In this case the distinction between $+1$ and -1 ceases to be relevant because -1 to an odd power is -1. So the equation under consideration becomes

$$x^p - y^q = 1, \tag{15.2}$$

where p and q are odd primes. From now on we call this the *Catalan equation*. The Catalan Conjecture is now equivalent to the assertion that for all odd primes p, q the Catalan equation has no whole number solutions.

Meanwhile, a different line of attack suggested that perhaps the Catalan Conjecture could be reduced to a computer calculation. Suppose a solution exists. Suppose moreover that the sizes of x, y, p and q are bounded; that is, if a solution exists, then there must be one for which these four numbers have a specific, limited size, called a bound. Then a computer could, in principle, try every possibility up to that bound. If no solutions were found in this range, the proof would be complete. Alternatively, a solution would appear, and that would prove the conjecture false.

Proving that such a bound exists, and working out what it is, would be difficult, but there were encouraging precedents. Bit by bit, pieces of the puzzle seemed to be slotting into place. In 1929 Karl Siegel proved a general theorem on Diophantine equations, and a direct consequence is that for fixed odd primes p and q, the Catalan equation has a finite number of solutions. The hope was that this finite number could be proved to be zero, but at least number theorists now knew that there cannot be infinitely many solutions. This, in turn, implies that for fixed p and q the sizes of x and y are bounded. However, Siegel's theorem does not say how big that bound is. It also fails to specify any bound on the sizes of p and q. So, while it is a step in the right direction, it does not open the way for a computer attack.

In 1955 Davenport and Klaus Roth proved another general theorem, and this one did provide an explicit bound on the sizes of x and y. However, the bound was so gigantic that a computer search would take much longer than the lifetime of the universe, so again the practical implications of their theorem were nil. Nonetheless, number theorists felt they were edging just a little bit closer to a proof, and further evidence for this belief soon showed up. In 1966–67 Alan Baker proved that in general, an integer combination of logarithms of rational numbers cannot be small; indeed, it has to be larger than some function of the integers and rationals concerned. In 1976 Robert Tidjeman [84] applied Baker's theorem to the Catalan equation, and proved that the exponents p and q must both be less than some explicitly computable bound.

Later work showed that the largest solution to (15.2) satisfies

$$|x| \leq \exp\exp\exp\exp 730.$$

At this point, a computer could in principle solve the problem, by checking all numbers in this range. But again, the range was much too great for any practical computation to be possible—even allowing for probable improvements in computer power. In 1999, Maurice Mignotte proved that if a nontrivial solution exists then $p < 7.15 \times 10^{11}$ and $q < 7.78 \times 10^{16}$. Modern computers can handle numbers that big—but those are the exponents, not the numbers x and y, let alone their powers x^p and y^q. So Mignotte's theorem, although an exciting theoretical result, did not reduce the problem to a feasible calculation.

Nevertheless, mathematicians were slowly whittling away at the Catalan Conjecture. A few more improvements of the same kind, and a computer solution would be within reach. But when a solution finally appeared, however, it came from a different direction altogether. And it made hardly any use of a computer. Instead, it went back to the tried and tested strategy of algebraic number theory—but with better tactics.

In 2002, Preda Mihăilescu startled the mathematical community by publishing a proof of the Catalan Conjecture [57]. His ingenious, highly technical, proof is based on cyclotomic integers. The methods come from his PhD thesis 'Cyclotomy of rings and primality testing', an area with no obvious connection to the Catalan Conjecture. An expanded description of the proof can be found in Schoof [72]. See also Bilu [5].

The first step is to rewrite the Catalan equation as $x^p - 1 = y^q$ and borrow a trick from Gauss. The left-hand side factorizes as $(x - 1)(x^{p-1} + x^{p-2} + ... + x + 1)$. The right-hand side is a qth power, therefore the left hand side is also a qth power. If the two numbers $x - 1$ and $x^{p-1} + x^{p-2} + ... + x + 1$ have no common factor, then each factor must also be a qth power. In 1960 John Cassels (usually known as 'Ian') proved that this assumption leads to a contradiction. He also showed that if the two numbers do have a common factor, it must be p. In fact, p divides the second term but p^2 does not. Cassels was unable to derive a contradiction from this result, but it convinced number theorists that further progress on the Catalan Conjecture might be possible by pursuing a similar strategy. There were some standard tricks that might apply, but none of them seemed to work.

Using the cyclotomic integer $\zeta = e^{2\pi i/p}$, the Catalan equation splits into linear factors:

$$y^q = x^{p-1} - 1 = (x - 1)(x - \zeta)(x - \zeta^2) \ldots (x - \zeta^{p-1})$$

If the factors on the right-hand side have no common factor, and prime factorization is unique in the cyclotomic integers for the prime p, then each term $x - \zeta^k$ must itself be a prime power. From this it is not too hard to derive a contradiction. However, neither of these statements is true. Undaunted, Mihăilescu decided to find out what happens instead. The resulting analysis is quite long, and too complicated to do more than summarize. It relied on a number of ideas about the ring of cyclotomic integers.

The current proof, a simplification of Mihăilescu's original version that includes more recent work—some by him—has several main components. The first step, which he completed in 2000, was to prove that the exponents p and q satisfy a very stringent condition: they are *double Wieferich primes*. This means that $p^{q-1} - 1$ is divisible by q^2 and $q^{p-1} - 1$ is divisible by p^2. Wieferich introduced a related condition in 1909 when working on Fermat's last theorem: a *Wieferich prime p* is one for which p^2 divides $2^{p-1} - 1$. The only known Wieferich primes are 1093 and 3511, and their rarity suggests that double Wieferich primes might also be rare. Indeed, if it could be proved that they do not exist, that would polish off the Catalan Conjecture. However, they do exist: an example is $p = 83, q = 4871$, and only five more such pairs are currently known. Nevertheless, this result is a good start.

The next step, accomplished in 2002, is to derive a much simpler condition. Namely, either $p - 1$ is divisible by q or $q - 1$ is divisible by p. A key idea is to analyse the structure of 'annihilators', a technical concept that turns statements about ideals back into useful statements about numbers. Another is to exploit known results about the units of cyclotomic numbers—cyclotomic integers whose reciprocal is also a cyclotomic integer. A deep and difficult theorem proved by Francisco Thaine in 1988 relates annihilators of units to ideals, and this is a key ingredient in proving the second step.

The third step relates the sizes of the primes p and q. Neither can be too large compared to the other; specifically, $p < 4q^2$ and $q < 4p^2$. Mihăilescu proved this in 2003, and it simplifies parts of his original proof. The method fails to work for some smallish values of p and q, so these have to be handled by other methods. In this manner, he proved that no new solutions of the Catalan equation exist if either p or q is 41 or less. The improved proof replaces 41 by 5.

By putting all these things together, Mihăilescu was able to prove that if the Catalan equation has a solution, then the two primes p and q that occur must satisfy the condition $p < q$. However, because p and q are odd, the equation is symmetric in p and q, in the sense that if $x^p - y^q = 1$ then $(-y)^q - (-x)^p = 1$. So we can swap p and q. Therefore, by the same argument, $q < p$. This is a contradiction, so no new solution to the Catalan equation exists.

Whenever someone solves a major problem in mathematics, an immediate reflex is to try the same method on other similar problems. However, Mihăilescu's proof uses so many special features of the Catalan equation that it is difficult to get any significant generalizations working. So, for the moment at least, it is a 'one-off', carefully steering its way through a host of difficulties. That makes Mihăilescu's achievement all the more remarkable.

The Pillai Conjecture

A stronger result was conjectured by Subbayya Sivasankaranarayana Pillai. The Pillai Conjecture states that each positive integer occurs only finitely many times as a difference of perfect powers. That is, for any integer $c > 0$ the equation $x^m - y^n = c$ has finitely many integer solutions. This problem remains open, but it would be a consequence of the ABC Conjecture 15.14 below.

Fermat–Catalan Conjecture

More generally, consider the Diophantine equation

$$x^a + y^b = z^c. \tag{15.3}$$

The 'surprising' solutions occur when a, b, c are 'large' in some sense. For the Pythagorean equation ($a = b = c = 2$), with its infinite family of solutions, the exponents a, b, c should clearly be considered small. We explain below why a sensible interpretation of the 'size' of a solution, in this context, is the number

$$s = \frac{1}{a} + \frac{1}{b} + \frac{1}{c}.$$

The *smaller* s is, the larger a, b, c must be. The crucial distinction is that 'large' solutions have $s < 1$ but 'small' ones have $s > 1$. The *only* known integer solutions of (15.3) for large a, b, c (see Mazur [55]) are:

$$
\begin{aligned}
1 + 2^3 &= 3^2, \\
2^5 + 7^2 &= 3^4 \\
7^3 + 13^2 &= 2^9 \\
2^7 + 17^3 &= 71^2 \\
3^5 + 11^4 &= 122^2 \\
17^7 + 76271^3 &= 21063928^2 \\
1414^3 + 2213459^2 &= 65^7 \\
9262^3 + 15312283^2 &= 113^7 \\
43^8 + 96222^3 &= 30042907^2 \\
33^8 + 159034^2 &= 15613^3.
\end{aligned}
\tag{15.4}
$$

By convention 1 is treated as 1^∞ and $\frac{1}{\infty} = 0$. So $s = \frac{5}{6}$ for the first solution above.

The first five of these solutions have been known for centuries; the last five are due to Frits Beukers and Zagier. The main conjecture here is:

Conjecture 15.9. (Fermat–Catalan Conjecture.) *In total, for all large* (a, b, c) *(that is, $s > 1$) there exists only a finite number of coprime integer solutions of Equation* (15.3).

The name of the conjecture is modern, due to Henri Darmon and Granville: it reflects the fact that a positive solution would imply both the Catalan Conjecture and Fermat's Last Theorem.

The main positive result is a recent theorem of Darmon and Loïc Merel, who prove that there are no solutions with $(a, b, c) = (g, g, 3)$ for $g > 3$. Darmon and Granville have proved that for each individual triple (a, b, c) with $s < 1$ there exist only finitely many coprime integer solutions x, y, z of (15.3). The Fermat–Catalan Conjecture says more than this: the number of triples (a, b, c) with $s < 1$ for which coprime integer solutions exist is also finite.

ABC Conjecture

The problems above lead to a far-reaching and potentially enormously powerful conjecture. Its proof would revolutionize number theory. In order to formulate the relevant conjectures and theorems with some precision, we require the following simple concepts:

Definition 15.10. Let N be an integer. Then the *radical* rad N is the product of all distinct prime factors of N.

If $N \neq 0, \pm 1$ then the *power function* of N is

$$P(N) = \frac{\log |N|}{\log \operatorname{rad} N}.$$

By convention,

$$P(\pm 1) = \infty.$$

Obviously $P(N) = 1$ if and only if N is squarefree. If N is a perfect kth power, then $P(N) \geq k$. We also define an *a-powered number* to be a number N for which $P(N) \geq a$. Roughly speaking, the larger a becomes, the rarer a-powered numbers are. For example:

Proposition 15.11. *As $x \to \infty$ the number of squarefree integers between 0 and x is of the form*

$$\frac{6}{\pi^2} x + O(\sqrt{x}).$$

Proof: See Exercise 8. □

Informally, this proposition tells us that roughly 60% of integers (up to a given size) are squarefree.

Fermat's Last Theorem asserts that a particular 'linear relation' between perfect nth powers is rare—indeed, so rare that it never happens. That is, the sum of two nth powers is never an nth power. More generally, we can look at linear relations between three a-powered numbers and ask how rare those are. To be precise, choose three real numbers $a, b, c \geq 1$ and a real number x. Let $S(a, b, c; x)$ be the set of all triples (A, B, C) of integers (assumed relatively prime and nonzero) such that

$$|A| \leq x \quad |B| \leq x \quad |C| \leq x,$$
$$A + B + C = 0,$$
$$P(A) \geq a \quad P(B) \geq b \quad P(c) \geq c.$$

Given a, b, c, how rapidly do we expect the cardinality of $S(a, b, c; x)$ to grow as $x \to \infty$? That is, how rare (or how common) are solutions to $A + B + C = 0$ in a, b, c-powered numbers A, B, C?

A heuristic argument leads to a striking guess. Ignore the condition that A, B, C be relatively prime, because this does not change the likely result much. Then there are roughly $x^{1/a}$ choices for A, $x^{1/b}$ choices for B, and $x^{1/c}$ choices for C. Since $|A + B + C| \leq 3x$, the probability (whatever that means here) that $A + B + C = 0$ is of the order $\frac{1}{a} + \frac{1}{b} + \frac{1}{c} - 1$, so the cardinality of $S(a, b, c; x)$ should be comparable to $x^{\frac{1}{a} + \frac{1}{b} + \frac{1}{c} - 1}$. This argument is very rough-and-ready, but it focuses attention on the *basic exponent*

$$d = \frac{1}{a} + \frac{1}{b} + \frac{1}{c} - 1 = s - 1$$

and leads us to consider three different cases:

$$d < 0 \quad d = 0 \quad d > 0$$

where we expect very different results. Roughly speaking, if $d > 0$ then we expect a proportion x^d of solutions with $|A|, |B|, |C| \leq x$, so that in particular if we allow x to take on all values, then we expect there to exist infinitely many solutions to $A + B + C = 0$. When $d < 0$, however, we expect there to be only finitely many solutions (allowing x to range over all values). When $d = 0$ we have a delicate transitional case and caution is needed even in formulating a sensible conjecture.

Suppose that $a \leq b \leq c \in \mathbb{N}$. Then we can tabulate all cases for which $d \geq 0$ (Table 15.1). This table is familiar from other areas of mathematics,

notably polyhedra and tilings in higher dimensions, which raises the hope
that the approach being adopted here is significant.

When $d > 0$ and $a, b, c \in N$ we can get large numbers of (a, b, c')
solutions with c' close to c from single Diophantine equations, such as
$x^a + y^b = Ez^c$ for some fixed integer $E \neq 0$. Thus, for instance, to obtain
lots of $(2, 2, c)$ solutions we might consider $x^2 + y^2 = z^c$. When $c = 2$
the problem is that of Pythagorean triples, and these occur in sufficient
profusion to establish the conjectured asymptotics, in the sense that the
number of solutions with $|A|, |B|, |C| < x$ is at least $x^{d-\epsilon}$ for any $\epsilon > 0$,
however small.

When $d < 0$ the problem becomes, if anything, more interesting. Extensive numerical experiments are consistent with:

Conjecture 15.12. ((a,b,c) Conjecture.) *If $\frac{1}{a} + \frac{1}{b} + \frac{1}{c} < 1$ then the number
of solutions A, B, C of the equation $A + B + C = 0$ with $P(A) \geq a$, $P(B) \geq$
b, $P(C) \geq c$ is finite.*

Indeed, there is a stronger conjecture:

Conjecture 15.13. (Uniform (a,b,c) Conjecture.) *Let $d_0 < 0$ be real. If
$\frac{1}{a} + \frac{1}{b} + \frac{1}{c} < 1$ then the number of solutions A, B, C of the equation $A +
B + C = 0$ with $P(A) \geq a$, $P(B) \geq b$, $P(C) \geq c$ and $\frac{1}{a} + \frac{1}{b} + \frac{1}{c} - 1 \leq d_0$ is
finite.*

The Uniform (a, b, c) Conjecture implies the (a, b, c) Conjecture. However, at the time of writing neither conjecture has been proved in any
case whatsoever. However, both would be consequences of the so-called
ABC-Conjecture of Masser and Oesterlé [50, 62], one of the biggest open

a	b	c	d
1	*	*	*
2	2	*	*
2	3	3	$\frac{1}{6}$
2	3	4	$\frac{1}{12}$
2	3	5	$\frac{1}{30}$
2	3	6	0
3	3	3	0

Table 15.1. Integer a, b, c for which $d \geq 0$.

questions in current number theory. We state this conjecture after setting up one necessary concept.

Define an *ABC solution* to be a triple (A, B, C) of nonzero coprime integers such that $A + B + C = 0$. Define the *power* of (A, B, C) to be

$$P(A, B, C) = \frac{\log \max(|A|, |B|, |C|)}{\log \operatorname{rad}(ABC)}.$$

Then the conjecture is:

Conjecture 15.14. (Masser and Oesterlé ABC Conjecture.) *For any real $\rho > 1$ there exist only finitely many ABC solutions with $P(A, B, C) \geq \rho$.*

In 2012 Shinichi Mochizuki announced a proof of the ABC Conjecture based on over 500 pages of work in four preprints. It developed a novel method, which he named 'inter-universal Teichmüller theory'. Experts have been analyzing this proof, which involves a substantial amount of new machinery and is therefore an intensely time-consuming task. So far one error has emerged, but Mochizuki has stated that this can be corrected so that the proof still works. He has not lectured on his ideas, but he has issued progress reports. The mathematical community had not reached a final decision about the proof's validity when this book went to press.

Beal Conjecture

Finally, we mention the Beal Conjecture: see also Mauldin [51]. Andrew Beal is a number theory enthusiast living in Dallas, Texas. When Fermat's Last Theorem was proved, he decided to follow the example of Paul Wolfskehl. He offered a prize of $5,000 (increasing annually by $5,000 up to a total of $50,000) for a proof of:

Conjecture 15.15. (Beal Conjecture.) *Let x, y, z, a, b, c be positive integers with $a, b, c > 2$. If $x^a + y^b = z^c$, then x, y, z have a common factor > 1.*

This conjecture is currently open, and the prize is now $1 million.

All ten 'large' solutions (15.4) of $x^a + y^b = z^c$ have one exponent equal to 2, so the Beal Conjecture is consistent with the known data related to the Fermat–Catalan Conjecture. It is clear that plenty of unsolved questions about sums of powers remain to keep the next generation of number theorists busy.

15.5 Exercises

1. Give a heuristic argument to show that the number of squarefree integers less than x is equal to $\frac{6}{\pi^2}x + O(\sqrt{x})$ for $x \to \infty$.

   (*Hint:* Let p_j be the primes in increasing order and consider the sequence of integers $1, 2, 3, ..., x$. Remove from this sequence all multiples of p_1^2, leaving approximately $(1 - \frac{1}{p_1^2})x$ integers. From these, remove all mutiples of p_2^2, then p_3^2, and so on. Continue until $p_j \sim \sqrt{x}$, so that the number of integers left (which are the squarefree ones) is approximately

   $$x \prod_{p_j \leq \sqrt{x}} \left(1 - \frac{1}{p_j^2}\right).$$

   Now use Euler's result that

   $$\frac{\pi^2}{6} = \sum_{n=1}^{\infty} \frac{1}{n^2} = \prod_{p_j} \left(1 - \frac{1}{p_j^2}\right)$$

   and estimate errors.)

2. If $f(z), g(z)$ are modular functions and

   $$F(z) = \frac{f'(z)}{g(z)}$$

   prove that

   $$F\left(\frac{az + b}{cz + d}\right) = (cz + d)^2 F(z).$$

3. Show that over the Gaussian integers $\mathbb{Z}[i]$ the equation

   $$(78 + 78i)^2 = (-23i)^3 + i$$

   is valid. That is, two perfect powers (other than 8 and 9) differ by a unit.

4. Prove that Table 15.1 lists all solutions of the inequality $d \geq 0$, where

   $$d = \frac{1}{a} + \frac{1}{b} + \frac{1}{c} - 1$$

   and a, b, c are positive integers.

5. Search the Internet to find the latest information on the status of the Fermat-Catalan, Beal, and ABC Conjectures.

IV

Appendices

A

Quadratic Residues

The theory of quadratic residues is one of the great triumphs of the classical period of number theory. An integer k that is prime to a positive integer m is said to be a *quadratic residue modulo* m if there exists $z \in \mathbb{Z}$ such that

$$z^2 \equiv k \pmod{m}.$$

Denoting the residue class of k modulo m by $\bar{k}$, this can be rephrased as: $\bar{k}$ is both a unit and a square in $\mathbb{Z}_m$.

We investigate the question of quadratic residues by determining the structure of the units in $\mathbb{Z}_m$. We also show how understanding quadratic residues leads to the solution of quadratic equations

$$ax^2 + bx + c = 0$$

in $\mathbb{Z}_m$.

The most remarkable theorem about quadratic residues is known as the *quadratic reciprocity law* which states:

If p, q are distinct odd primes, at least one of which is congruent to 1 modulo 4, then p is a quadratic residue of q if and only if q is a quadratic residue of p; otherwise precisely one of p, q is a quadratic residue of the other.

The reciprocal nature of the relationship between p and q in the first case gives rise to the name of the law. Gauss first proved it in 1796 when he was eighteen years old. The result had been conjectured earlier by Euler and Legendre, though Gauss said that he did not know this at the time. He thought so highly of this theorem that he called it 'the gem of higher arithmetic' and developed six different proofs. In the 19th century quadratic

reciprocity continued to arouse interest, and more than fifty different methods of proof were found by mathematicians such as Cauchy, Eisenstein,
Jacobi, Leopold Kronecker, Kummer, Liouville, and Karl Zeller. In fact it
was when Kummer was studying higher reciprocity laws that he devised
his partial proof of Fermat's Last Theorem. In 1850 Kummer referred to
the higher reciprocity laws as the 'the pinnacle of contemporary number
theory', regarding Fermat's Last Theorem as a 'curiosity', Edwards [24]. It
is only right and proper, therefore, that any text on Fermat's Last Theorem
should include a description of the result that so fascinated number theorists, and whose study led to Kummer's proof of a special case of Fermat's
Last Theorem as a by-product.

A.1   Quadratic Equations in $\mathbb{Z}_m$

An obvious topic of number-theoretic study is the solution of polynomial
equations modulo a positive integer m.

A linear equation

$$ax + b \equiv 0 \pmod{m} \tag{A.1}$$

can clearly be solved when a is prime to m, because there exist integers
c, d such that

$$ac + dm = 1,$$

so $ac \equiv 1 \pmod{m}$. Multiply (A.1) by c and simplify to get the solution

$$x \equiv -bc \pmod{m}.$$

If, on the other hand, a and m have highest common factor $h > 1$, then
(A.1) can have a solution only if h divides b, in which case, writing $a = a_0 h$,
$b = b_0 h$, $m = m_0 h$, (A.1) reduces to

$$a_0 x + b_0 \equiv 0 \pmod{m_0}$$

where once more a_0, m_0 are coprime. Thus the solution of linear equations
modulo m is straightforward.

Quadratic equations are more interesting. The equation

$$ax^2 + bx + c \equiv 0 \pmod{m} \tag{A.2}$$

where m does not divide a may be simplified by multiplying throughout
(including the modulus) by $4a$ and completing the square to get the equivalent equation

$$4a^2 x^2 + 4abx + 4ac \equiv 0 \pmod{4am}$$

or

$$(2ax + b)^2 \equiv b^2 - 4ac \pmod{4am}.$$

Now substituting $4am = m_0$, $2ax + b \equiv z \bmod m_0$ and $b^2 - 4ac \equiv k \bmod m_0$, we replace (A.2) by the two equations:

$$z^2 \equiv k \pmod{m_0} \tag{A.3}$$

$$2ax + b \equiv z \pmod{m_0} \tag{A.4}$$

If we find z from (A.3) we can then attack (A.4) by the given method for linear congruences, so the solution of the general quadratic (A.2) reduces to solving (A.3). We can further reduce this if k, m_0 are not coprime by supposing they have highest common factor h where $k = k_0 h$ and $m_0 = m_1 h$, and then factorizing h as

$$h = e^2 r$$

where e^2 is the largest square factor of h. If (A.3) has a solution for z then er is a factor of z. Let $z = erw$; then

$$e^2 r^2 w^2 \equiv k_0 e^2 r \pmod{m_1 e^2 r}$$

so

$$rw^2 \equiv k_0 \pmod{m_1}. \tag{A.5}$$

Now suppose that the highest common factor of r, m_1 is s. For (A.5) to have a solution, s must be a factor of k_0. But k_0, m_1 are coprime, so $s = 1$ and r, m_1 are also coprime. Thus there exists an integer d prime to m_1 such that

$$dr \equiv 1 \pmod{m_1},$$

so multiplying (A.5) by d and simplifying gives

$$w^2 \equiv dk_0 \pmod{m_1}. \tag{A.6}$$

But d and k_0 are both prime to m_1, so putting $dk_0 = k_1$, the solution of the general quadratic equation (A.2) reduces to linear congruences and the congruence

$$w^2 \equiv k_1 \pmod{m_1}$$

where k_1, m_1 are coprime.

If $m > 1$ and k are integers, recall that k is a quadratic residue modulo m if

(1) k, m are coprime
(2) There exists $w \in \mathbb{Z}$ such that

$$w^2 \equiv k \pmod{m}.$$

If $\bar{k}$ is the residue class of k in $\mathbb{Z}_m$, then these conditions are equivalent to
    (1)' $\bar{k}$ is a unit in $\mathbb{Z}_m$,
    (2)' $\bar{w}^2 = k$ in $\mathbb{Z}_m$,
so $\bar{k}$ is both a unit and a square in $\mathbb{Z}_m$. We attack the problem of finding quadratic residues by first computing the structure of the units in $\mathbb{Z}_m$.

A.2   The Units of $\mathbb{Z}_m$

An element $\bar{k} \in \mathbb{Z}_m$ is a unit if and only if k, m are coprime, so the units $U(\mathbb{Z}_m)$ of $\mathbb{Z}_m$ are

$$U(\mathbb{Z}_m) = \{\bar{k} \in \mathbb{Z}_m | 1 \le k < m; k, m \text{ coprime}\}.$$

The number of elements in $U(\mathbb{Z}_m)$ is called the *Euler function* $\phi(m)$ and is equal to the number of positive integers k less than m and prime to it. For example $\phi(10) = 4$ since 1, 3, 7, 9 are the integers between 1 and 10 and prime to 10, and there are four of them. For later reference we record:

Lemma A.1. *If p is prime, then $\phi(p^e) = p^{e-1}(p - 1)$, and in particular $\phi(p) = p - 1$.*

Proof: There are $p^e - 1$ elements satisfying $1 \le k < p^e$, and of these, if $k = rp$, then

$$1 \le rp < p^e$$

implies

$$1 \le r < p^{e-1},$$

so there are $p^{e-1} - 1$ elements not prime to p, giving

$$\phi(p^e) = (p^e - 1) - (p^{e-1} - 1). \qquad \Box$$

    The units $U(\mathbb{Z}_m)$ form a group under multiplication whose structure we compute. First we factorize $m = p_1^{e_1} \ldots p_r^{e_r}$ where $p_1, \ldots, p_r$ are distinct primes, and reduce the problem to considering each prime separately. For typographical reasons we write $p_i^{e_i} = P_i$.

Lemma A.2. *If $m = p_1^{e_1} \ldots p_r^{e_r}$ where $p_1, \ldots, p_r$ are distinct primes, then, writing $p_i^{e_i} = P_i$, there is a ring isomorphism*

$$\mathbb{Z}_m \cong \mathbb{Z}_{P_1} \times \ldots \times \mathbb{Z}_{P_r}.$$

Proof: Define $\pi : \mathbb{Z} \to \mathbb{Z}_{P_1} \times \ldots \times \mathbb{Z}_{P_r}$ by $\pi(k) = (k_1, \ldots, k_r)$ where k_i is the residue class of k modulo $P_i = p_i^{e_i}$. Clearly π is a ring homomorphism and $k \in \ker \pi$ if and only if P_i divides k for every i, which implies $\ker \pi = \langle m \rangle$ (the ideal generated by m). Thus π induces a monomorphism

$$\bar{\pi} : \mathbb{Z}_m \to \mathbb{Z}_{P_1} \times \ldots \times \mathbb{Z}_{P_r}.$$

But $\mathbb{Z}_{P_1} \times \ldots \times \mathbb{Z}_{P_r}$ has $P_1 \times \ldots \times P_r = m$ elements, so $\bar{\pi}$ is in fact an isomorphism. $\qquad \square$

Lemma A.3. *With the above notation,*

$$U(\mathbb{Z}_m) \cong U(\mathbb{Z}_{P_1}) \times \ldots \times U(\mathbb{Z}_{P_r}).$$

Proof: Under the ring isomorphism $\bar{\pi}$, an element $\bar{k} \in \mathbb{Z}_m$ is a unit if and only if $\bar{\pi}(k)$ is a unit, which holds if and only if each of its components is a unit. $\qquad \square$

Lemma A.3 reduces the study of units in $\mathbb{Z}_m$ to the case of $\mathbb{Z}_{p^e}$ where p is prime. To tackle this case we begin with $e = 1$. Here we see that $U(\mathbb{Z}_p)$ is a cyclic group of order $p - 1$. It might seem that the easiest way to show this would be to find a generator, but that turns out not to be feasible in general. So we attack the problem indirectly by introducing an auxiliary notion, which will prove useful in several ways: the *exponent h* of a finite group G, which is defined as the smallest positive integer such that (in multiplicative notation) $x^h = 1$ for all $x \in G$.

By Lagrange's Theorem, $x^n = 1$ where n is the order of G, so clearly $h \leq n$. Also, for any $x \in G$, if x has order k, then k divides h. An alternative definition of the exponent is, therefore, the least common multiple of all the orders of the elements in G.

We claim that in any finite abelian group of exponent h, there exists an element x_0 of order h. This applies in particular to $U(\mathbb{Z}_p)$. Two facts follow. First, by Lagrange's Theorem, the exponent h divides the order of G. Second, if we can demonstrate that the exponent *equals* the order, then G is cyclic with generator x_0.

To demonstrate the existence of such an x_0, we begin with:

Lemma A.4. *In an abelian group G, if a, b have finite orders q, r which are coprime, then ab has order qr.*

Proof: $(ab)^{qr} = a^{qr} b^{qr} = 1$. If $(ab)^s = 1$ then

$$a^s = b^{-s}$$

so the elements a^s and b^{-s} have the same order k. However, the order of a^s divides the order of a, so k divides q, similarly k divides r. Since q, r are coprime, $k = 1$, which implies

$$a^s = b^{-s} = 1,$$

whence q divides s and r divides s. Since q, r are coprime, qr divides s. $\square$

Lemma A.5. *If the finite abelian group G has exponent h, then there exists $x_0 \in G$ such that the order of x_0 is h.*

Proof: Let p^k be the highest power of a prime p dividing h. It is easy to see that G must have an element x of order $p^k r$ where r is prime to p, for if not, the highest power of p dividing the order of every element is p^{k-1} or less, contrary to the definition of h. The element $y = x^r$ is then of order p^k. Find elements y for all distinct primes p dividing h and then use Lemma A.4. $\square$

Proposition A.6. *$U(\mathbb{Z}_p)$ is a cyclic group of order $p - 1$.*

Proof: The order of $U(\mathbb{Z}_p)$ is $\phi(p) = p - 1$ by Lemma A.1. To show $U(\mathbb{Z}_p)$ is cyclic by Lemma A.5, we verify that the exponent h of $U(\mathbb{Z}_p)$ is $p - 1$. Certainly $h \leq p - 1$. Conversely, every element of $U(\mathbb{Z}_p)$ satisfies $x^h = 1$ by definition, and interpreting $x^h - 1 = 0$ as a polynomial equation over $\mathbb{Z}_p$, this has, at most, h roots, hence $p - 1 \leq h$. $\square$

Examples A.7.
$U(\mathbb{Z}_2) = \{\bar{1}\}$, generator $\bar{1}$.
$U(\mathbb{Z}_3) = \{\bar{1}, \bar{2}\}$, generator $\bar{2}$.
$U(\mathbb{Z}_5) = \{\bar{1}, \bar{2}, \bar{3}, \bar{4}\}$, generators $\bar{2}, \bar{3}$.
$U(\mathbb{Z}_7) = \{\bar{1}, \bar{2}, \bar{3}, \bar{4}, \bar{5}, \bar{6}\}$, generators $\bar{3}.\bar{5}$.

If $\bar{s}$ is a generator of $U(\mathbb{Z}_p)$, then $s \in \mathbb{Z}$ is called a *primitive root modulo* p. Primitive roots play a central part in our computations, since every element of $U(\mathbb{Z}_p)$ is of the form $\bar{s}^r$ where s is a primitive root modulo p. If we can find a primitive root, then because the order of $U(\mathbb{Z}_p)$ is even, the even powers $\bar{s}^{2r}$ are clearly quadratic residues and the odd powers $\bar{s}^{2r+1}$ are not. In general we do not attack the problem this way, because we do not know the value of s, but it illustrates the theoretical importance of primitive roots. We isolate two properties which will prove essential later:

Lemma A.8.

(a) If s is a primitive root modulo p, then so is s^r if and only if r, $p-1$ are coprime.

(b) If s is a primitive root modulo p and k is a positive integer, then there is another primitive root λ modulo p such that $\bar{s} = \bar{\lambda}^{p^k}$.

Proof: (a) If r, $p-1$ are coprime, then there exist integers a, b such that $ar + b(p-1) = 1$, so

$$\bar{s} = \bar{s}^{ar+b(p-1)} = (\bar{s}^r)^a,$$

and $\bar{s}^r$ generates $U(\mathbb{Z}_p)$.

Conversely, if r, $p-1$ have a common factor $d > 1$, where $p-1 = dq$, $r = dc$, then

$$(\bar{s}^r)^q = \bar{s}^{dcq} = \bar{s}^{(p-1)c} = 1,$$

so $\bar{s}^r$ cannot generate $U(\mathbb{Z}_p)$.

(b) Since p^k, $p-1$ are coprime, there exist integers a, b where a is prime to $p-1$ such that

$$ap^k + b(p-1) = 1.$$

Hence $\lambda = s^a$ is a primitive root modulo p by part (a) and

$$\bar{\lambda}^{p^k} = \bar{s}^{ap^k} = \bar{s}^{ap^k+b(p-1)} = \bar{s}. \qquad \square$$

We are now in a position to describe the structure of $U(\mathbb{Z}_{p^e})$ for prime p, which we do by explicit computation, first for an odd prime.

Proposition A.9. *If p is a prime, $p \neq 2$, $e \geq 2$, then $U(\mathbb{Z}_{p^e})$ is a cyclic group of order $p^{e-1}(p-1)$ with generator $\bar{s}(\overline{1+p})$, where $\bar{s}$ is a primitive root modulo p.*

Proof: Since the order of $U(\mathbb{Z}_{p^e})$ is $p^{e-1}(p-1)$ by Lemma A.1, and $p-1$, p^{e-1} are coprime, then using Lemma A.4 it is sufficient to show that $\bar{s}$ has order $p-1$ and $\overline{1+p}$ has order p^{e-1} in $U(\mathbb{Z}_{p^e})$. For $\bar{s}$,

$$\bar{s}^{p-1} = \bar{\lambda}^{p^{e-1}(p-1)} \quad \text{using Lemma A.7 (ii)}$$
$$= 1 (\bmod\, p^e) \quad \text{by Lagrange's theorem.}$$

On the other hand

$$s^r \equiv 1 \quad (\bmod\, p^e)$$

implies

$$s^r \equiv 1 \quad (\bmod\, p)$$

and since s is a primitive root modulo p we have

$$s^r \not\equiv 1 \quad (\bmod \, p^e) \quad \text{for } 1 < r < p - 1,$$

demonstrating that $\bar{s}$ is of order $p - 1$ in $U(\mathbb{Z}_{p^e})$. To prove that $\overline{1 + p}$ is of order p^{e-1}, we establish by induction on $e \geq 2$ that

$$(1 + p)^{p^{e-2}} \equiv 1 + kp^{e-1} \quad (\bmod \, p^e) \tag{A.7}$$

where k depends on e, but $k \not\equiv 0 \pmod{p}$. For $e = 2$ this is true with $k = 1$. Assume it true for some $e \geq 2$. Then

$$\begin{aligned}
(1 + p)^{p^{e-2}} &= 1 + kp^{e-1} + rp^e \\
&= 1 + sp^{e-1}
\end{aligned}$$

where $s = k + rp \not\equiv 0 \pmod{p}$. Hence

$$\begin{aligned}
(1 + p)^{p^{e-1}} &= (1 + sp^{e-1})^p \\
&= 1 + psp^{e-1} + \binom{p}{2} s^2 p^{2(e-1)} + \ldots + s^p p^{p(e-1)}.
\end{aligned}$$

For $e \geq 2$ and prime $p \neq 2$ this is of the form

$$1 + sp^e + bp^{e+1}.$$

(For $p = 2$ this breaks down only when $e = 2$.) Therefore

$$(1 + p)^{p^{e-1}} \equiv 1 + sp^e \quad (\bmod \, p^{e+1}) \tag{A.8}$$

where $s \not\equiv 0 \pmod{p}$, completing the induction proof of (A.7).
  Then (A.7) implies

$$(\overline{1 + p})^{p^{e-2}} \neq \bar{1} \text{ in } \mathbb{Z}_{p^e}$$

and (A.8) implies

$$(\overline{1 + p})^{p^{e-1}} = \bar{1} \text{ in } \mathbb{Z}_{p^e},$$

which together show $\overline{1 + p}$ is of order p^{e-1} in $U(\mathbb{Z}_{p^e})$. $\qquad \square$

Since the proof in Proposition A.9 breaks down for $p = 2$, we must treat this case separately. We find:

Proposition A.10. $U(\mathbb{Z}_4) = \{\bar{1}, -\bar{1}\}$ *is cyclic with generator* $-\bar{1}$. *For* $e \geq 3$, $U(\mathbb{Z}_{2^e})$ *is not cyclic, but* $-\bar{1}$ *is of order* 2, $\bar{5}$ *is of order* 2^{e-2} *and* $U(\mathbb{Z}_{2^e})$ *is the direct product of the cyclic groups generated by* $-\bar{1}, \bar{5}$.

Proof: The assertion concerning $U(\mathbb{Z}_4)$ is trivial. For $e \geq 3$, by Lemma A.1, the order of $U(\mathbb{Z}_{2^e})$ is 2^{e-1}. Clearly the order of $-\bar{1}$ in $U(\mathbb{Z}_{2^e})$ is 2. For the element $\bar{5}$ we note that by induction on $e \geq 3$ we may establish

$$5^{2^{e-3}} = (1 + 2^2)^{2^{e-3}} \equiv 1 + 2^{e-1} \pmod{2^e}.$$

Hence $\bar{5}^{2^{e-3}} \neq \bar{1}$ in $\mathbb{Z}_{2^e}$, but

$$5^{2^{e-2}} \equiv (1 + 2^{e-1})^2 \equiv 1 \pmod{2^e}$$

so $\bar{5}$ is of order 2^{e-2} in $U(\mathbb{Z}_{2^e})$.

Now $-\bar{1}$ is not a power of $\bar{5}$ in $U(\mathbb{Z}_{2^e})$ since

$$-1 \not\equiv 5^r \pmod 4,$$

so certainly

$$-1 \not\equiv 5^r \pmod{2^e}.$$

Hence if C is the cyclic subgroup generated by $\bar{5}$, the cosets C, $-\bar{1}C$ are disjoint. But the index of C in $U(\mathbb{Z}_{2^e})$ is $2^{e-1}/2^{e-2} = 2$, so these two cosets exhaust $U(\mathbb{Z}_{2^e})$. Thus every element of $U(\mathbb{Z}_{2^e})$ is uniquely of the form $(-\bar{1})^a \bar{5}^b$ where $a = 0$ or 1 and $0 \leq b < 2^{e-2}$. Since multiplication is commutative, $U(\mathbb{Z}_{2^e})$ is the direct product of the cyclic subgroups generated by $-\bar{1}, \bar{5}$. The exponent of $U(\mathbb{Z}_{2^e})$ is 2^{e-2} which is less than the order, so $U(\mathbb{Z}_{2^e})$ cannot be cyclic. $\qquad\square$

Having described the structure of $U(\mathbb{Z}_{p^e})$, we are now in a position to investigate quadratic residues.

A.3 Quadratic Residues

As in the last section we can speedily reduce the problem of finding residues modulo m to the case of prime powers:

Proposition A.11. *If* $m = p_1^{e_1} \ldots p_r^{e_r}$ *where* $p_1, \ldots, p_r$ *are distinct primes and* k *is relatively prime to* m, *then* k *is a quadratic residue modulo* m *if and only if it is a quadratic residue of* $p_i^{e_i}$ *for* $1 \leq i \leq r$.

Proof: Using the isomorphism $\bar{\pi} : U(\mathbb{Z}_m) \to U(\mathbb{Z}_{P_1}) \times \ldots \times U(\mathbb{Z}_{P_r})$ of Lemma A.3, $\bar{k}$ is a square if and only if each component of $\bar{\pi}(\bar{k})$ is a square, and the ith component is the residue class of k modulo $p_i^{e_i}$. ☐

This reduces the general problem of finding quadratic residues to the simpler problem of finding quadratic residues modulo a prime power. Following the last section we distinguish between the case of an odd prime and $p = 2$ first, because this can be given an immediate answer:

Proposition A.12. *The odd integer k is a quadratic residue modulo 4 if and only if $k \equiv 1 (\mathrm{mod}\, 4)$, and is a quadratic residue modulo 2^e for $e \geq 3$ if and only if $k \equiv 1 (\mathrm{mod}\, 8)$.*

Proof: Since $U(\mathbb{Z}_4) = \{\bar{1}, \bar{3}\}$ the only square in $U(\mathbb{Z}_4)$ is $\bar{1}$. For $e \geq 3$, if $\bar{z}^2 = \bar{k}$ in $U(\mathbb{Z}_{2^e})$, we use Proposition A.10 to write

$$\bar{z} = (-\bar{1})^a \bar{5}^b, \quad \bar{k} = (-\bar{1})^c \bar{5}^d,$$

and then

$$(-\bar{1})^{2a} \bar{5}^{2b} = (-\bar{1})^c \bar{5}^d,$$

whence c is even and $2b \equiv d \pmod{2^{e-2}}$. Given d, the congruence can be solved for b if and only if d is even. Thus $\bar{k}$ is a quadratic residue modulo 2^e if and only if

$$\bar{k} = \bar{5}^d$$

in $\mathbb{Z}_{2^e}$ where d is even. Putting $d = 2r$ this implies

$$k \equiv 5^{2r} \pmod{2^e}$$

for $e \geq 3$, hence

$$k \equiv 25^r \pmod 8$$
$$\equiv 1 \pmod 8.$$

Conversely if $k \equiv 1 \pmod 8$ and $k \equiv (-1)^c 5^d \pmod{2^e}$, then

$$(-1)^c 5^d \equiv 1 \pmod 8.$$

This happens only when c, d are even, and then $(-\bar{1})^c \bar{5}^d$ is a square in $U(\mathbb{Z}_{2^e})$. ☐

In the case p odd we first characterize the quadratic residues modulo p by using a primitive root modulo p:

Lemma A.13. *If s is a primitive root modulo p, then $\bar{k} = \bar{s}^a$ is a quadratic residue if and only if a is even.*

Proof: If $a = 2b$, then $\bar{k} = (\bar{s}^b)^2$. Now $\bar{s}$ has even order $p - 1$, it cannot be a square, nor can $\bar{s}^a$ for a odd. $\qquad\square$

This characterization of quadratic residues modulo p immediately gives:

Proposition A.14. *If p is an odd prime, then k is a quadratic residue modulo p^e for $e \geq 2$ if and only if k is a quadratic residue modulo p.*

Proof: If $z^2 \equiv k \pmod{p^e}$, then clearly $z^2 \equiv k \pmod{p}$, so a quadratic residue modulo p^e also serves modulo p. Conversely, suppose k is a quadratic residue modulo p. By Proposition A.8 we can write $k \equiv s^a(1 + p)^a \pmod{p^e}$, and reducing this modulo p gives $k \equiv s^a \pmod{p}$, so Lemma A.12 implies that $a = 2b$ for an integer b, so $k \equiv [s^b(1 + p)^b]^2 \pmod{p^e}$ and k is a quadratic residue modulo p^e. $\qquad\square$

This leaves the central core of the problem: to determine quadratic residues modulo any odd prime p. Legendre, who published two volumes on number theory in 1830, introduced a deceptively simple notation which is ideally suited to the task. He defined the symbol (k/p) for an odd prime p and an integer k not divisible by p as

$$(k/p) = \begin{cases} +1 & \text{if } k \text{ is quadratic residue modulo } p. \\ -1 & \text{otherwise.} \end{cases}$$

As remarked in the introductory chapter, the Legendre symbol is commonly written

$$\left(\frac{k}{p}\right)$$

but this is less convenient in print.

The value of this notation can be seen by writing $\bar{k} = \bar{s}^a$ where s is a primitive root modulo p. By Lemma A.12, k is a quadratic residue modulo p if and only if a is even, hence

$$(k/p) = (-1)^a.$$

From this it is easy to deduce the following useful properties:

Proposition A.15.
(a) $k \equiv r \pmod{p}$ implies $(k/p) = (r/p)$.
(b) $(kr/p) = (k/p)(r/p)$.

Proof: (a) is immediate, and (b) follows by writing $\bar{k} = \bar{s}^a$, $\bar{r} = \bar{s}^b$, whence $\overline{kr} = \bar{s}^{a+b}$ and

$$(kr/p) = (-1)^{a+b} = (-1)^a(-1)^b = (k/p)(r/p). \qquad \square$$

It is now possible to give a computational test for quadratic residues:

Proposition A.16. (Euler's Criterion.) *For an odd prime p and an integer k not divisible by p,*

$$(k/p) \equiv k^{(p-1)/2} \pmod{p}.$$

Proof: For a primitive root s modulo p we have $s^{p-1} \equiv 1 \pmod{p}$, and since $p - 1$ is even,

$$(s^{(p-1)/2} - 1)(s^{(p-1)/2} + 1) = (s^{p-1} - 1) \equiv 0 \pmod{p}.$$

Because $s^{(p-1)/2} \not\equiv 1 \bmod p$, we deduce

$$s^{(p-1)/2} \equiv -1 \pmod{p}.$$

Hence, writing $\bar{k} = \bar{s}^a$ as before,

$$\begin{aligned}
(k/p) &= (-1)^a \\
&\equiv (s^{(p-1)/2})^a \pmod{p} \\
&= (s^a)^{(p-1)/2} \\
&\equiv k^{(p-1)/2} \pmod{p}.
\end{aligned}$$

$$\square$$

Example A.17. k is a quadratic residue mod 5 if $k^2 \equiv 1 \pmod 5$, giving $k = 1, 4$.

We soon see the weakness in this criterion if we attempt to find the quadratic residues modulo a larger prime, for example $p = 19$. In this case k is a quadratic residue if and only if $k^9 \equiv 1 \pmod{19}$, and the calculations concerned involve more work than just calculating all the squares of elements in $U(\mathbb{Z}_{19})$ and solving the problem by inspection. However, Euler's criterion can be used to deduce a much more useful test, due to Gauss.

What Gauss did was to partition the units modulo p by writing them in the form

$$\begin{aligned}
U(\mathbb{Z}_p) &= \{-\overline{(p-1)/2}, \ldots, -\bar{2}, -\bar{1}, \bar{1}, \bar{2}, \ldots, \overline{(p-1)/2}\} \\
&= N \cup P
\end{aligned}$$

where
$$N = \{-\overline{(p-1)/2}, \ldots, -\bar{2}, -\bar{1}\}$$
and
$$P = \{\bar{1}, \bar{2}, \ldots, \overline{(p-1)/2}\}.$$

For instance, if $p = 7$, then
$$N = \{-\bar{3}, -\bar{2}, -\bar{1}\}, \quad P = \{\bar{1}, \bar{2}, \bar{3}\}.$$

Using the usual multiplicative notation $aS = \{as | s \in S\}$, we can write $N = (-\bar{1})P$. To find out whether k is a quadratic residue, Gauss computed the set $\bar{k}P$ and proved:

Proposition A.18. (Gauss's Criterion.) *With the above notation, if $\bar{k}P \cap N$ has ν elements then $(k/p) = (-1)^{\nu}$.*

Proof: Since $\bar{k}$ is a unit, the elements of $\bar{k}P$ are distinct so $|\bar{k}P| = |P|$. Furthermore if $\bar{a}$, $\bar{b}$ are distinct elements of P, then we may take $0 < a < b \le (p-1)/2$. We cannot have $\bar{k}\bar{a} = \bar{r}$ and $\bar{k}\bar{b} = -\bar{r}$, for that implies $k(a+b)$ is divisible by p, hence $a+b$ is divisible by p, contradicting the inequalities satisfied by a, b. Thus the elements $\bar{k}, \bar{k}\bar{2}, \ldots, \bar{k}(p-1)/2$ of $\bar{k}P$ consist precisely of the elements $\pm\bar{1}, \pm\bar{2}, \ldots, \pm\overline{(p-1)/2}$, possibly in a different order, where the number of minus signs is the number of elements of $\bar{k}P$ in N. Hence
$$\bar{k} \cdot \bar{k}\bar{2} \ldots \bar{k}\overline{(p-1)/2} = (\pm\bar{1})(\pm\bar{2}) \ldots (\pm\overline{(p-1)/2})$$
so
$$\bar{k}^{(p-1)/2} = (-1)^{\nu}$$
where ν is the number of elements in $\bar{k}P \cap N$. Thus
$$k^{(p-1)/2} \equiv (-1)^{\nu} \pmod{p}$$
and Euler's criterion gives
$$(k/p) = (-1)^{\nu}.$$

$\square$

Example A.19. Is 3 a quadratic residue modulo 19? To answer this we calculate $\bar{3}P = \{3, 6, 9, 12, 15, 18, 2, 5, 8\}$
$$= \{3, 6, 9, 2, 5, 8\} \cup \{-7, -4, -1\},$$

so $\nu = 3$ and Gauss's criterion tells us that 3 is not a quadratic residue modulo 19.

These two criteria take us further in the search for quadratic residues k modulo an odd prime p, for by factorizing

$$k = (-1)^a 2^b p_1^{e_1} \cdots p_r^{e_r}$$

then k is a square if $a, b, e_1, \ldots, e_r$ are even; moreover, it is a quadratic residue if the factors with odd exponent are quadratic residues. Thus the question of quadratic residues is finally reduced to determining whether -1, 2 or an odd prime q (distinct from p) are quadratic residues modulo an odd prime p.

The given criteria solve the question for -1, 2:

Proposition A.20.
(a) $(-1/p) = (-1)^{(p-1)/2}$, so -1 is a quadratic residue modulo p if and only if $p \equiv 1 \pmod 4$.
(b) $(2/p) = (-1)^{(p^2-1)/8}$, so 2 is a quadratic residue modulo p if and only if $p \equiv \pm 1 \pmod 8$.

Proof: (a) is a trivial consequence of Euler's criterion.
     (b) $2P = \bar{2}, \bar{4}, \ldots, \overline{p-1}$, so $|\bar{2}P \cap N|$ is $\nu = (p-1)/2 - r$ where r is the largest integer such that $2r \leq (p-1)/2$. The proof now splits into two cases.
     Case 1. $(p-1)/2$ is even and $2r = (p-1)/2$, whence $\nu = (p-1)/2 - (p-1)/4 = (p-1)/4$. Thus $(2/p) = (-1)^{(p-1)/4}$.
     Case 2. $(p-1)/2$ is odd and $2r = (p-1)/2 - 1$, so that $\nu = (p-1)/2 - (p-1)/4 + \frac{1}{2} = (p+1)/4$. Thus $(2/p) = (-1)^{(p+1)/4}$. We can put these two cases together by noting that in the first case $(p-1)/2$ is even if and only if $(p+1)/2$ is odd. Raising $(-1)^n$ to an odd power does not change it, so in case 1

$$(2/p) = [(-1)^{(p-1)/4}]^{(p+1)/2} = (-1)^{(p^2-1)/8}.$$

Case 2 gives the same result by raising to the odd power $(p-1)/2$.   □

We see from the work required in case (b) that Gauss's criterion is still not subtle enough to decide easily whether an odd prime q is a quadratic residue modulo p. To complete the solution we refine the criterion further to obtain Gauss's 'gem of higher arithmetic':

Theorem A.21. (Quadratic Reciprocity Law.) *If p, q are distinct odd primes,*
then

$$(p/q)(q/p) = (-1)^{(p-1)(q-1)/4}.$$

Proof: By Gauss's criterion $(q/p) = (-1)^{\nu}$, where ν is the number of
integers a in $1 \le a \le (p-1)/2$ such that there exists an integer b satisfying

$$aq = bp + r, \quad -p/2 < r < 0.$$

There can be at most one such integer b for each a, so we can rephrase
the requirement: ν is the number of ordered pairs (a, b) of integers
satisfying

$$1 \le a \le (p-1)/2, \tag{A.9}$$

$$-p/2 < aq - bp < 0. \tag{A.10}$$

From (A.9) and (A.10) we deduce

$$bp < aq + p/2 \le (p-1)q/2 + p/2 < pq/2 + p/2 = p(q+1)/2.$$

Hence $b < (q+1)/2$, and (A.10) implies $b \ge 1$, so

$$1 \le b \le (q-1)/2. \tag{A.11}$$

Since (A.9) and (A.10) imply (A.11), it does no harm to add (A.11) to the
list of requirements to be satisfied by the ordered pair (a, b), so ν is the
number of pairs (a, b) of integers satisfying (A.9)–(A.11). It actually does
a lot of good because of the symmetry of (A.9) and (A.11). Interchanging
p, q and a, b, we also have

$$(p/q) = (-1)^{\mu}$$

where μ is the number of ordered pairs of integers (a, b) satisfying (A.11),
(A.9) and

$$-q/2 < bp - aq < 0 \tag{A.10}'$$

which can be written

$$0 < aq - bp < q/2. \tag{A.12}$$

Since p, q are distinct primes, (A.9) and (A.11) imply $aq - bp \ne 0$, so $\nu + \mu$
is the number of ordered pairs of integers (a, b) satisfying (A.9), (A.11)
and

$$-p/2 < aq - bp < q/2. \tag{A.13}$$

Now

$$(p/q)(q/p) = (-1)^{\nu+\mu}, \tag{A.14}$$

so the problem reduces to finding the value of $\nu + \mu$ mod 2.

To do this, let

$$R = \{(a,b) \in \mathbb{Z}^2 | 1 \le a \le (p-1)/2,\ 1 \le b \le (q-1)/2\}.$$

Then R has $(p-1)(q-1)/4$ elements. Partition R into three subsets

$$\begin{aligned} R_1 &= \{(a,b) \in R | aq - bp \le -p/2\} \\ R_2 &= \{(a,b) \in R | -p/2 < aq - bp < q/2\} \\ R_3 &= \{(a,b) \in R | q/2 \le aq - bp\}. \end{aligned}$$

Then R_2 is the set of solutions of (9), (11), (13) as required. The map $f : \mathbb{Z}^2 \to \mathbb{Z}^2$ given by $f(a,b) = ((p+1)/2 - a, (q+1)/2 - b)$ is easily seen to restrict to a bijection from R_1 to R_3, so $|R_1| = |R_3|$. This implies that

$$|R| = |R_1| + |R_2| + |R_3| \equiv |R_2| \quad (\text{mod } 2)$$

so

$$(p-1)(q-1)/4 = \nu + \mu \quad (\text{mod } 2).$$

From (A.14)

$$(p/q)(q/p) = (-1)^{(p-1)(q-1)/4}. \qquad \Box$$

An immediate deduction is the quadratic reciprocity law in the form stated by Gauss:

Theorem A.22. *If p and q are distinct odd primes, at least one of which is congruent to 1 modulo 4, then p is a quadratic residue of p if and only if q is a quadratic residue of p; otherwise if neither is congruent to 1 modulo 4 then precisely one is a quadratic residue of the other.*

Proof: If at least one of p, q is congruent to 1 modulo 4, then $(p-1)(q-1)/4$ is even, so

$$(p/q)(q/p) = 1,$$

whence $(p/q) = +1$ if and only if $(q/p) = +1$. If neither is congruent to 1 modulo 4, then $(p-1)(q-1)/4$ is odd,

$$(p/q)(q/p) = -1,$$

so precisely one of (p/q), (q/p) is $+1$ and the other is -1. $\qquad \Box$

We can imagine Gauss's intense pleasure at discovering this remarkable result. To see its power, we only have to compare the easy way this resolves problems involving quadratic residues compared with the two criteria given earlier or with *ad hoc* methods. Its use is even more clear when allied with Legendre's clever symbol.

Example A.23. Is 1984 a quadratic residue modulo 97?

$$(1984/97) = (44/97) = (2/97)^2(11/97) = (\pm1)^2(11/97)$$
$$= (11/97) = (97/11) = (9/11) = (3/11)^2 = 1$$

because $97 \equiv 1 \bmod 4$. Hence 1984 is a quadratic residue modulo 97.

By putting together the appropriate results of this chapter the question of whether a specific number k is a quadratic residue modulo m may be completely solved by a succession of reductions of the problem:

(a) By factorizing m and using Proposition A.11 which says k is a quadratic residue modulo m if and only if it is a quadratic residue modulo each prime factor of m.

(b) If 2 is a prime factor, that part of the problem may be solved by Proposition A.12: if 2^e is the largest power of 2 dividing m, for $e = 1$, any odd k is a quadratic residue, for $e = 2$ we must check that $k \equiv 1 \pmod 4$ and for $e \geq 3$ we must check that $k \equiv 1 \pmod 8$.

(c) For an odd prime factor p of m, we calculate (k/p). First, reduce k modulo p to assume that $1 \leq k < p$. Then factorize $k = q_1^{f_1} \ldots q_s^{f_s}$ where the q_i are primes and write

$$(k/p) = (q_1/p)^{f_1} \ldots (q_s/p)^{f_s}.$$

We need consider only (q_i/p) for f_i odd, and since $q_i < p$, we can use quadratic reciprocity to obtain

$$(q_i/p) = (p/q_i)^2(q_i/p) = (p/q_i)(-1)^{(p-1)(q_i-1)/4}$$

This reduces the problem to calculating (p/q_i) where $p > q_i$. Reducing p modulo q_i, we have to calculate Legendre symbols for smaller primes. Successive reductions of this nature lead to a complete solution.

Example A.24. Is 65 a quadratic residue modulo 124?

Since $124 = 2^2 \times 31$, we must check whether 65 is a quadratic residue modulo 2^2 and 31. Modulo 2^2 we have $65 \equiv 1 \pmod 4$, so the answer is yes by Proposition A.12(a). Modulo 31

$$(65/31) = (3/31) = (31/3)(-1)^{30 \times 2/4} = -(31/3) = -(1/3),$$

and 1 is a quadratic residue modulo 3, so $(65/31) = -1$. Thus 65 is not a quadratic residue modulo 124.

It is the reduction of the seemingly complicated problem of quadratic residues to simple arithmetic such as this which highlights the brilliance of the jewel in Gauss's number-theoretic crown.

A.4 Exercises

1. Solve the following congruences (where possible)

    (1) $3x \equiv 14 \pmod{17}$,

    (2) $6x \equiv 3 \pmod{35}$,

    (3) $3x \equiv 13 \pmod{18}$,

    (4) $20x \equiv 60 \pmod{80}$.

2. Solve the quadratic congruences (where possible):

    (1) $3x^2 + 6x + 5 \equiv 0 \pmod{7}$,

    (2) $x^2 + 5x + 3 \equiv 0 \pmod{4}$,

    (3) $x^2 \equiv 1 \pmod{12}$,

    (4) $x^2 \equiv 0 \pmod{12}$,

    (5) $x^2 \equiv 2 \pmod{12}$.

3. Let $a_1, a_2, \ldots, a_n$ be the complete set of residues modulo n (not in any specific order), and let b be an integer relatively prime to n and c any integer. Show that

$$a_1 b + c, a_2 b + c, \ldots, a_n b + c$$

    is also a complete set of residues.

4. Calculate the Euler function $\phi(n) = |U(\mathbb{Z}_n)|$ for $n = 4, 6, 12, 18$.

5. Determine all the generators of $U(\mathbb{Z}_7)$, $U(\mathbb{Z}_{11})$, $U(\mathbb{Z}_{17})$.

6. Show that there exists a primitive root s modulo p such that

$$s^{p-1} \not\equiv 1 \pmod{p^2}.$$

7. Calculate the exponents of the following groups: $U(\mathbb{Z}_4)$, $U(\mathbb{Z}_6)$, $U(\mathbb{Z}_8)$, $U(\mathbb{Z}_{10})$. Which of these groups are cyclic?

8. Show that for an odd prime p there are as many square residue classes as non-squares in $U(\mathbb{Z}_{p^e})$ for $e \geq 1$.

9. Show that in $U(\mathbb{Z}_{2^e})$ there are exactly 2^{e-3} squares and $3 \cdot 2^{e-3}$ non-squares for $e \geq 3$.

10. Determine the squares in the following groups: $U(\mathbb{Z}_7)$, $U(\mathbb{Z}_{12})$, $U(\mathbb{Z}_{49})$.

11. Use Euler's criterion to check whether 7 is a quadratic residue modulo 23. Answer the same question using Gauss's criterion. Now calculate $(7/23)$ using Gauss reciprocity.

12. Compute the following Legendre symbols: $(-1/179)$, $(6/11)$, $(2/97)$.

13. Compute $(97/1117)$, $(2437/811)$, $(23/97)$.

14. Is 1984 a quadratic residue modulo 365?

15. Is 2001 a quadratic residue modulo 1820?

16. Find the primes for which 11 is a quadratic residue.

17. Define the Jacobi symbol (k/m) for relatively prime integers k, m $(m > 0)$ by factorizing $m = p_1^{e_1} \ldots p_r^{e_r}$ and writing

$$(k/m) = (k/p_1)^{e_1} \ldots (k/p_r)^{e_r}.$$

If k, r are prime to m and $k \equiv r \pmod{m}$, show

$$(k/m) = (r/m).$$

18. If (k/m) is the Jacobi symbol of Exercise 17, for m positive and odd, prove that

(1) $(-1/m) = (-1)^{(m-1)/2}$,

(2) $(2/m) = (-1)^{(m^2-1)/8}$.

For k, m positive and relatively prime, prove that

$$(k/m)(m/k) = (-1)^{(k-1)(m-1)/4}.$$

Dirichlet's Units Theorem

In this appendix we look a little more deeply at properties of the units in the integers of a number field. These properties are significant for the general theory, but are not essential to our development of Fermat's Last Theorem. Units are important because, while ideals are best suited to technicalities, there may come a point at which it is necessary to return to elements. But the generator of a principal ideal is ambiguous up to multiples by a unit. To translate results about ideals to their corresponding generators, we therefore need to know about units in the ring of integers. The most fundamental and far-reaching theorem on units is that of Dirichlet, which gives an almost complete description, in abstract terms, of the group of units of the ring of integers of any number field. In particular it implies that this group is finitely generated. We prove Dirichlet's theorem in this appendix. The methods are 'geometric' in that we use Minkowski's theorem, together with a 'logarithmic' variant of the space $\mathbb{L}^{st}$.

B.1 Introduction

We have already described the units in the integers of $\mathbb{Q}(\sqrt{d})$ for negative squarefree d in Proposition 4.2. For $d = -1$ the units are $\{\pm 1, \pm i\}$, for $d = -3$ they are $\{\pm 1, \pm \omega, \pm \omega^2\}$ where $\omega = e^{2\pi i/3}$, and for all other $d < 0$, the units are just $\{\pm 1\}$.

In all cases U is a finite cyclic group of even order (2, 4, or 6) whose elements are roots of unity. It is in any case obvious that every unit of finite order is a root of unity in any number field.

For other number fields the structure of U is more complicated. For example in $\mathbb{Q}(\sqrt{2})$

$$(1 + \sqrt{2})(-1 + \sqrt{2}) = 1$$

so $\epsilon = 1 + \sqrt{2}$ is a unit. Now ϵ is not a root of unity since $|\epsilon| = 1 + \sqrt{2} \neq 1$. It follows that ϵ has infinite order, all the elements $\pm\, \epsilon^n$ ($n \in Z$) are distinct units, and U is an infinite group. In fact, though we do not prove it here, the $\pm\epsilon^n$ are all the units of $\mathbb{Q}(\sqrt{2})$, so U is isomorphic to $\mathbb{Z}_2 \times \mathbb{Z}$.

After we have proved Dirichlet's theorem it will emerge that this more complicated structure of U is in some sense typical.

B.2   Logarithmic Space

Let K be a number field of degree $n = s + 2t$, as in Chapter 8, and let $\mathbb{L}^{st}$ be as described there. We use the notation of Chapter 8 in what follows. Define a map

$$l : \mathbb{L}^{st} \to \mathbb{R}^{s+t}$$

as follows. For $x = (x_1, \ldots, x_s; x_{s+1}, \ldots, x_{s+t}) \in \mathbb{L}^{st}$ put

$$l_k(x) = \begin{cases} \log|x_k| & \text{for } k = 1, \ldots, s \\ \log|x_k|^2 & \text{for } k = s + 1, \ldots s + t. \end{cases}$$

Then set

$$l(x) = (l_1(x), \ldots, l_{s+t}(x)).$$

The additive property of the logarithm leads at once to the property

$$l(xy) = l(x) + l(y) \tag{B.1}$$

for $x, y \in \mathbb{L}^{st}$. The set of elements of $\mathbb{L}^{st}$ with all coordinates non-zero is a group under multiplication, and l is a homomorphism from this group into $\mathbb{R}^{s+t}$. By (8.1) in Chapter 8,

$$\sum_{h=1}^{s+t} l_k(x) = \log|\mathrm{N}(x)|. \tag{B.2}$$

For $\alpha \in K$ define

$$l(\alpha) = l(\sigma(\alpha))$$

where $\sigma : K \to \mathbb{L}^{st}$ is the standard map. This ambiguity in the use of l causes no confusion, and is tantamount to an identification of α with $\sigma(\alpha)$. Explicitly,

$$l(\alpha) = (\log|\sigma_1(\alpha)|, \ldots, \log|\sigma_s(\alpha)|, \log|\sigma_{s+1}(\alpha)|^2, \ldots, \log|\sigma_{s+t}(\alpha)|^2).$$

The map $l : K \to \mathbb{R}^{s+t}$ is the *logarithmic representation* of K, and $\mathbb{R}^{s+t}$ is the *logarithmic space*.

By (8.2) of Chapter 8 and (B.1),

$$l(\alpha\beta) = l(\alpha) + l(\beta) \quad (\alpha, \beta \in K)$$

so l is a homomorphism from the multiplicative group $K^* = K \setminus \{0\}$ of K to the additive group of $\mathbb{R}^{s+t}$. Further we have, setting $l_k(\alpha) = l_k(\sigma(\alpha))$,

$$\sum_{k=1}^{s+t} l_k(\alpha) = \log |N(\alpha)| ,$$

using (8.3) of Chapter 8 and (B.2).

B.3 Embedding the Unit Group in Logarithmic Space

Why all these logarithms? Because the group of units is multiplicative, whereas Minkowski's theorem applies to lattices, which are additive. We must pass from one context to the other, and it is just for this purpose that logarithms were created.

Let U be the group of units of $\mathfrak{D}$, the ring of integers of K. By restriction we obtain a homomorphism

$$l : U \to \mathbb{R}^{s+t}.$$

It is not injective, but the kernel is easily described:

Lemma B.1. *The kernel W of $l : U \to \mathbb{R}^{s+t}$ is the set of all roots of unity belonging to U. This is a finite cyclic group of even order.*

Proof: We have $l(\alpha) = 0$ if and only if $|\sigma_i(\alpha)| = 1$ for all i. The field polynomial

$$\prod_i (t - \sigma_i(\alpha))$$

lies in $\mathbb{Z}[t]$ by Theorem 2.6 (a) combined with Lemma 2.13. We can therefore appeal to Lemma 11.6 to conclude that all the $\sigma_i(\alpha)$ are roots of unity, in particular α itself.

The image $\sigma(\mathfrak{D})$ in $\mathbb{L}^{st}$ is a lattice by Corollary 8.3, so it is discrete by Theorem 6.1. Since the unit circle in $\mathbb{C}$ maps to a bounded subset in $\mathbb{L}^{st}$ it follows that $\mathfrak{D}$ contains only finitely many roots of unity, so W is finite.

But any finite subgroup of K^* is cyclic (see Stewart [78], Theorem 16.7, p. 171). Finally, W contains -1 which has order 2, so W has even order. $\square$

Obviously the next thing to find out is the image E of U in $\mathbb{R}^{s+t}$:

Lemma B.2. *The image E of U in $\mathbb{R}^{s+t}$ is a lattice of dimension $\leq s+t-1$.*

Proof: The norm of any unit is ± 1, so for any unit ϵ

$$\sum_{k=1}^{s+t} l_k(\epsilon) = \log |N(\epsilon)| = \log 1 = 0.$$

Hence all points of E lie in the subspace V of $\mathbb{R}^{s+t}$ consisting of those elements $(x_1, \ldots, x_{s+t})$ such that

$$x_1 + \ldots + x_{s+t} = 0.$$

This has dimension $s + t - 1$.

To prove E is a lattice it is sufficient to prove it discrete by Theorem 6.1. Let $\| \ \|$ be the usual length function on $\mathbb{R}^{s+t}$. Suppose that $0 < r \in \mathbb{R}$, and

$$\|l(\epsilon)\| < r.$$

Now $|l_k(\epsilon)| \leq \|l(\epsilon)\| < r$, so

$$|\sigma_k(\epsilon)| < e^r \quad (k = 1, \ldots, s)$$
$$|\sigma_{s+j}(\epsilon)|^2 < e^r \quad (j = 1, \ldots, t).$$

Hence the set of points $\sigma(\epsilon)$ in $\mathbb{L}^{st}$ corresponding to units with $\|l(\epsilon)\| < r$ is bounded, so finite by Corollary 8.3. Hence E intersects each closed ball in $\mathbb{R}^{s+t}$ in a finite set, so E is discrete. Therefore E is a lattice. Since $E \subseteq V$ it has dimension $\leq s + t - 1$. $\square$

Already we know quite a lot about U. In particular, U is finitely generated, because W is finite and $U/W \cong E$ is a lattice, so free abelian, with rank $\leq s + t - 1$. All that remains is to find the exact dimension of the lattice E. In fact it is $s + t - 1$, as we prove in the next section.

B.4   Dirichlet's Theorem

The main thing we lack is a topological criterion for deciding whether a lattice L in a vector space V has the same dimension as V. We remedy this lack with:

Lemma B.3. *Let L be a lattice in $\mathbb{R}^m$. Then L has dimension m if and only if there exists a bounded subset B of $\mathbb{R}^m$ such that*

$$\mathbb{R}^m = \bigcup_{x \in L} (x + B).$$

Proof: If L has dimension m then we may take B to be a fundamental domain of L and appeal to Lemma 6.2.

Suppose conversely that B exists but, for a contradiction, L has dimension $d < m$. An intuitive argument goes like this: the quotient $\mathbb{R}^m/L$ is, by Theorem 6.6, the direct product of a torus and $\mathbb{R}^{m-d}$. The condition on B says that the image of B under the natural map $\nu : \mathbb{R}^m \to \mathbb{R}^m/L$ is the whole of $\mathbb{R}^m/L$. But because B is bounded this contradicts the presence of a direct factor $\mathbb{R}^{m-d}$ which is unbounded. By taking more account of the topology than we have done hitherto, this argument can easily be made rigorous. Alternatively, we operate in $\mathbb{R}^m$ instead of $\mathbb{R}^m/L$ as follows.

Let V be the subspace of $\mathbb{R}^m$ spanned by L. If L has dimension less than m then $\dim V < \dim \mathbb{R}^m$. Hence we can find an orthogonal complement V' to V in $\mathbb{R}^m$. The condition on B implies that $\mathbb{R}^m = \cup_{v \in V} (v + B)$, so V' is the image of B under the projection $\pi : \mathbb{R}^m \to V'$. But π is distance-preserving, so V' is bounded, contradiction. $\qquad\qquad\square$

In fact, what we are saying topologically is that L has dimension m if and only if the quotient topological group $\mathbb{R}^m/L$ is *compact*. This can profitably be compared with Theorem 1.17. In fact there is some kind of analogy between free abelian groups and sublattices of vector spaces; as witness to which the reader should compare Lemma 9.3 and Theorem 1.17.

Before proving that E has dimension $s+t-1$ it is convenient to extract one computation from the proof:

Lemma B.4. *Let $y \in \mathbb{L}^{st}$ and let $\lambda_y : \mathbb{L}^{st} \to \mathbb{L}^{st}$ be defined by $\lambda_y(x) = yx$. Then λ_y is a linear map and*

$$\det \lambda_y = \mathrm{N}(y).$$

Proof: It is obvious that λ_y is linear. To compute $\det \lambda_y$ we use the basis (8.4) of Chapter 8. If

$$y = (x_1, \ldots, x_s; y_1 + iz_1, \ldots, y_t + iz_t)$$

then we obtain for $\det \lambda_y$ the expression

$$\begin{vmatrix} x_1 & & & & & \\ & \ddots & & & \mathbf{0} & \\ & & x_s & & & \\ & & & y_1 & -z_1 & \\ & & & z_1 & y_1 & \\ & \mathbf{0} & & & & \ddots \\ & & & & & y_t & -z_t \\ & & & & & z_t & y_t \end{vmatrix}$$

which is
$$x_1 \ldots x_s(y_1^2 + z_1^2) \ldots (y_t^2 + z_t^2) = \mathrm{N}(y). \qquad \square$$

The way is now clear for the proof of:

Theorem B.5. *The image E of U in $\mathbb{R}^{s+t}$ is a lattice of dimension $s+t-1$.*

Proof: As before let V be the subspace of $\mathbb{R}^{s+t}$ whose elements satisfy

$$x_1 + \ldots + x_{s+t} = 0.$$

Then $E \subseteq V$, and $\dim V = s + t - 1$. To prove the theorem we appeal to Lemma B.3: it is sufficient to find in V some bounded subset B such that

$$V = \bigcup_{e \in E} (e + B).$$

This additive property translates into a multiplicative property in $\mathbb{L}^{st}$. Every point in $\mathbb{R}^{s+t}$ is the image under l of some point in $\mathbb{L}^{st}$, so every point in V is the image of some point in $\mathbb{L}^{st}$. In fact, for $x \in \mathbb{L}^{st}$, we have $l(x) \in V$ if and only if $|\mathrm{N}(x)| = 1$. So if we let

$$S = \{x \in \mathbb{L}^{st} : |\mathrm{N}(x)| = 1\}$$

then $l(S) = V$. If $X_0 \subseteq S$ is bounded, then so is $l(X_0)$, as may be verified easily. If $x \in S$ then the multiplicativity of the norm implies that $xX_0 \subseteq S$ if $X_0 \subseteq S$. In particular if ϵ is a unit then $\sigma(\epsilon)X_0 \subseteq S$. So if we can find a bounded subset X_0 of S such that

$$S = \bigcup_{\epsilon \in U} \sigma(\epsilon)X_0 \qquad (\text{B.3})$$

then $B = l(X_0)$ will do what is required in V.

Now we find a suitable X_0. Let M be the lattice in $\mathbb{L}^{st}$ corresponding to $\mathfrak{O}$ under σ. Consider the linear transformation $\lambda_y : \mathbb{L}^{st} \to \mathbb{L}^{st}(y \in \mathbb{L}^{st})$

of Lemma B.4. If $y \in S$ then the determinant of λ_y is $N(y)$ which is ± 1. Therefore λ_y is unimodular. By the remark after Lemma 9.3, this implies that any fundamental domain for the lattice $yM(= \lambda_y(M))$ has the same volume as a fundamental domain for M. Call this volume v.

Choose real numbers $c_i > 0$ with

$$Q = c_1 \ldots c_{s+t} > \left(\frac{4}{\pi}\right)^t v.$$

Let X be the set of $x \in \mathbb{L}^{st}$ for which

$$|x_k| < c_k \quad (k = 1, \ldots, s)$$
$$|x_{s+j}|^2 < c_{s+j} \quad (j = 1, \ldots, t).$$

By Lemma 9.2 there exists in yM a non-zero point $x \in X$. We have

$$x = y\sigma(\alpha) \quad (0 \neq \alpha \in \mathfrak{O}).$$

Now

$$N(x) = N(y)N(\alpha) = \pm N(\alpha)$$

so

$$|N(\alpha)| < Q.$$

By Theorem 5.17 (c) only finitely many ideals of $\mathfrak{O}$ have norm $< Q$. Consider principal ideals and recall that their generators are ambiguous up to unit multiples. Then there exist in $\mathfrak{O}$ only finitely many pairwise non-associated numbers

$$\alpha_1, \ldots, \alpha_N$$

whose norms are $< Q$ in absolute value. Thus for some $i = 1, \ldots, N$ we have $\alpha\epsilon = \alpha_i$ for a unit ϵ. Therefore

$$y = x\sigma(\alpha_i^{-1})\sigma(\epsilon). \tag{B.4}$$

Now define

$$X_0 = S \cap \left(\bigcup_{i=1}^N \sigma(\alpha_i^{-1})X\right). \tag{B.5}$$

Since X is bounded so are the sets $\sigma(\alpha_i^{-1})X$, and since N is finite X_0 is bounded. Obviously X_0 does not depend on the choice of $y \in S$.

But now, since y and $\sigma(\epsilon) \in S$, we have $x\sigma(\alpha_i^{-1}) \in S$, hence $x\sigma(\alpha_i^{-1}) \in X_0$. Then (B.4) shows that

$$y \in \sigma(\epsilon)X_0.$$

Hence (B.3) holds for an arbitrary element $y \in S$. $\square$

We put this result into a more explicit form, obtaining the *Dirichlet Units Theorem*:

Theorem B.6. **(Dirichlet Units Theorem.)**   *The group of units of $\mathfrak{O}$ is isomorphic to*

$$W \times \mathbb{Z} \times \ldots \times \mathbb{Z}$$

where W is as described in Lemma B.1 and there are $s + t - 1$ direct factors $\mathbb{Z}$.

Proof: By Theorem B.5, $U/W \cong \mathbb{Z} \times \ldots \times \mathbb{Z} = \mathbb{Z}^{s+t-1}$. Since W is finite, U is a finitely generated abelian group, hence a direct product of cyclic groups, see Fraleigh [28] Theorem 9.3, p. 90. Since W is finite and U/W is torsion-free, W is the set of elements of U of finite order, which is the product of all the finite cyclic factors in the direct decomposition. The other factors are all infinite cyclic; looking at U/W tells us there are exactly $s + t - 1$ of them. $\square$

In more classical terms, Dirichlet's theorem asserts the existence of a system of $s + t - 1$ *fundamental units*

$$\eta_1, \ldots, \eta_{s+t-1}$$

such that every unit of $\mathfrak{O}$ is representable *uniquely* in the form

$$\zeta \cdot \eta_1^{r_1} \ldots \eta_{s+t-1}^{r_{s+t-1}}$$

for a root of unity ζ and rational integers r_i.

We return briefly to $\mathbb{Q}(\sqrt{2})$, which we looked at in Section B.1. For this field, $s = 2$, $t = 0$, so $s + t - 1 = 1$. Hence U is of the form $W \times \mathbb{Z}$ where W consists of the roots of unity in $\mathbb{Q}(\sqrt{2})$. These are just ± 1, so we get $U \cong \mathbb{Z}_2 \times \mathbb{Z}$ as asserted in Section 1. Note, however, that we have still not proved that $1 + \sqrt{2}$ is a fundamental unit. In fact, this is true in general of Dirichlet's theorem: it does not allow us to find any *specific* system of fundamental units. Other methods can be developed to solve this problem, and the Dirichlet theorem is still needed to tell us when we have found sufficiently many units.

B.5   Exercises

1. Find units, not equal to 1, in the rings of integers of the fields $\mathbb{Q}(\sqrt{d})$ for $d = 2, 3, 5, 6, 7, 10$.

2. Use Dirichlet's theorem to prove that for any squarefree positive integer d there exist infinitely many integer solutions x, y to the *Pell equation*
$$x^2 - dy^2 = 1.$$
(Really this should not be called the Pell equation, since Pell did not solve it. It was mistakenly attributed to him by Euler, and the name stuck.)

3. Prove that $1 + \sqrt{2}$ is a fundamental unit for $\mathbb{Q}(\sqrt{2})$.

4. Let $\eta_1, \ldots, \eta_{s+t-1}$ be a system of fundamental units for a number field K. Show that the *regulator*
$$R = |\det(\log|\sigma_i(\eta_j)|)|$$
is independent of the choice of $\eta_1, \ldots, \eta_{s+t-1}$.

5. Show that the group of units of a number field K is finite if and only if $K = \mathbb{Q}$ or K is an imaginary quadratic field.

6. Show that a number field of odd degree contains only two roots of unity.

Bibliography

[1] H. Anton. *Elementary Linear Algebra* (5th ed.), Wiley, New York 1987.

[2] T.M. Apostol. *Mathematical Analysis*, Addison–Wesley, Reading MA 1957.

[3] A. Baker. Linear forms in the logarithms of algebraic numbers, *Mathematika* **13** (1966) 204–216.

[4] K. Barner. Paul Wolfskehl and the Wolfskehl Prize, *Notices Amer. Math. Soc.* **44** (1997) 1294–1303.

[5] Y. Bilu. Catalan's conjecture (after Mihăilescu), *Astérisque* **294** vii (2004) 1–26.

[6] B.J. Birch. Diophantine analysis and modular functions, *Proc. Conf. Algebraic Geometry*, Tata Institute, Bombay 1968, 35–42.

[7] S. Bloch. The proof of the Mordell Conjecture, *Math. Intelligencer* **6** No.2 (1984) 41–47.

[8] Z.I. Borevič and I.R. Šafarevič. *Number Theory*, Academic Press, New York 1966.

[9] E. Brieskorn and H. Knörrer. *Plane Algebraic Curves*, Birkhäuser, Basel 1986.

[10] J. Brillhart, J. Tonascia, and P. Weinberger. On the Fermat quotient, in *Computers in Number Theory*, Academic Press, New York 1971, 213–222.

[11] J. Buhler, R. Crandall, R. Ernvall, and T. Metsankyla. Irregular primes and cyclotomic invariants to four million, *Math. Comp.* **60** (1993) 161–153.

[12] R.P. Burn. *Groups: a Path to Geometry*, Cambridge University Press, Cambridge 1985.

[13] A. Cauchy. *Oeuvres 1(X)*, Gauthier–Villars, Paris 1897, 276–285 and 296–308.

[14] A. Cayley. *An Elementary Treatise on Elliptic Functions*, Dover, New York 1961.

[15] H. Chatland and H. Davenport. Euclid's algorithm in real quadratic fields, *Canad. J. Math.* **2** (1950) 289–296.

[16] D.A. Clark. *Manuscripta Mathematica* **83** (1994) 327–330.

[17] H. Cohen. *A Course in Computational Algebraic Number Theory*, Springer–Verlag, Berlin 1993.

[18] David A. Cox. Introduction to Fermat's Last Theorem, *Amer. Math. Monthly* **101** (1994) 3–14.

[19] H.S.M. Coxeter. *Introduction to Geometry* (2nd ed.), Wiley, New York 1969.

[20] H. Darmon. A proof of the full Taniyama–Shimura Conjecture is announced, *Notices Amer. Math. Soc.* **46** (1999) 1397–1401.

[21] H. Darmon, F. Diamond, and R. Taylor. Fermat's Last Theorem, in *Current Developments in Mathematics 1995*, International Press, Cambridge MA 1994, 1–154.

[22] V.A. Demjanenko. L. Euler's Conjecture, *Acta Arithmetica* **25** (1973/4) 127–135.

[23] M. Deuring. Imaginäre quadratischen Zahlkörper mit der Klassenzahl Eins, *Invent. Math.* **5** (1968) 169–179.

[24] H.M. Edwards. The background of Kummer's proof of Fermat's Last Theorem for regular primes, *Arch. Hist. Exact Sci.* **14** (1975) 219–236.

[25] H.M. Edwards. *Fermat's Last Theorem*, Springer–Verlag, New York 1977.

[26] N. Elkies, On $A^4 + B^4 + C^4 = D^4$, *Math. Comp.* **51** (1988) 825–835.

[27] M.H. Fenrick. *Introduction to the Galois Correspondence*, Birkhäuser, Boston 1992.

[28] J.B. Fraleigh. *A First Course in Abstract Algebra*, Addison–Wesley, Reading MA 1989.

[29] G. Frey. Rationale Punkte auf Fermatkurven und gewisteten Modularkurven, *J. Reine Angew. Math.* **331** (1982) 185–191.

[30] G. Frey. Links between stable elliptic curves and certain diophantine equations, *Ann. Univ. Sarav.* **1** (1986) 1–40.

[31] D.J.H. Garling. *A Course in Galois Theory*, Cambridge University Press, Cambridge 1986.

[32] C.F. Gauss. *Disquisitiones Arithmeticae* (translated by A.A. Clarke), Yale University Press, New Haven CT 1966.

[33] D.M. Goldfeld. Gauss's class number problem for imaginary quadratic fields, *Bull. Amer. Math. Soc* **13** (1985) 23–37.

[34] J.R. Goldman. *The Queen of Mathematics*, A.K. Peters, Wellesley MA 1998.

[35] E.S. Golod and I.R. Šafarevič. On class field towers, *Izv. Akad. Nauk SSSR ser. Mat.* **28** (1964) 261–272; *Amer. Math. Soc. Trans., 2nd ser.* **48** (1965) 91–102.

[36] H. Hancock. *Theory of Elliptic Functions*, Dover, New York 1958.

[37] G.H. Hardy. *A Course of Pure Mathematics*, Cambridge University Press, Cambridge 1960.

[38] M. Harper. $\mathbf{Z}(\sqrt{14})$ is Euclidean, *Canad. J. Math.* **56** (2004) 55–70.

[39] K. Heegner. Diophantische Analysis und Modulfunktionen, *Math. Zeit.* **56** (1952) 227–253.

[40] H. Heilbronn and E.H. Linfoot. On the imaginary quadratic corpora of class-number one, *Quart. J. Math. (Oxford)* **5** (1934) 293–301.

[41] Y. Hellegouarch. Points d'ordre $2p^h$ sur les courbes elliptiques, *Acta Arith.* **26** (1975) 253–263.

[42] J.F. Humphreys. *A Course in Group Theory*, Oxford University Press, Oxford 1996.

[43] K. Inkeri. Über den Euklidischen Algorithmus in quadratischen Zahlkörpern, *Ann. Acad. Scient. Fennicae* **41** (1947) pp.35.

[44] N. Jacobson. *Basic Algebra* vol. 1, Freeman, San Francisco 1980.

[45] R.B. King. *Beyond the Quartic Equation*, Birkhäuser, Boston 1996.

[46] L.J. Lander and T.R. Parkin. Counterexamples to Euler's Conjecture on sums of like powers, *Bull. Amer. Math. Soc.* **72** (1966) 1079.

[47] S. Lang. *Algebraic Number Theory*, Addison–Wesley, Reading MA 1970.

[48] S.L. Loney. *The Elements of Coordinate Geometry*, Macmillan, London 1960.

[49] I.D. Macdonald. *The Theory of Groups*, Oxford University Press, Oxford 1968.

[50] D. Masser. Open problems, *Proc. Symp. Analytic Number Thy.* (ed. W.W.L. Chen), Imperial College, London 1985.

[51] R.D. Mauldin. A generalization of Fermat's Last Theorem: the Beal Conjecture and prize problem, *Notices Amer. Math. Soc.* **44** (1997) 1436–1437.

[52] B. Mazur. Modular curves and the Eisenstein ideal, *Publ. Math. IHES* **47** (1977) 33–186.

[53] B. Mazur. Rational isogenies of prime degree, *Invent. Math.* **44** (1978) 129–162.

[54] B. Mazur. Number theory as gadfly, *Amer. Math. Monthly* **98** (1991) 593–610.

[55] B. Mazur. Questions about powers of numbers, *Notices Amer. Math. Soc.* **47** (2000) 195–202.

[56] H. McKean and V. Moll. *Elliptic Curves.* Cambridge University Press, Cambridge 1999.

[57] Preda Mihăilescu. Primary Cyclotomic Units and a Proof of Catalan's Conjecture, *J. Reine Angew. Math.* **572** (2004) 167–195.

[58] L.J. Mordell. *Diophantine Equations*, Academic Press, New York 1969.

[59] W. Narkiewicz. Class number and factorization in quadratic number fields, *Colloq. Math.* **17** (1967) 167–190.

[60] P.M. Neumann. *The Mathematical Writings of Évariste Galois*, European Mathematical Society, Zürich 2011.

[61] P.M. Neumann, G.A. Stoy, and E.C. Thompson. *Groups and Geometry*, Oxford University Press, Oxford 1994.

[62] J. Oesterlé. Nouvelles approches du 'theorème' de Fermat, *Astérisque* **161/162** (1988) 165–186.

[63] C. Reid. *Hilbert*, Springer–Verlag, Berlin 1970.

[64] P. Ribenboim. *13 Lectures on Fermat's Last Theorem*, Springer–Verlag, New York 1979.

[65] K. Ribet. On modular representations of $\mathrm{Gal}(\overline{\mathbb{Q}}/\mathbb{Q})$ arising from modular forms, *Invent. Math.* **100** (1990) 431–476.

[66] K. Ribet. Galois representations and modular forms, *Bull. Amer. Math. Soc.* **32** (1995) 375–402.

[67] B. Riemann. Grundlagen für eine allgemeine Theorie der Funktionen einer veränderlichen complexen Grosse, *Werke* (2nd ed.) 3–48.

[68] J. Roe. *Elementary Geometry*, Oxford University Press, Oxford 1993.

[69] J.J. Rotman. *An Introduction to the Theory of Groups*, Allyn & Bacon, Boston 1984.

[70] K. Rubin and A. Silverberg. A report on Wiles' Cambridge lectures, *Bull. Amer. Math. Soc.* **31** (1994) 15–38.

[71] P. Samuel. About Euclidean rings, *J. Algebra* **19** (1971) 282–301.

[72] R. Schoof. *Catalan's Conjecture*, Springer–Verlag, New York 2008.

[73] J.L. Selfridge and B.W. Pollock. Fermat's Last Theorem is true for any exponent up to 25,000, *Notices Amer. Math. Soc.* **11** (1967) 97, abstract no. 608–138.

[74] D. Sharpe. *Rings and Factorization*, Cambridge University Press, Cambridge 1987.

[75] C.L. Siegel. Zum Beweise des Starkschen Satzes, *Invent. Math.* **5** (1968) 169–179.

[76] H.M. Stark. A complete determination of the complex quadratic fields of class-number one, *Michigan Math. J.* **14** (1967) 1–27.

[77] I. Stewart. *The Problems of Mathematics*, Oxford University Press, Oxford 1987.

[78] I. Stewart. *Galois Theory* (4rd ed.), Chapman and Hall/CRC, Boca Raton FL 2015.

[79] I. Stewart and D.O. Tall. *The Foundations of Mathematics* (2nd ed.), Oxford University Press, Oxford 2015.

[80] I. Stewart and D.O. Tall. *Complex Analysis*, Cambridge University Press, Cambridge 1983.

[81] D.J. Struik. *A Concise History of Mathematics*, Bell, London 1962.

[82] R.L. Taylor and A. Wiles. Ring theoretic properties of certain Hecke algebras, *Ann. of Math.* **141** (1995) 553–572.

[83] H. te Riele and H. Williams. New computations concerning the Cohen-Lenstra heuristics, *Experimental Math.* **12** (2003) 99–113.

[84] R. Tijdeman. On the equation of Catalan, *Acta Arith.* **29** (1976) 197–209.

[85] E.C. Titchmarsh. *The Theory of Functions*, Oxford University Press, Oxford 1960.

[86] S. Wagstaff. The irregular primes to 125000, *Math. Comp.* **32** (1978) 583–591.

[87] A. Weil. Sur un théorème de Mordell, *Bull. Sci. Math.* **54** (1930) 182–191.

[88] A. Weil. *Number Theory: an Approach Through History from Hammurapi to Legendre*, Birkhäuser, Boston 1984.

[89] R.S. Westfall. *Never at Rest*, Cambridge University Press, Cambridge 1980.

[90] A. Wiles. Modular elliptic curves and Fermat's Last Theorem, *Ann. of Math.* **141** (1995) 443–551.

Further Reading on Algebraic Number Theory

1. S. Alaca and K.S. Williams. *Introductory Algebraic Number Theory*, Cambridge University Press, Cambridge 2003.

2. J.W.S. Cassels and A. Fröhlich. *Algebraic Number Theory*, Academic Press, New York 1967.

3. Ph. Cassou-Noguès and M.J. Taylor. *Elliptic Functions and Rings of Integers*, Birkhäuser, Basel 1987.

4. H. Cohn. *A Classical Invitation to Algebraic Numbers and Class Fields*, Springer–Verlag, New York 1978.

5. P.M. Cohn. *Algebraic Numbers and Algebraic Functions*, Chapman & Hall, London 1991.

6. G.H. Hardy and E.M. Wright. *An Introduction to the Theory of Numbers*, Oxford University Press, Oxford 1954.

7. F. Jarvis. *Algebraic Number Theory*, Springer-Verlag, New York 2014.

8. S. Lang. *Algebraic Number Theory*, Springer-Verlag, New York 2000.

9. O.T. O'Meara. *Introduction to Quadratic Forms*, Springer–Verlag, New York 1973.

10. H. Pollard. *The Theory of Algebraic Numbers*, Math. Assoc. America, Buffalo NY 1950.

11. P. Ribenboim. *Algebraic Numbers*, Wiley–Interscience, New York 1972.

12. P. Samuel and A.J. Silberger. *Algebraic Theory of Numbers*, Dover, New York 2008.

13. H. P. F. Swinnerton-Dyer. *A Brief Guide to Algebraic Number Theory*, Cambridge University Press, Cambridge, 2001.

14. A. Weil. *Basic Number Theory*, Springer–Verlag, New York 1967.

15. E.Weiss. *Algebraic Number Theory*, Dover, New York 1999.

16. H. Weyl. *Algebraic Theory of Numbers*, Princeton University Press, Princeton NJ 1940.

Index

Printed in the United States
by Baker & Taylor Publisher Services

Printed in the United States
by Baker & Taylor Publisher Services